AF873062

Climate Alert
Climate Change Monitoring and Strategy

Edited by
Yuzhu You and Ann Henderson-Sellers

SYDNEY UNIVERSITY PRESS

Published 2010 by SYDNEY UNIVERSITY PRESS
Fisher Library, University of Sydney
sydney.edu.au/sup

National Library of Australia Cataloguing-in-Publication entry

Title:	Climate alert : climate change monitoring and strategy / editors Yuzhu You, Ann Henderson-Sellers.
ISBN:	9781920899417 (pbk.)
Notes:	Includes index.
	Bibliography.
Subjects:	Climatic changes--Australia.
	Greenhouse effect, Atmospheric--Australia.
	Environmental monitoring--Australia
	Environmental indicators--Australia.
	Australia--Climate.
Other Authors/Contributors:	
	You, Yuzhu.
	Henderson-Sellers, Ann.
Dewey Number:	551.5253

Cover design by Miguel Yamin, the University Publishing Service

Contents

Preface

If you smoke you risk lung cancer. The world is smoking. As societies develop, their energy needs increase and the burning of fossil fuels accelerates. It is not always easy for individuals to give up smoking because the addiction is real. The world's addiction to burning carbon is also very real. How can we help the world decide to stop smoking?

Several things are necessary. Firstly, the damage caused by smoking has to be communicated beyond doubt. Climate change monitoring can help here and the chapters in this book discuss the compelling evidence that the climate is changing, and changing at an alarming rate. The chapters also consider the broad alliance between researchers, politicians and all citizens that is required to mandate governments with the power to take positive steps.

It is also important to minimise the impact of withdrawing from the burning of fossil fuels. Alternatives exist and their use must be demonstrated and encouraged.

Everyone has a part to play. Researchers play a role in identifying evidence, communicating evidence, and helping find alternatives and solutions. But every citizen has a role. This century is the time when the world will discover whether reason, intellect and foresight can triumph over ignorance and inertia, and whether or not the world that we know will continue to flourish or be irreparably damaged.

Professor Merlin Crossley[1]
Director of the Centre for the Human Aspects of Science and Technology
University of Sydney
August 2009

---

[1] Former Acting Deputy Vice-Chancellor for Research, University of Sydney, who opened the National Climate Change Monitoring Symposium on 12 December 2008 at the Webster Theatre, Veterinary Science Conference Centre.

About the authors

Scott Bainbridge, of the Australian Institute of Marine Science, is Project Manager for the Great Barrier Reef Ocean Observing System (GBROOS), a part of the Australian Integrated Marine Observing System (IMOS) project. With background in coral reef ecology, computing, informatics and data systems, Scott is currently working on real-time data systems and their use to address issues such as climate change and the long-term sustainability of the Great Barrier Reef.

Ray Berkelmans is a Senior Research Scientist at the Australian Institute of Marine Science. He is a coral ecologist interested in how climate change impacts on coral reefs and the links between ecological responses and environmental forcing factors. Ray also runs a long-term temperature logger program on Australia's tropical and subtropical reefs.

Wylie Bradford is a senior lecturer and member of the Burma Economic Watch team in the Department of Economics at the Macquarie University. His research interests cover the Burmese economy and the relationship between political philosophy and economic theory, as well as issues pertaining to sustainability and the economics of the environment.

Maria Byrne is professor of marine and developmental biology at the University of Sydney, and Director of the University of Sydney's One Tree Island Research Station in the southern Great Barrier Reef. Maria specialises in the biology of marine invertebrates. As part of her recent research on the interactive effects of climate change stressors on marine life histories, she is investigating the impact of climate change on the marine biota of eastern Australia.

Russ Davis is a professor emeritus of physical oceanography at the Scripps Institution of Oceanography specialising in long-duration ocean observations. He is a member of the National Academy of Science and a fellow of the American Geophysical Union and the American Meteorological Society.

Rana Fine is a professor at the Rosenstiel School of Marine and Atmospheric Science at the University of Miami. The objective of her research is to understand the role of the oceans in climate change occurring on decadal time scales. The research involves understanding the physical processes that determine the capacity of the oceans to take up atmospheric constituents, such as carbon-dioxide.

Masao Fukasawa is the General Director at the Research Institute of Global Change of Japan Agency of Marine and Earth Science and Technology. His main interest is the global meridional overturning circulation in the context of terrestrial climate variability. He is a member of the Japan Earth Observation Committee, the council of the Japan Oceanographic Society, and co-chair of Japan GEOSS working group under the Japanese cabinet.

Albert Gabric is an associate professor in the School of Environment at the Griffith University. His research interests are in oceanography, with particular emphasis on the role of the oceans in climate change.

Hiroyasu Hasumi is an associate professor at Center for Climate System Research, University of Tokyo. He specialises in numerical modeling of the ocean, especially in the physics of the global thermohaline circulation and its controlling processes.

Ann Henderson-Sellers is an ISI 'most highly cited' author of over 500 publications on climate and climatic change. Prior to her current position as an Australian Research Council Professorial Fellow at Macquarie University, she was Director of the United Nations' World Climate Research Programme and Deputy Vice-Chancellor of RMIT. Ann is an elected Fellow of Australia's Academy of Technological Sciences and Engineering and was awarded the Centenary Medal of Australia for Service to Australian Society in 2003.

Neil Holbrook is convenor of the Adaptation Research Network for Marine Biodiversity and Resources; associate professor of climatology and climate change at the University of Tasmania; Executive Secretary of the International Commission on Climate of IAMAS/IUGG; and Associate Editor of the *Australian Meteorological and Oceanographic Journal*. Neil is an international expert in the regional- to large-scale ocean and climate dynamics of the South Pacific, oceanic Rossby waves, climate predictability, extremes and climate change risks. He is interested in the complex feedbacks in climate change science, impacts and adaptation.

Dunxin Hu is Chinese Academy of Sciences Academician, professor of the Key Laboratory of Ocean Circulation and Waves, Institute of Oceanography, Chairman of the Joint Center for Ocean Circulation and Climate/Environment Studies, and Editor-in-chief of the *Chinese Journal of Oceanology and Limnology*. His research interests include ocean circulation and global climate change. He is a pioneer in the research of China Ocean Flux Research (JGOFS) and has been awarded various prizes and medals in China for excellence of scientific performance including the Scientific Progress Award and Science Award of the Chinese Academy of Sciences, Zhu Kezhen Field Science Award and National Defense Technology Progress Award.

Abdul Gani Ilahude is an oceanographic researcher, retired since 2002 from the Oceanographic Research Center of The Indonesian Institute of Sciences, Jakarta, where he worked from 1960. During his tenure at the Institute, he conducted surveys on the hydrography (properties, distribution and circulation) of water masses of the Indonesian and adjacent (Pacific and Indian) waters. He also carried out more than 75 large and small oceanographic cruises in these waters. Last year he was honoured with the Life Time Award conferred by the Indonesian Association of Oceanographer, Jakarta.

Momoki Koga is an associate professor of physical oceanography in the Department of Physics and Earth Sciences, Faculty of Science at the University of the Ryukyus. His current research interests include the variability of the Kuroshio, the ocean surface heat flux and air-sea interaction in the East China Sea and surrounding upstream area.

Jae Hak Lee is Director of Climate Change and Coastal Disaster Research Department at Korea Ocean Research and Development Institute. He is a physical oceanographer with research interests in ocean mixing and the volume transports variability at choke points such as the Drake Passage and Indonesian seas gap.

Tong Lee is a principal scientist at NASA Jet Propulsion Laboratory in Pasadena, California. His area of research is physical oceanography, with emphasis on the analysis of ocean circulation variability, upper-ocean heat balance, and their relationships to climate using satellite observations, ocean modeling and assimilation.

Cho-Teng Liu is a professor of physical oceanography at National Taiwan University. His research focuses on Kuroshio-related phenomena that include upwelling, volume transport, meandering, deep counter-current, interaction with impinging eddies and optical properties of Kuroshio water. His future project is on acoustic tomographic measurement of Kuroshio structure.

Janice Lough is a Senior Principal Research Scientist at the Australian Institute of Marine Science. She is a climatologist with research interests in the nature and impacts of climate change on tropical coral reef ecosystems and the development of high-resolution palaeoclimate records from massive coral skeletons.

Yukio Masumoto is an associate professor of physical oceanography at the University of Tokyo and a Program Director at the Research Institute for Global Change, Japan Agency for Marine-Earth Science and Technology. His research focuses on the ocean and climate variability in the Indo-Pacific sector. He is a co-chair of CLIVAR/GOOS Indian Ocean Panel.

Supriya Mathew is a PhD student in the Department of Environment and Geography, Macquarie University. She also holds a scholarship from CSIRO's climate adaptation flagship through which she collaborates with researchers in the sustainable ecosystems group. Her research interests are in local government prioritisation of adaptation strategies to reduce climate extremes.

Kendal McGuffie is an associate professor in the Department of Physics and Advanced Materials at the University of Technology, Sydney. His research interests are climate analysis and climate change, particularly in modelling the effects of climate change at the land surface and the effect of climate change on the hazard of tropical cyclones.

Gary Meyers was formerly a Chief Research Scientist in CSIRO Marine and Atmospheric Research and was the founding Director of the Australian Integrated Marine Observing System. He is now retired and an Honorary

Fellow in CSIRO. He has avidly promoted sustained ocean observing throughout his career, and is at currently researching the role of the tropical Pacific and Indian Oceans in the climate system.

Robert Molcard was a senior scientist at the Centre National de la Recherche Scientifique. Now retired, he has an honorary position at the Laboratoire d'Océanographie et du Climat: Expérimentation et Approches Numériques.

Wahyu W. Pandoe is currently the Head of Program Division at the Technology Center for Marine Survey of the government Agency for the Assessment and Application of Technology (BPPT), Indonesia. His research interests include the study of the Indonesian Throughflow. His office operates four Baruna Jaya Research Vessels and supports the marine and oceanographic researches as well as offshore engineers within Indonesian waters.

Tangdong Qu is a senior researcher of the International Pacific Research Center, SOEST, University of Hawaii at Manoa. As a physical oceanographer, Qu has particular interest in upper ocean dynamics and its impact on the ocean's heat and fresh water budget.

Joachim Ribbe is an associate professor of climatology and head of the Department of Biological and Physical Sciences at the University of Southern Queensland. Joachim is interested in the physical processes that drive the oceanic circulation. In his current research, he aims to better understand the impact of Australian climate drivers upon rainfall using field observations, data from climate models, and models of the ocean and climate system.

Philip Richardson is Scientist Emeritus at the Woods Hole Oceanographic Institution and researches issues concerning large scale ocean circulation. His recent study looked at the measurement of inter-ocean flow of water from the Indian Ocean around Southern Africa into the Atlantic Ocean as part of the meridional overturning circulation (KAPEX), and the importance of the Agulhas Current rings in transporting this water into the Atlantic.

Stephen Riser is a professor of physical oceanography at the University of Washington in Seattle. His research focuses on observing large-scale ocean and atmosphere interactions, the ocean's effect on climate, and large-scale ocean circulation. He is actively involved in programs dealing with the development of in-situ ocean observing systems, such as the Argo program.

Robin Robertson is a senior lecturer at School of Physical, Environmental and Mathematical Sciences, UNSW@ADFA, Canberra. Robin is a physical oceanographer and her research interests include internal tides and internal waves, vertical mixing mechanisms in the ocean, Antarctic seas, and long-term variability of the deep waters.

Moninya Roughan is a senior lecturer in the Coastal and Regional Oceanography Laboratory at the University of New South Wales. She specialises in oceanographic processes and biophysical coupling in the East Australian Current. She leads the physical oceanography component of the NSW-IMOS program and is the deputy node leader for NSW-IMOS.

Tom Rossby is a professor of physical oceanography at the University of Rhode Island with a strong interest in ocean observation and particularly the Gulf Stream. He is a member of the National Academy of Engineering and the American Academy of Arts and Sciences. He has been awarded several medals and is a Fellow of the American Geophysical Union and the American Meteorological Society.

Jennifer Scott has an adjunct role with the University of Western Sydney and is a partner in a climate adaptation research project with Macquarie University. For the past three years Jennifer has been working within local government to provide policy advice on complex trans-disciplinary issues such as climate change. She has published in business, law and science journals on a wide range of sustainability topics. In her current role she is acting Sustainability Program Leader at Ku-ring-gai Council.

Peter Sigray is a senior scientist and professor at the Department of Meteorology at Stockholm University. His primary research interest is on ocean electrodynamics. His focus has been on transport measurements in the Baltic and the North Atlantic utilizing in-use cables.

Craig Steinberg is a physical oceanographer at the Australian Institute of Marine Science. He has 25 years experience researching the tropical marine environment in northern Australia. He currently manages sub-facilities of the

Integrated Marine Observing System: Satellite Remote Sensing and the Queensland and Northern Territory Australian National Mooring Network.

Margery Street is a research assistant within Macquarie University's Faculty of Science and recently completed a Masters degree in Environmental Studies.

Dwi Susanto is a scientist and Director for Indonesian Research Coordination at Lamont-Doherty Earth Observatory, Columbia University. Dwi is a physical oceanographer, currently working on the INSTANT (International Nusantara Stratification and Transport) program, analysis of submesoscale in the Indonesian seas; upwelling along the coast of Java and Sumatra and its relation to throughflow, climate phenomena and productivity, Indo-Pacific sea level and sea surface temperature variability and climate impacts to fisheries and primary productivity variability in the Indonesia Seas.

Iain Suthers is leader of NSW-IMOS, one of five national nodes of the Integrated Marine Observing System based at the Sydney Institute of Marine Science. He is the lead proponent for the deployment of ocean gliders into eddies of the East Australian Current, and is researching the ecology and increasing abundance of salps in response to climate change.

Ros Taplin is a professor and Director of the Environmental Management and Carbon Management programs at Bond University, Gold Coast. Her research background and qualifications are in environmental management and her primary research interests are in climate change including impacts and adaptation, international policy on climate change, and Australian climate change and carbon management policy. Ros has a decade of consulting experience involving a broad range of sectors of industry (insurance, manufacturing, energy, mining and resources) and government (state and federal).

Stefan Trück is an associate professor in the Economics Department, Macquarie University. He has held positions at Queensland University of Technology and University of Karlsruhe in Germany where he received a PhD in Statistics. His current research interests focus on risk management, quantitative finance, energy and environmental economics. He has many years of consulting experience for financial institutions in Europe.

Cesar Villanoy is a professor at the Marine Science Institute, University of the Philippines. He is a physical oceanographer and his main interests are in the application of remote sensing techniques and numerical hydrographic models to problems related to coastal pollution and coastal habitat connectivities. Cesar has extensive experience and has developed tools to map out variations in water residence times in coastal areas used for aquaculture.

David Walland is a senior climatologist at the Bureau of Meteorology and is responsible for development of climate considerations in a range of projects and policies. Having spent time as the Manager of the National Climate Centre, his interests revolve around developing the best climate service possible for the Australian community. His education reflects his interests with a mix of scientific and management qualifications following on from a major in Physics at the University of Melbourne (Hons) with a PhD in Meteorology and an MBA.

Yuzhu You is a senior research associate on ocean climate monitoring and palaeoclimate modeling with the University of Sydney's Institute of Marine Science. After completing his PhD on physical oceanography in 1991 at the University of Sydney, he worked as a visiting professor and fellow in many top oceanographic institutions around the world including the USA, France, Germany, Japan, China, Korea, South Africa and Taiwan. In 2004 he was awarded an ARC grant on palaeoclimate modeling.

Walter Zenk was a senior scientist at the Leibniz Institute of Marine Sciences IFM-GEOMAR in Kiel, one of Europe's leading marine research institutions. After retirement in 2005 he worked as a guest investigator in the research division Physical Oceanography. His professional interests include spreading and mixing of water masses at mid-depths and in the abyssal. He has extensive experience studying the in-situ flow field by Eulerian and Lagrangian methods.

1

Monitor climate, monitor change

Yuzhu You

*Abstract**

At the first National Climate Change Monitoring symposium held in the Webster Theatre, Veterinary Science Conference Centre, University of Sydney, on 12 December 2008, "Monitor climate, monitor change" was urged to meet the present challenge of climate change. This book, *Climate alert: climate change monitoring and strategy*, is the result of that symposium and provides a parallel effort to ensure that outcomes are made available for a wider audience including policy makers and the general public. The symposium presentations amalgamated a wide range of evidence of global and regional climate change in the Antarctic, Southern Ocean, Great Barrier Reef, coastal regions and the Australian land mass. This book therefore complements the recently released Garnaut Report (2008) on climate change evidence and monitoring. Overall, the symposium and book aim to promote long-term climate change monitoring and strategies for Australia.

1. Introduction

1.1 Evidence of rapid climate change

Climate change is real and is happening now. Since the Industrial Revolution in the late 18th century, the global mean surface temperature has increased by 0.8°C. The upward trend has increased since the 1980s. In the last 30 years, the instrumental surface temperature has increased by about 0.2°C per

---

* Chapter originally submitted in August 2009.

decade (Hansen et al. 2006). At the same time, atmospheric carbon dioxide has increased with a doubling time of about 30 years since the Industrial Revolution. The growth rate has doubled since measurement began at the Mauna Loa Observatory in 1958 from 1 ppmv (parts per million by volume) per year prior to 1970 to more than 2 ppmv per year in recent years (Hofmann et al. 2009). Over this time period, the net increase of carbon dioxide has been 120 ppmv, the present (2008) CO_2 level having reached 387 ppmv.

These measures appear to have coincided with the increase of human population since the Industrial Revolution. The Intergovernmental Panel on Climate Change Fourth Assessment Report (2007; referenced as IPCC AR4 [2007] hereafter) has unequivocally concluded that the observed rapid warming is primarily due to human activities with at least a 90% certainty. The current level of atmospheric carbon dioxide is the highest since the peak value attained 650,000 years ago.

In response to global warming, sea level has risen at a rate of 2 mm yr^{-1} in the last century, a total rise of 20 cm since the 1880s (Douglas 1997). The rate has recently increased to 2.8±0.4 mm yr^{-1} in 1993–2003 (Chambers 2003). Sea-level rise occurs mainly because of thermal expansion resulting from the warming of sea water as well as from water-mass exchange. However, melting of the polar ice sheets has not been sufficiently accounted for in the calculation of sea-level rise due to their unpredictable future state as noted in the IPCC AR4 (2007). Melting of polar ice sheets can contribute from a few metres to several tens of metres to sea-level rise over a period of hundreds or thousands of years.

The IPCC AR4 (2007) predicts a temperature increase of 1.4°–5.8°C and a sea-level rise of 19–59 cm by the year 2100 based on mid-range scenarios. A temperature rise of 0.2°C per decade is projected for the next two decades. Using an increase in temperature for 0.2°C, 20 ppmv in CO_2 and 30 mm in sea-level rise per decade from the year 2000 levels, the global mean surface temperature, atmospheric CO_2 and sea level will increase, respectively, by 2°C (since the Industrial Revolution), 140 ppmv (to a total 510 ppmv) and 20 cm (a total rise of 40 cm since 1880s) by the year 2070.

However, there is well-recognised uncertainty about climate variability. Recently, Knight et al. (2009) noted that global warming was flat for the just past decade due to the interference by other natural force mechanisms such as the El Niño – Southern Oscillation (ENSO) cycle, pollutant aerosols, volcanic aerosols and solar variability. In 10 climate model runs for 21st century climate projection totalling 700 years of simulation, they discovered 17 separate decade-long intervals showing warming trends resembling that of the past decade with no greenhouse warming. The overall simulated trend for the 700 years shows no surprise, indicating greenhouse warming as expected.

1.2 A view of global climate change

In addition to the evidence provided above, there are numerous reports about global climate changes around the world such as the Greenland and land-based glacial retreats, polar ice sheet melting, sea-ice reduction, sea and land surface temperature increases, changes in severe weather rates and patterns and the increase of greenhouse gases. There are several features considered relevant to our changing climate. These are plate movement, solar constant and orbital parameters, volcanism, ocean circulation and human effects. The present global warmth is mainly attributed to human-activity-induced influences such as burning fossil fuel for power, manufacturing (e.g. cement), transportation, agriculture and deforestation (IPCC AR4 2007). The polar region has been warming strongly as a result of human-contributed climate change (Gillett et al. 2008). In 2007, the summer minimum sea ice in the Arctic was an unprecedented 40% below the minimum sea ice extents of the 1980s. Autumn temperatures were 6°C higher than the 1958–98 mean. The dramatic sea ice drop is caused by temperature increase (Overland 2008). The Greenland ice sheet has been melting, with a 16% increase from 1979–2002. Based on satellite gravity data between 2003–08, a more recent study reports a melting rate of 195 cubic kilometres per year, representing a contribution of 0.5 ± 0.1 mm yr^{-1} to global sea-level rise (Wouters 2008).

In Antarctica, a significant temperature anomaly of more than +2°C has been found in the Antarctic Peninsula. In response to the warming, a collapse of the Antarctic ice shelf has been reported both in the east and west of the Antarctic Peninsula in recent years; the Larsen B ice shelf in 2002 following the Larsen A ice shelf disintegration in 1995 (Scambos et al. 2004) and the Wilkins ice shelf in 2008 (National Snow and Ice Center 2008). However, no significant temperature rise and loss of ice of any kind have been found in the rest of Antarctica. On the other hand, sea ice from satellite observations shows an increase for the last two decades. From an oceanic viewpoint using a global multicategory thickness and enthalpy distribution sea ice model coupled to an ocean model, a cause is found in the enhanced thermohaline stratification (due to atmospheric warming) leading to a decrease of the upward ocean heat transport available to melt sea ice (Zhang 2007). From atmospheric model runs, Turner et al. (2009) discovered that stratospheric ozone depletion causes sea ice growth due to enhanced autumn wind speeds around the Antarctic continent, deepening the Amundsen Sea Low. As it is hoped that the ozone depletion will recover, sea ice growth is expected to cease and more sea ice will be melted under the overall global warming environment.

Climate change in the 20th century has forced a poleward contraction of the southern hemisphere subpolar westerly winds. England (2008) analysed the implications of this wind shift for the ocean's thermohaline circulation in his ocean general circulation models (OGCMs) compared with available observations. He found that substantial heat content anomalies can be linked to changes in the latitude and strength of the shifted westerly winds. Effects are also found in the sea surface temperature (SST), thermohaline overturning circulation and Antarctic water-mass formation such as Subantarctic Mode Water (SAMW) and Antarctic Intermediate Water (AAIW).

Climate-change-induced changes in water properties, ocean circulation, ventilation and respiration may affect the oxygen distribution of the interior ocean and thus the oxygen minimum zone (Karstensen 2008). Indeed, the entire global ocean ecosystem is expected to change in the future.

More evidence of global climate change has been provided in the IPCC AR4 (2007) including changes in atmospheric composition, global average temperature, ocean condition etc. Except for the aforementioned increase in carbon dioxide, other greenhouse gases have also increased markedly. For example, the amount of methane in the atmosphere reached 1774 ppb (parts per billion) in 2005, far exceeding the natural range of 320–790 ppb in the last 650,000 years. The increase of atmospheric methane is partly attributed to permafrost thaw in the Siberian and Arctic shelf (Zimov et al. 2006; Walter et al. 2006). Another greenhouse gas, nitrous oxide, has risen from the preindustrial level of 270 ppb to 319 ppb in 2005.

The primary source of increase in greenhouse gases is due to human activities such as burning fossil fuel and agriculture. Chlorofluorocarbons (CFC-11, CFC-12, CFC-113, CFC-114 and CFC-115) are typical human-contributed greenhouse gases resulting from human usage of foams, packaging materials and refrigerants. CFCs are also known to cause stratospheric ozone depletion.

Global warming is likely to lead to less frequent cold days, cold nights and frost and more frequent hot days, hot nights and heat waves. As a consequence, wind intensity increases. The North Atlantic hurricanes have been observed increasing their intensity since the 1970s associated with the SST increase. Mountain glaciers, snow cover and permafrost have declined on average in both hemispheres while drought and heavy precipitation events have increased.

At the National Climate Change Monitoring (NCCM) symposium, Church (2008) reported recent results from his research group on sea-level rise. From an historical viewpoint, sea level has risen over120 m since the last glacial maximum 21,000 years ago at an average rate of ~1 m per century and peak rates of several metres per century. The rate of sea-level rise has increased since last century with an upward trend of 1.8±0.3 mm yr^{-1} from 1960 to 2000. The melting of the Greenland ice sheet has contributed about 0.2 mm yr^{-1} to sea-level rise for 1993–2003 while the melting of the Antarctic ice sheet is estimated as having contributed the same to sea-level rise for that period of time (Lemke et al. 2007). The best estimate for total sea-level rise for 1961–2003 is ~60 mm. Since coastal city societies have developed during a relatively low sea-level rise period (the last 3000 years of human civilisation), current enhanced sea-level rise has already increased the frequency of flooding events from one in five years to one in two years during the 20th century at both Fremantle and Sydney. For a 0.5 m rise, a one in 100 year event could occur several times per year at many Australian locations (Church 2008).

The consequence of sea-level rise is serious since about 100 million people worldwide live within a metre of the present sea level. Over the years, Pacific Island countries have been calling on the world community for help. The same message was received from the recent 40th Pacific Island Forum leaders' meeting in Cairns, 5–6 August 2009. Australia and New Zealand have been urged to cut greenhouse gas emissions by at least 50% by 2050. Sea-level rise causes many islands and coastal cities around the world to become submerged and will potentially result in millions of climate refugees.

1.3 A view of Australia's climate change

Australia crosses almost an entire latitudinal band in the southern hemisphere from the tropics to Antarctica and is bounded by three oceans: the Pacific Ocean in the east, the Indian Ocean in the west and the Southern Ocean in the south. Therefore, Australia is exposed to both global and regional climate regimes and is affected more severely than other countries by climate change. Australia's mainland is a vast continent, hot and dry, and is the only land in the world with a positive annual mean E-P (evaporation minus precipitation) value. Its habitat is therefore highly vulnerable to climate change.

1.3.1 The Antarctic and Southern Ocean

Ice cores reproduce detailed climate records dating back over 800,000 years, revealing a significant influence of the Antarctic on the global climate and the regional climate of Australia (Press 2008). As Australian Antarctic stations are mostly located in the eastern Antarctica, away from the Antarctic Peninsula, global warming is felt to a lesser degree. However, the coastal Antarctic stations of Casey and Davis show a slight warming of less than 0.1°C per decade in the last 50 years although Mawson station shows a slight cooling (Jacka et al. 2004). A similar scale of warming is conjectured in the interior of the Antarctic continent but with high spatial variability. The general warming of the Antarctic is consistent with the reported average growth rates of Antarctic mosses for different species ranging from 0.6 to 1.3 mm yr^{-1} over the last 50 years (Clarke et al. 2008).

At Mawson station, Hei et al. (1996) found that ice growth is slowed by oceanic heat sources. The bottom water in the world's ocean, called Antarctic Bottom Water (AABW), is known to be formed around Antarctica, especially in the Weddell Sea and the Ross Sea. AABW is a limb of the global thermohaline or great ocean conveyor belt circulation. Using a historical data set, Robertson (2008) found a warming trend of ~0.007°C±0.004°C yr^{-1} from 1960 to 2005 in the Weddell Sea, when the Weddell Polynyas bias is considered in the mid-1970s and mid-to-later 1990s. Between the 1970s and 1990s, the warming trend is higher at ~0.012°C±0.007°C yr^{-1}.

Climate change has also affected the acidification of the Antarctic marine ecosystem. At the NCCM symposium, Riddle et al. (2009) reported the recent discovery of a diverse, cold-water coral community in a canyon system on the Antarctic continental slope off East Antarctica. The aragonite corals were spatially dominant at 600–800 m, well below the reach of modern icebergs. The corals were sparse at 900–1000 m and apparently absent at 1400 m. Water column CO_2 chemistry indicates that the aragonite saturation horizon in the region occurred at about 900 m. With increasing atmospheric CO_2, the continued uptake of CO_2 will drive the saturation state of carbonate minerals lower, causing the saturation horizon to rise towards the surface. They predict that these cold-water coral communities in Antarctica will be among the first globally to show calcification effects due to the crossing of saturation thresholds, possibly within 5–10 years.

The Australian sector of the Southern Ocean appears to be changing physically and biologically in ways that were considered to be only theoretical a decade ago (Press 2008). On Macquarie Island, the annual mean surface temperature increased by over 1°C with a warming rate of 0.042°C yr^{-1} in the austral summer and autumn and 0.018°C yr^{-1} in spring since instrumental records started in 1949 (Adamson et al. 1988). The Heard Island ice sheet has retreated by over 12% since 1947 from 288 km^2 to 253 km^2 in 2000 (Ruddell 2005).

1.3.2 The Great Barrier Reef

The Great Barrier Reef (GBR) is World Heritage listed, being thus declared in 1981, and is economically important to Queensland and Australia in general. The GBR extends for ~2000 km, along the Queensland coast (9°–24°S) and up to 200 km from the shore, spanning an area of about 35 million hectares. This vast area includes nearly 3000 coral reefs, home to 360 species of hard coral, 1500 species of fishes, 4000 species of molluscs, 400 species of sponges and the list goes on (Lough 2008). However, the world's best managed and protected coral reef ecosystem is in danger due to climate change as a result of increasing greenhouse gases. Lough (2008) reported that significant impact of climate change on the GBR has already occurred. The GBR SST has increased by 0.4°C between 1871 and 2007. The climate zone has shifted southward with a SST anomaly of about 0.6°C that is higher south of 23°S than north of this latitude. Significant reef bleaching has been caused by thermal stress that occurred in 1998, 2002 and 2006, and has occurred more frequently since the 1970s. The increase of atmospheric greenhouse gases results in an extra 30% of the CO_2 being absorbed by the ocean, causing ocean acidification, resulting in a difficulty for corals to form their skeletons. Furthermore, the temperature increase of sea water increases coral disease. Climate change also increases the frequency and intensity of tropical cyclones, causing strong physical destruction of coral reefs. Other damaging factors include extreme weather events such as more rainfall and freshwater extending on to the reef, storm surges, erosion and river discharges.

1.3.3 Coastal impact

Except for the aforementioned southward shift of the coastal high SST anomaly water along the east coast, the dry and hot climate in Australia has generally impacted its coastal water and the environment surrounding its continent. Ribbe (2008) observed hypersalinity in estuaries during recent decades, which is due to long-term drought and evaporation in land and coastal areas.

The change in the coastal water environment, especially coastal water warming, has a strong impact on coastal ecosystems. Byrne et al. (2008) reported the different adaptation ability of marine species, especially abalone and sea urchin, to the warming coastal water. They found that the distribution of these species overlaps with the eastern Australian climate change hot spot where disproportionate ocean warming and ocean acidification make regional marine biota particularly vulnerable to climate change. In the context of scenarios projected for local waters, they determined synergistic effects of ocean warming and acidification in fertilisation, development and calcification. With experimental treatments (20°–26°C, pH: 7.8–8.2 and pCO_2: 370–500 ppmv) including all combinations for the 2070 business-as-usual scenario, the percentage of fertilisation for both species did not differ across treatments. For sea urchins, there was also no difference in the percentage of normal cleavage and gastrulation at 20°C and 24°C across all pH treatments. For abalone, however, cleavage was more sensitive to the increased temperature. Thus, Byrne et al. (2008) concluded that +4°C above ambient SST and pH-0.4 below ambient did not impact the early development of sea urchins but did impact abalone development. The higher temperature at 26°C was the major teratogen causing morbidity regardless of pH. For embryos that developed to the larval stage, increased pCO_2 ($pH<8.0$) had a major impact on calcification at all temperatures, with abalone being more sensitive than sea urchin.

The impacts of climate-change-induced sea-level rise are of great concern to the public since most capital cities and the majority of the Australian population are located in coastal regions. Holbrook (2008) demonstrated recently that changes in the baroclinic representation of East Australian Current intensity using a linear reduced-gravity ocean model provides an impressive surrogate for the observed interannual to multi-decadal time scale sea-level changes recorded at the long-record Fort Denison tide-gauge in Sydney Harbour. Cowell and Goodwin (2008) presented their study of coastal impacts of sea-level rise at the NCCM symposium. Evidence for the signal of geomorphic impacts on coasts due to sea-level rise at present is hidden amidst noise from other well-recognised drivers including the effects of historical climate variability and direct human impacts. Their model results show that the effects of sea-level rise in driving coasts inland account for only about half of the potential response under a range of conditions. Sedimentation rates are equally important on the climate change time scale.

Their modelled effect of increased storms on the coastal geomorphology is minor compared to other drivers.

1.3.4 The Australian land

According to the CSIRO, Bureau of Meteorology (CSIRO & BOM 2007) and the Garnaut Report (2008), Australia's annual mean temperature has increased by about 0.9°C since 1910, about 0.1°–0.2°C more than the global mean temperature increase between 0.7°–0.8°C since 1900. Warming is not uniform spatially with the greatest warming occurring in central Australia (Murphy & Timbal 2008). In southeast Australia, average maximum temperatures have increased, the climate becoming drier and hotter (Nicholls 2004). Bush fires in Victoria, such as the black Saturday on 7 February 2009 that claimed 176 lives, could be a consequence of the changed environment by climate change in the region. A single bushfire may not simply be attributed to climate change; however, on the black Saturday a historical record of temperature at 46.7°C was observed following a 50-year warming trend and a record low humidity of 5% following a 12-year below average rainfall (more facts following). Extreme weather events such as hot days and nights have increased in Australia since 1955 and heat waves have become increasingly common (CSIRO & BOM 2007). Nearly two-thirds of the continent recorded maximum temperatures over 39°C with peaks at 48.5°C in western NSW during the period from 1955 (Lynch et al. 2008).

Human-induced climate change has resulted in up to 50% reduction in rainfall in the southwest of Western Australia and the decline of snow cover in the southeast Australian alpine region (Cai & Cowan 2006). Since 1950, rainfall across eastern Australia has been in strong decline and, correspondingly, since 1977 the frequency of El Niño events has increased in the region (Power & Smith 2007). A decline in the annual average rainfall has also occurred in southeast Australia as seen in historical records (Murphy & Timbal 2008). At the NCCM symposium, Walland (2008) presented evidence of the magnitude of the decline in rainfall over the last decade or so. The last five to 10 years mark one of the most severe drought periods in Australia's history with the worst declines centred over Australia's most productive agricultural zones in the Murray-Darling Basin and in the capital cities of Perth, Adelaide, Hobart, Melbourne and Sydney, with Brisbane also being adversely affected. Among these cities, some areas around Perth, Melbourne and Brisbane are noted as being the driest on record. Almost all of Victoria has rainfall measures that are very much below average. Walland (2010) presented evidence that associated the change in rainfall with a significant shift in the mean position of the subtropical ridge across Australia and extra-tropical cyclones and rainfall bearing systems occurring more frequently in wintertime. The frequency of anticyclones (high pressure systems) between 1950 and 2006 shows a clear increase across southern Australia. High pressure systems tend to be associated with stable, fine weather and reduced rainfall.

The reduction in rainfall has, as a consequence, decreased streamflows affecting agriculture and the urban freshwater supply. In the Murray-Darling Basin, a 10% change in rainfall has been found to result in a 35% change in streamflow (Jones et al. 2001). The decreased streamflows have resulted in a shortfall of urban freshwater supply. For example, the water storage in Perth has been in marked decline since 1970s, from 336 GL averaged over 1911–1975 to 170 GL averaged over 1976–2000 and only 87 GL over 2001–07 (Western Australia Water Corporation 2008).

When considering the impact of climate change on tropical cyclones, Henderson-Sellers et al.'s (1998) study concluded that the broad geographical areas of cyclogenesis are unlikely to change significantly under enhanced greenhouse gases, but the changed thermodynamic environment would likely lead to enhanced storm intensity of around 10–20%. With limited data, a substantial increase in the number of tropical cyclones occurred over the eastern coast of Australia in the 1950s but decreased in the 1970s (CSIRO & BOM 2007). This decrease is linked to an increased number of El Niño events. On the Australian west coast, the proportion of severe cyclones has increased from 29% in 1974–88 to 41% in 1989–98.

Bushfires show a general increase across east and southeast Australia between 1973 and 2007. From 23 measuring locations, Lucas et al. (2007) found that the three years with the highest Forest Fire Danger Index occurred after the year 2000, and that 50 out of 69 bushfires occurred after 2000.

2. Climate change monitoring and strategy

The current round of United Nations Framework Convention on Climate Change (UNFCCC) negotiations for a new global climate treaty aims to limit global warming to no more than 2°C above pre-industrial temperatures. Achieving no more than a 2°C temperature increase requires that global greenhouse gas emissions be cut by at least 50% below the 1990 levels by the year 2050. In the long term, CO_2 levels need to be stabilised well below 450 ppmv. In order to stay below 2°C, emissions must drop rapidly over the next 10 to 15 years. Due to limited scientific knowledge about the total climate system, effective monitoring of climate change in a wide range of topics is crucial for policy making and adaptation and mitigation.

As a historical mirror, You et al. (2008, 2009) compared present global warming with the geologically most recent warming: the Middle Miocene Climate Optimum (MMCO) that occurred 15 Ma BP. This high quality compiled proxy data set indicated that the global surface temperature at the MMCO was about 3°C higher than the present and is equivalent to the warming predicted for the next century. Applying the latest National Center for Atmospheric Research (NCAR) Community Atmosphere Model CAM3.1 and Land Model CLM3.0 coupled to a slab ocean model and a suitable set of integrated meridional SSTs, they simulated a range of CO_2 levels at 460–580 ppmv to drive the MMCO 3°C warming. This lower threshold of 460 ppmv for the MMCO is consistent with the 450 ppmv threshold urged for stabilizing the CO_2 level.

At the NCCM symposium, some proposals for climate change monitoring and strategy were made. Henderson-Sellers (2008) raised problems in the media and made suggestions on how to improve communication about present climate change in order to take better action on climate change. She emphasises the necessity of monitoring climate change communication in future.

Lyster (2008) raised an interesting and important issue about climate law and pointed out that monitoring the impacts of climate change is equally important where law is used as an instrument to drive adaptation response. Measuring and monitoring of climate change lie at the heart of climate law.

Under the rubric of climate law, two important goals must be achieved: the mitigation of greenhouse gas emissions and appropriate adaptation responses to the impacts of climate change. Legal responses to greenhouse gas mitigation are now pervasive in both developed and developing countries attendant upon the insertion of the *United Nations Framework Convention on Climate Change* and the *Kyoto Protocol* into International Environmental Law. Regional, national, state and local governments are enacting legislation to halt these emissions primarily through the use of emissions trading schemes and other complementary measures. All of these legal regimes require the measurement of greenhouse gas emissions and, to have any credibility, the performance of countries and corporations against their legal obligations must be continually monitored by the regulator.

Climate change monitoring for the Australian ocean and land areas was proposed at the NCCM symposium. Suthers and Myers (2008) introduced the ongoing Integrated Marine Observing System (IMOS) which is a $92M project established from the National Collaborative Infrastructure Strategy (NCRIS) and co-investments from 10 operators including university and government agencies. It provides a nationally distributed set of equipment established and maintained at sea, providing streams of oceanographic data and information services that collectively will meet the needs of marine research in open oceans and coastal oceans around Australia. Over the long term, the IMOS will permit identification and management of climate change in the marine environment, an area of research that is as yet almost a blank page. In complying with the IMOS, the GBR is now monitored by an extended program called the GBR Ocean Observing System (GBROOS). Details of GBROOS are given in Lough et al. (2010).

The Indonesian Throughflow (ITF) is, both globally and regionally, an important current and water drainage system from the equatorial Pacific to the eastern Indian Ocean. The ITF has a significant impact on both the global and Australian climate and coastal ecosystems on the west and south coasts of Australia. You (2008) introduced an international ocean climate program, PACSWIN (ITF: PACific Source Water INvestigation), to the NCCM symposium to pursue an international coordinated, cost-effective, long-term monitoring program of the ITF. One of the PACSWIN components is the usage of submarine cables in the Indonesian seas, which can satisfy the decadal to centennial time scales of climate monitoring. A comprehensive description of the PACSWIN program is given in the You et al. (2010) chapter.

Australia's atmosphere and hydrosphere are mainly monitored by the Bureau of Meteorology (BOM) and CSIRO as government agencies. National meteorological and hydrological services such as the BOM take lead responsibility for the efforts to effectively monitor Australia's atmosphere and hydrosphere although they rely on other agencies' support as well. Australia has 103 reference climate stations covering the continent. Typically, these stations measure air temperature, humidity, pressure, wind direction/speed, rainfall and, in addition, observe weather, visibility, cloud type/amount etc. Data output include surface data, upper air, atmospheric chemistry, solar radiation, and hydrology. In some cases, monitoring of the atmosphere has occurred for over 100 years. As reported by Walland (2008), BOM and CSIRO are determined to continue monitoring climate change with the highest possible quality in spite of challenges in the monitoring of the climate system in Australia.

Adam and McBratney (2008) presented the topic of monitoring ecosystem and carbon fluxes and the carbon pool. Monitoring of the biosphere-atmosphere fluxes of greenhouse gases and biosphere-stored carbon is becoming a critical part of the worldwide scientific effort. Soil carbon is particularly susceptible to changes in climate, especially changes in temperature and moisture regimes, and may provide a key feedback to changes in atmosphere CO_2 and rising temperature. Currently, Australia has no soil carbon monitoring scheme in operation but CSIRO has plans for 3000 future sites. The Australian contribution to the study of the effects of climate change on vegetation is mainly from controlled environment and laboratory studies. The study by Adam and McBratney highlights that soil carbon content and turnover, such as the rates of respiration and its temperature dependence, are frequently species, and thus fire-regime, dependent. Their models suggest that changes in phenology and species composition play major roles in carbon and hydrological consequences of changing climates.

An Australian Climate Change and Earth Observatory (ACCEO) is proposed by Cleugh et al. (2008), which aims to provide an early warning system of large or rapid shifts in the climate system; data for improving climate and earth system models; information to support and evaluate mitigation and adaptation policies and a foundation for international engagement. To build, deliver and sustain such a program requires a nationally coordinated approach that builds on existing observation-based programs. The current urgent scientific issues for the observatory to resolve include the aerosol/cloud processes, extreme events, ice sheet dynamics, the hydrological cycle (ocean-atmosphere-cryosphere), biogeochemical feedbacks and abrupt climate change.

An overview of a number of climate change issues was given by Pearman (2008) at the NCCM symposium. He listed reasons for climate change as a "diabolical" issue as stated in the Garnaut Report: uncertain in its format and extent, insidious rather than (as yet) confrontational, long-term rather than immediate, international as well as national; in the absence of effective mitigation there is a risk of dangerous consequences. He emphasises that there is an ongoing urgency for adaptation and mitigation actions that relate to observed changes, the potential impacts and the timescales associated with the climate system.

3. *Introduction to the book chapters*

This book contains rich sources of evidence that climate change is happening now, together with the necessary recommended monitoring. A wide range of themes is covered spatially from the Antarctic to tropical Indonesian seas, the GBR, open ocean, coast and the land. Topics include the monitoring of communication, ecological response of the GBR to climate change, water-mass formation in the Weddell Sea, the atmosphere, hydrology, the impact of climate change stressors on marine invertebrate life histories, the impact of climate change on the hypersalinity in coastal bays, the Indonesian Throughflow in an international ocean climate program, ocean circulation, integrated marine observing systems, the South Pacific Ocean climate dynamics, future cyclone season characteristics under a changing climate and assessing climate change adaption for local government.

In the second chapter 'How seriously are we taking climate change: monitoring climate change communication', Henderson-Sellers (2010) analyses three possible reasons to have climate change deniers' voices still heard in the media and proposes improved climate change communication in our society. One reason for the lack of effective climate change communication may be that the terminology we have developed may slow our ability to galvanise action. The second probable reason is the extreme difficulty of separating personal viewpoints from the scientific voice. And the third reason is that there is an unhealthy polarisation into 'believers' and, the only alternative, 'non-believers', which excludes scientific scepticism of a type not previously recognised – including concern that the IPCC consensus is too conservative. These reasons are partly responsible for the lack of urgency on greenhouse action.

Lough et al. (2010) present evidence of the impact of climate change on the GBR and propose physical monitoring of the GBR in the third chapter entitled 'Physical monitoring of the Great Barrier Reef to understand ecological responses to climate change'. Significant warming of the GBR has already occurred. Assessing biological responses to warming requires high quality, ongoing observations of the physical environment over a range of spatial and temporal scales. These authors describe the GBR monitoring programs developed in the Australian Institute of Marine Science (AIMS) through automatic weather stations, remote sensing, ocean moorings and an intensive water temperature logger program. These monitoring programs are now being enhanced and expanded through the GBR Ocean Observing System (GBROOS).

The Weddell Sea is the major location for AABW formation, and is an important limb of the global thermohaline circulation and a sink of greenhouse gas uptake by ocean bottom water. Global warming may indeed slow down AABW formation. In the fourth chapter 'Are the deep waters of the Weddell Sea still warming?', Robertson (2010) uses historical conductivity-temperature-depth (CTD) data and bottle data and examines if the Warm Deep Water (WDW) is still warming since it was warmed from

1970s to 1990s and then cooled starting in the mid-1990s. From the extended dataset, she finds that a general warming trend was present in the WDW from 1960s to the present of ~0.007±0.004°C yr^{-1}. Two significant cooling episodes occurred during this period associated with the Weddell Polynyas of the mid-1970s and the mid- to late 1990s. Without the cooling episodes, the warming trend was higher at ~0.012 ±0.007°C yr^{-1} from the 1970s to the 1990s. The continued monitoring of the WDW is crucial in monitoring global climate change.

Walland (2010) reported on the climate monitoring programs in the Bureau of Meteorology (BOM) in the fifth chapter 'Monitoring the atmosphere and hydrology'. BOM has been collecting observations of the atmospheric and hydrological cycle for large parts of the past century. The data underpin our knowledge about Australian climate, climate variability and change. The climate record incorporates data from a number of networks including surface, upper air, atmospheric chemistry, surface radiation and of course hydrology networks. The resulting climate record gives us a wonderful resource to examine trends in our climate and assist in the process of attribution of these trends to climate change.

How does marine biota respond to warming and acidification caused by global warming and increased atmospheric CO_2? Byrne (2010) examines the thermo- and pH tolerance of fertilisation and development of sea urchins and bivalves to address their future vulnerabilities in her chapter 'Impact of climate change stressors on marine invertebrate life histories with a focus on the Mollusca and Echinodermata' (chapter 6). She uses data from single stressor studies of estuarine, coastal and shallow water species and places them within the setting of projected ocean change. She finds that fertilisation in many urchin and bivalve species as well as a number of other invertebrate taxa exhibits a broad tolerance and therefore may not be a suitable indicator to discern the effects of climate change on marine life histories. However, their development stages do exhibit a greater sensitivity to thermal and pH stress. Early embryos of some species are vulnerable within the range of future warming. Multifactorial experiments with marine embryos using methods adopted from ecotoxicology provide a framework to determine the potential effects of climate change on marine biota.

In the seventh chapter 'Climate impact on hypersalinity in an Australian coastal bay', Ribbe (2010) monitors a longer episode of hypersalinity as a consequence of a drying and warming climate in a subtropical east Australian bay: Hervey Bay off the central Queensland coast. He observed the response of the coastal ocean to changes in the hydrological cycle in several Australian coastal systems, including the response to persistent drought conditions resulting in increased salinity and more frequent hypersaline conditions. His first modelling experiments indicate the frequency of hypersaline conditions has increased during the last 20 years. Future routine monitoring of coastal salinity is essential in order to evaluate future impacts of climate change.

For cost-effective long-term climate monitoring of the Indonesian Throughflow (ITF), You et al. (2010) propose an internationally coordinated climate program called PACSWIN in the Indonesian seas and the adjacent region in the eighth chapter 'Indonesian throughflow: PACific Source Water INvestigation (PACSWIN): An international ocean climate program'. The ITF, draining water from the tropical Pacific into the Indian Ocean, plays an important role in both the Australian and global climate. It is reasonable to expect that the potential dramatic climate change associated with the weakening of the ocean circulation in the northern and tropical North Atlantic could also affect the ITF and cause subsequent regional climate change in southeast Asia and Australia because the ITF is a choke point of the global ocean circulation affecting the Earth's climate. The International Nusantara Stratification and Transport (INSTANT) program from 2004–06 measured the ITF simultaneously at several of the primary straits and showed a 20–30% increase of the ITF transport under relatively weak El Niño – Southern Oscillation (ENSO) conditions. This suggests that prior ITF transports could be underestimated. Further simultaneous observations encompassing a complete ENSO cycle are required for a more accurate estimate of the ITF transport. Climate monitoring requires cost-effective, long-term measurement of the ITF on decadal or even centennial time scales. Currently, no such sustainable monitoring program exists for the Indonesian seas and the adjacent region. To address this need, a coordinated international program is necessary.

The IMOS was established as part of the Australian Government's National Collaborative Research Infrastructure Strategy (NCRIS) initially for a five year period to June 2011, with an extension through to June 2013 announced in 2009. In their chapter 'The Australian Integrated Marine Observing System (IMOS) and the regional implementation in New South Wales' (chapter 9), Roughan et al. (2010) introduce the IMOS and its strategic research goals to assemble and provide free, open and timely access to streams of data that support research on the role of the oceans in the climate system and the impact of major boundary currents on continental shelf environments, ecosystems and biodiversity. These authors describe the implementation of IMOS for the NSW node as an example, with some preliminary results.

From a basin view of the South Pacific, Holbrook (2010) discusses the role of baroclinic ocean Rossby waves in sea level and climate in his chapter 'South Pacific Ocean dynamics: potential for enhancing sea level and climate forecasts' (chapter 10). The southwest Pacific boundary converts impinging Rossby wave energy into Kelvin wave transfers to the equatorial zone, contributing to the timing of equatorial wave dynamics and ENSO climate variability. This complex boundary also hosts the poleward-flowing East Australian Current (EAC), which transports heat to the mid-latitudes and moderates and modulates the regional climate of the southwest Pacific. Holbrook (2010) uses wind stress-forced upper ocean dynamics in the South Pacific to investigate interannual to decadal changes in sea level along the New South Wales coast and the relationship with summer rainfall totals across NSW. He found that there is a highly significant relationship between modelled first baroclinic-mode ocean dynamics, the baroclinic component of EAC intensity at 34°S and observed sea level at Fort Denison. Holbrook concludes with a discussion of the large-scale ocean observing and monitoring of extra-tropical Rossby waves and the potential for enhancing future ENSO climate forecasts.

In his chapter 'Future cyclone season characteristics from a thermodynamic model driven by coupled model climate simulations' (chapter 11), McGuffie (2009) uses coupled model climate simulations to examine future cyclone season characteristics under a changing climate.

When he examines an enhanced greenhouse condition with 550 ppmv, results from five CMIP3 models show an overall tendency for more severe cyclone seasons with either an increased length of season or increased maximum potential intensity, although it varies from basin to basin. On a mean case and worst case basis for a 500 ppmv greenhouse scenario, analysis indicates changes in maximum potential intensity of tropical cyclones of roughly half a category and one category respectively. Increases in maximum potential intensity are found to occur differentially across basins and it seems likely that no simple global conclusion can be made on overall intensity from this study, but a consistent prediction for increases in months of most intense cyclone activity of around one category is found.

Climate change requires mitigation and adaptation, which are challenging tasks, and needs input from all levels of government. Climate projections and economic evaluation need to be available at the local level (scales of a few tens of kilometres at most) to ensure optimal local climate investment actions. In their chapter 'Assessing climate change adaption options for local government' (chapter 11), Trück et al. (2010) describe the early stages of their study on adaption options for a local government authority, based on available climate information and perception of the local community. These authors find that, to date, there have been surprisingly few studies at the local level on the economic cost of mitigation and adaptation to climate change effects. They suggest that valuing the non-monetary impacts of climate change via Quadruple Bottom Line analysis that considers the social, economic, environment and governance dimensions as a useful method for identifying the appropriate importance of each impact category. The authors further suggest the use of the so-called Loss Distribution Approach (LDA) in combination with expert judgments and discounted present value analysis as a potential framework for modelling the losses from climate change-impacted hazards at the local scale. For the choice of the optimal discount rate, these authors suggest for the financial dimension to use 'cost of capital' for local governments, while for the social discount rate, it might be appropriate to use clearly expressed lower discount rates as suggested, for example, in the reports by Stern (2006) and Garnaut (2008).

References

Adam M & McBratney A (2008). Monitoring soil carbon and emissions: natural and anthropogenic influences. National Climate Change Monitoring Symposium program, University of Sydney, an unpublished abstract, p14.

Adamson DA, Whetton P & Selkirk PM (1988). An analysis of air temperature records for Macquarie Island: decadal warming, ENSO cooling and southern hemisphere circulation patterns. *Papers and Proceedings of the Royal Society of Tasmania*, 122, 107–12.

Byrne M, Ho M, Nguyen H, Wong E, Soars N, Davis A, Dworjanyn S & Selvakumaraswamy P (2008). Naked abalone and corrupted urchins, impact of climate change on benthic marine invertebrates. National Climate Change Monitoring Symposium program, University of Sydney, an unpublished abstract, p20.

Byrne M (2010). Impact of climate change stressors on marine invertebrate life histories with a focus on the Mollusca and Echinodermata. In Y You & A Henderson-Sellers (Eds), *Climate alert: climate change monitoring and strategy*, (pp169–212). Sydney: Sydney University Press.

Cai W & Cowan T (2006). SAM and regional rainfall in IPCC AR4 models: can anthropogenic forcing account for southwest Western Australian winter rainfall reduction? *Geophysical Research Letters*, 33, L24708, doi:10.1029/2006GL028037.

Chambers DP, Ries JC & Urban TJ (2003). Calibration and verification of Jason-1 using Global Along-Track Residuals with TOPEX. *Marine Geodesy*, 26, 305, doi:10.1080/714044523.

Church JA (2008). Sea-level variability and rise: understanding the past, implications for the future. National Climate Change Monitoring Symposium program, University of Sydney, an unpublished abstract, p18.

Clarke LJ, Robinson SA, Hua Q & Fink D (2008). The influence of climate change on the growth rate of east Antarctic mosses since the 1960s.

National Climate Change Monitoring Symposium program, University of Sydney, an unpublished abstract, p12.
Cleugh H, Wijffels S & May P (2008). An Australian climate change and Earth observatory. National Climate Change Monitoring Symposium program, University of Sydney, an unpublished abstract, page 15.
Cowell P & Goodwin I (2008). Coastal impacts of sea-level rise. National Climate Change Monitoring Symposium program, University of Sydney, an unpublished abstract, p25.
CSIRO & BOM (2007). *Climate change in Australia: technical report 2007*. Melbourne: CSIRO.
Douglas BC (1997). Global sea rise: a redetermination. *Surveys in Geophysics*, 18, 279–92.
England M (2008). Ocean's role in climate and climate change. National Climate Change Monitoring Symposium program, University of Sydney, an unpublished abstract, p23.
Garnaut R (2008). *Garnaut climate change review: final report*. Melbourne: Cambridge University Press.
Gillett NP, Stone DA, Stott PA, Nozawa T, Karpechko AY, Hegerl GC, Wehner MF & Jones PD (2008). Attribution of polar warming to human influence. *Nature Geoscience*, 1, 750–54.
Hansen J, Sato M, Ruedy R, Lo K, Lea WD & Medina-Elizade M (2006). Global temperature change. *PNAS*, 103, 14288–293.
Heil P, Allison I & Lytle VI (1996). Seasonal and interannual variations of the oceanic heat flux under a landfast Antarctic sea ice cover. *Journal of Geophysical Research*, 101, 741–52.
Henderson-Sellers A, Zhang H, Berz G, Emanuel K, Gary W, Landsea C, Holland G, Lighthill J, Shieh S-L, Webster P & McGuffie K (1998). Tropical cyclones and global climate change: a post-IPCC assessment. *Bulletin of American Meteorological Society*, 79, 19–38.
Henderson-Sellers A (2008). How seriously are we talking climate change: monitoring climate change communication. National Climate Change Monitoring Symposium program, University of Sydney, an unpublished abstract, p6.

Henderson-Sellers A (2010). How seriously are we taking climate change: monitoring climate change communication. In Y You & A Henderson-Sellers (Eds), *Climate alert: climate change monitoring and strategy*, (pp28–65). Sydney: Sydney University Press.

Hofmann DJ, Butler JH & Tans PP (2009). A new look at atmospheric carbon dioxide. *Atmospheric Environment*, 43, 2084–86.

Holbrook N (2008). South Pacific Ocean climate dynamics. National Climate Change Monitoring Symposium program, University of Sydney, an unpublished abstract, p14.

Holbrook NJ (2010). South Pacific Ocean dynamics: potential for enhancing sea level and climate forecasts. In Y You & A Henderson-Sellers (Eds), *Climate alert: climate change monitoring and strategy*, (pp313–42). Sydney: University of Sydney Press

IPCC AR4 (2007). Climate change 2007: the physical science basis (summary for policy makers), IPCC.

Jacka TH, Budd WF & Holder A (2004). A further assessment of surface temperature changes at stations in the Antarctic and Southern Ocean, 1949–2002. *Annals of Glaciology*, 39: 331–38.

Jones R, Whetton P, Walsh K & Page C (2001). Future impact of climate variability, climate change and land use change on water resources in the Murray-Darling Basin. Overview and Draft Program of Research, CSIRO.

Knight J, Kennedy JJ, Folland C, Harris G, Jones GS, Palmer S, Parker D, Scaife A & Stott P (2009). Do global temperature trends over the last decade falsify climate prediction? Meteorological Office, UK. Available at: www.metoffice.gov.uk.

Lemke P, Ren J, Alley RB, Allison I, Carrasco J, Flato G, Fujii Y, Kaser G, Mote P, Thomas RH & Zhang T (2007). Observations: changes in snow, ice and frozen ground. In S Solomon, D Qin, M Manning, Z Chen, M Marquis, KB Averyt, M Tignor & HL Miller (Eds), *Climate change 2007: the physical science basis. Contribution of working group I to the Fourth Assessment Report of the Intergovernmental Panel on*

Climate Change. Cambridge and New York: Cambridge University Press.
Lough J (2008). A changing climate for the Great Barrier Reef: physical observations and biological responses. National Climate Change Monitoring Symposium program, an unpublished abstract, page 16.
Lough J, Bainbridge S, Berkelmans R & Steinberg C (2010). Physical monitoring of the Great Barrier Reef to understand ecological responses to climate change. In Y You & A Henderson-Sellers (Eds), *Climate alert: climate change monitoring and strategy*, (pp66–110). Sydney: Sydney University Press.
Lucas C, Hennessey K, Mills G & Bathols J (2007). Bushfire weather in southeast Australia: recent trends and projected climate change impacts, consultancy report prepared for the Climate Institute of Australia.
Lynch A, Nicholls N, Alexander L & Griggs D (2008). Defining the impacts of climate change on extreme events, report prepared for the Garnaut Climate Change Review.
Lyster R (2008). Measuring and monitoring: at the heart of climate law. National Climate Change Monitoring Symposium program, University of Sydney, an unpublished abstract, p9.
Karstensen J (2008). Why do we care about oxygen minimum zone? National Climate Change Monitoring Symposium program, University of Sydney, an unpublished abstract, p24.
McGuffie K (2010). Future cyclone season characteristics from a thermodynamic model driven by coupled model climate simulations. In Y You & A Henderson-Sellers (Eds), *Climate alert: climate change monitoring and strategy*, (pp343–61). Sydney: Sydney University Press.
Murphy BF & Timbal B (2008). A review of recent climate variability and climate change in southeastern Australia. *International Journal of Climatology*, 28(7), 859–80.
National Snow and Ice Center (2008). Antarctic ice shelf disintegration underscores a warming world. Press Room, 25 March 2008.

Nicholls N (2004). The changing nature of Australian droughts. *Climatic Change*, 63, 323–36.

Overland J, Turner J, Francis J, Gillett N, Marshall G & Tjernstrom M (2008). The Arctic and Antarctic: two faces of climate change. *EOS*, 89, 177–84.

Power S & Smith IN (2007). Weakening of the walker circulation and apparent dominance of El Niño both reach record levels, but has ENSO really changed? *Geophysical Research Letters*, 34, L18702, doi:10.1029/2007GL030854.

Press AJ (2008). Antarctic climate science: Australia's efforts in the icy frontier and what we can learn about the future. National Climate Change Monitoring Symposium program, University of Sydney, an unpublished abstract, p5.

Ribbe J (2008). Observing the impact of climate variability on Hervey Bay. National Climate Change Monitoring Symposium program, University of Sydney, an unpublished abstract, p26.

Ribbe J (2010). Climate impact on hypersalinity in an Australian coastal bay. In Y You & A Henderson-Sellers (Eds), *Climate alert: climate change monitoring and strategy*, (pp213–37). Sydney: Sydney University Press.

Riddle MJ, Tilbrook B, Snape I, Paterson K, Berry K & Gore DB (2008). Deep, cold-water coral communities in Antarctica may be early indicators of the wider impacts of ocean acidification on coral-based ecosystems. National Climate Change Monitoring Symposium program, University of Sydney, an unpublished abstract, p7.

Robertson R (2008). Are the deep waters of the Weddell Sea still warming? National Climate Change Monitoring Symposium program, University of Sydney, an unpublished abstract, p8.

Robertson R (2010). Are the deep waters of the Weddell Sea still warming? In Y You & A Henderson-Sellers (Eds), *Climate alert: climate change monitoring and strategy*, (pp111–41). Sydney: Sydney University Press.

Roughan M, Suthers I & Meyers G (2010). The Australian Integrated Marine Observing System (IMOS) and the regional implementation in New South Wales. In Y You & A Henderson-Sellers (Eds), *Climate alert: climate change monitoring and strategy*, (pp299–312). Sydney: Sydney University Press.

Ruddell AR (2006). An inventory of present glaciers on Heard Island and their historical variation. In K Green & EJ Woehler (Eds), *Heard Island: Southern Ocean sentinel*, (pp28–51). Sydney: Surrey Beatty & Sons.

Scambos TA, Bohlander JA, Shuman CA & Skvarca P (2004). Glacier acceleration and thinning after ice shelf collapse in the Larsen B embayment, Antarctica. *Geophysical Research Letters*, 31, L18402, doi:10.1029/2004GL020670.

Suthers I & Meyers G (2008). The Australian Integrated Marine Observing System. National Climate Change Monitoring Symposium program, University of Sydney, an unpublished abstract, p19.

Trück S, Bradford W, Henderson-Sellers A, Mathew S, Scott J, Street M & Taplin R (2010). Assessing climate change adaption options for local government. In Y You & A Henderson-Sellers (Eds), *Climate alert: climate change monitoring and strategy*, (pp362–99). Sydney: Sydney University Press.

Turner J, Comiso JC, Marshall GJ, Lachlan-Cope TA, Bracegirdle T, Maksym T et al. (2009). Non-annular atmospheric circulation change induced by stratospheric ozone depletion and its role in the recent increase of Antarctic sea ice extent. *Geophysical Research Letters*, 36, L08502, doi:10.1029/2009GL037524.

Walland D (2008). Monitoring the atmosphere and hydrology. National Climate Change Monitoring Symposium program, University of Sydney, an unpublished abstract, p22.

Walland D (2010). Monitoring the atmosphere and hydrology. In Y You & A Henderson-Sellers (Eds), *Climate alert: climate change monitoring and strategy*, (pp142–68). Sydney: Sydney University Press.

Walter KM, Zimov SA, Chanton JP, Verbyla D & Chapin FS (2006). Methane bubbling from Siberian thaw lakes as a positive feedback to climate warming. *Nature*, 443, 71–75, doi:10.1038/nature05040.
West Australia Water Corporation (2008). Annual streamflows into Perth's dams. Data supplied by the West Australia Water Corporation to the Garnaut Climate Change Review.
Wouters B, Chambers D & Schrama EJO (2008). GRACE observes small-scale mass loss in Greenland. *Geophysical Research Letters*, 35, L20501, doi:10.1029/2008GL034816.
You Y, Huber M, Müller D, Poulsen CJ & Ribbe J (2008). The previous global warming. National Climate Change Monitoring Symposium program, University of Sydney, an unpublished abstract, p13.
You Y, Huber M, Muller RD, Poulsen CJ & Ribbe J (2009). Simulation of the Middle Miocene Climate Optimum. *Geophysical Research Letters*, 36, L04702, doi:10.1029/2008GL036571.
You Y (2008). Introducing an international ocean climate program, PACSWIN, and using submarine cable for cost-effective long-term climate monitoring. National Climate Change Monitoring Symposium program, University of Sydney, an unpublished abstract, p28.
You Y, Rossby T, Zenk W, Ilahude AG, Fukasawa M, Davis R et al. (2010). Indonesian Throughflow: PACific Source Water INvestigation (PACSWIN): An international ocean climate program. In Y You & A Henderson-Sellers (Eds), *Climate alert: climate change monitoring and strategy*, (pp238–98). Sydney: Sydney University Press.
Zhang J (2007). Increasing Antarctic sea ice under warming atmospheric and oceanic conditions. *Journal of Climate*, 20, 2515–29.
Zimov SA, Schuur EA & Chapin FS (2006). Climate change: permafrost and the global carbon budget. *Science*, 312, 1612–13, doi:10.1126/science.1128908.

2
How seriously are we taking climate change? Monitoring climate change communication

Ann Henderson-Sellers

*Abstract**

Scientists and now most of our politicians know that climate change is real and requires action but, despite the award of the 2007 Nobel Peace Prize jointly to the IPCC and Al Gore for their communication efforts, global warming still seems to suffer from a credibility shortfall. Those of us who try to disseminate strong messages of urgency and certainty into public arenas can be dismayed by, for example, our national newspaper, *The Australian*, running anti-greenhouse stories day after day and Robyn Williams, an excellent science journalist and ABC editor, giving two Ockham's Razor programs to a 'should-know-better' warming debater. Here I try to explore some of the reasons why human-caused climate change is still to be acted upon in Australia. I ask – why do greenhouse 'nay-sayers' still get airtime in our mass media and apparently credibility with the general public. One part of the reason for the lack of effective climate change communication may be the 'code' we use: the very terminology we have developed to talk about this diabolical challenge for humanity may be slowing our ability to galvanise action. Another aspect is the extreme difficulty of separating a personal, or citizen's, viewpoint from the scientific voice. Finally, I believe there is an unhealthy polarisation into 'believers' and, the only alternative of 'non-believers', which is excluding scientific scepticism of a type not previously recognised – concern that the IPCC consensus is too conservative. In the

* Chapter originally submitted in December 2008.

light of these reasons for the lack of urgency on greenhouse action, I conclude with a new proposal for monitoring and hence, hopefully, improving climate change communication in Australia.

1. Mainstreaming human-induced climate change – here, now we now hear

From the "Eighties" to "Noughties" global warming transitioned from a scientific possibility to being termed 'unequivocal'. Throughout this quarter of a century development, weather services, UN agencies and, most importantly, the world's media have sought for 'balance' giving equal platforms to advocates and greenhouse dissenters. Now there is agreement: the Intergovernmental Panel on Climate Change (IPCC) has become mainstream. This was clear at the end of 2007 when, within two weeks the IPCC jointly with Al Gore received the, well-deserved, Nobel Peace Prize for 2007 and presented the results of its Fourth Assessment to the government delegations at the Bali Conference of Parties to the UN Framework Convention on Climate Change (UNFCCC).

The 2007 IPCC Fourth Assessment Report (AR4), compiled by several thousand climate scientists and representing a consensus view of the best available peer-reviewed science, has unequivocally concluded that our climate is warming rapidly, and that there is now at least 90% certainty that this is primarily due to human activities. The amount of carbon dioxide in our atmosphere now far exceeds the natural range of the past 650,000 years, and it is rising alarmingly quickly, increasing by 2 ppm (parts per million by volume) per year, due to human activity (Global Carbon Project 2008). If this trend is not halted soon, many millions of people will be at risk from extreme events such as heat waves, drought, fire, floods and storms; our coasts and cities will be threatened by rising sea levels; vector, water and food borne diseases will spread rapidly; and many ecosystems, plants and animal species will be extinct or in very serious danger of extinction (IPCC WG2 SPM 2007).

The current round of UNFCCC negotiations for a new global climate treaty is underway. The prime goal of this new regime must be to limit global warming to no more than 2°C above the pre-industrial temperature, a limit that has already been identified by many as our best estimate of the threshold above which "dangerous anthropogenic interference with the climate" occurs (e.g. Schellnhuber et al. 2006). Therefore, the European Union, South Africa and a number of other countries have formally adopted this 2°C maximum allowed temperature rise. Based on current scientific understanding, achieving no more than a 2°C temperature increase requires that global greenhouse gas emissions be reduced by at least 50% below their 1990 levels by the year 2050. In the long run, greenhouse gas concentrations need to be stabilised at a level well below 450 ppm CO_2-e (CO_2 equivalent concentration). In order to stay below 2°C, the peak of global emissions must be very soon and then emissions must drop rapidly i.e. already in decline in the next 10 to 15 years. Recent reports from the UK Met Office and other climate modelling centres estimated that even stabilisation at 450 ppm CO_2-e gives us between 20–40% chance of avoiding a 2°C warming. *There is no time to lose.*

1.1 Why are global warming warnings weak?

In the final months of the Intergovernmental Panel on Climate Change's Fourth Assessment reporting in 2007, the world's three leading climate science agencies (the Global Climate Observing System Programme [GCOS], the World Climate Research Programme [WCRP], and the International Geosphere-Biosphere Programme [IGBP]) asked people directly and intimately involved with the report for their views on how the process had gone and some of the key issues it raised. These agencies also convened a workshop in Sydney in October 2007 on 'Learning from the IPCC Fourth Assessment Report', for which I drafted an outline of a workshop paper, based entirely on responses to the survey.

Recently, I went back to check whether the first glimpse we received of the "real gut feeling" of some of the IPCC authors and coordinators as they finished up the massive task of completing the Fourth Assessment Report (AR4) held steadfast throughout the meeting process and the subsequent re-writings of views and clarification of expressed opinions. I did this because I believe it is essential for the climate change research community to be transparent and honest about what it can and cannot deliver and how, if ever, current inadequacies can be resolved.

Although the full details of the questionnaire and the replies have now been restricted an early summary can still be found online.[1] In a web article (Henderson-Sellers 2008) I report what the IPCC éminences grises said under the following headings:

- Urgent policy issues that climate change research must tackle
- Serious inadequacies in climate change prediction that are of real concern
- Climate change research topics identified for immediate action
- International organisation issues, especially regarding the IPCC
- Institutional/ infrastructure issues hamper climate change information delivery.

It seems to me, even one year later, that the urgency and forthrightness of many of these comments still hold and need to be shared, discussed and acted upon. Those that concern me most include the simple, but dastardly, statement that "until and unless major oscillations in the Earth System can be predicted to the extent that they are predictable, regional climate is not a well defined problem. *It may never be.* The rush to emphasise regional climate does not have a scientifically sound basis." This is underscored in another quote: "adding complexity to models, when some basic elements are not working right (e.g. the hydrological cycle) *is not sound science.*" (The quotes are from IPCC Lead Authors, see Henderson-Sellers [2008] but the italics were added for emphasis here).

There are other, more organisational but still very pertinent views such as, "future IPCC reports must be more focused, shorter and timely"; "human-ware is very much a depleted commodity"; and "The Fourth Assessment Report is rather weak at including the latest research and thereby

---

[1] wcrp.ipsl.jussieu.fr/Documents/ACC/Sydney2007/Sydney07_SurveySummary_v1.5_240807.pdf (30/7/09)

is losing credibility in the science community."[2] Finally there is a cry from the heart – remember these quotes all come from IPCC lead authors, at the end of a five-year process of painstaking assessment, "there are too many committees and working groups and way too much time spent talking!"

The GCOS-WCRP-IGBP Workshop in October 2007 specifically tasked its committee and discussion leaders to tackle the tough questions about the IPCC and especially how AR4 might have been improved. These leaders were told to focus on the following climate change questions:

- What important processes/elements are missing from existing knowledge?
- What are the most uncertain of the important processes that are currently included?
- What modifications in current models will be needed to address these gaps?
- What process studies are needed to reduce the uncertainties?
- What observations are needed to reduce the uncertainties?
- What additional observations are needed, and what improvements to current observations are most critical?
- Are existing observations used effectively? If not, what mechanisms need to be established to increase the use and usefulness of observational data?

Leaders were asked to achieve an order of priority on the five most important recommendations about the IPCC and its processes by means of which improvements for future IPCC assessments might be achieved.

Did the final workshop outputs genuinely deliver these messages as clearly as they were stated in the responses to our questionnaire and at the Workshop? Did we say: "The IPCC was designed 20 years ago ... we absolutely need a complete re-think"; and we need "more time to learn from each other how to do integrative research to answer key questions"? My personal view is that we did not (see Doherty et al. 2009). Perhaps due to the

---

[2] The word "weak" in this quote is worthy of some clarification: it is IPCC's rules of assessment that are problematic. These mean that "latest research" gathered over the preceding five years excludes about the "last 12 months".

natural desire to tone down criticism in written documents and perhaps because some of us wonder what else can be said – e.g. WG I is "becoming a bit repetitive and may have little original to say other than 'we told you so last time and mean it even more now!'"

In some parts of the climate change community there is an unhealthy fear of misquoting by global warming 'deniers'. We are hesitant to stress a comment such as "The Fourth Assessment Report missed doing this owing essentially to the timelines that were arranged." Another intriguing example of this fear is that the original suggestion was to entitle the Sydney workshop, "What did the IPCC get wrong?" This proposal was quickly squashed in the corridors of the World Meteorological Organisation lest the anti-greenhouse lobby picked it up and repeated it as criticism of the IPCC.

Climate change communication entered a new and different regime with the publication of the *IPCC Fourth Assessment Report*. There is no longer any question about "whether" human activities are changing the climate; instead research must tackle the urgent questions of: "how fast?"; "with what impacts?" and "what responses are needed?" Climate change researchers cannot hide behind the need to improve models and observations any longer. Answers are now being demanded faster than, and at higher resolutions than, research can deliver. In my view we must quickly overcome our *reluctance to state the magnitude of the challenges to be overcome* before we can reliably address "how fast", "with what regional scale impacts" and "what responses are needed".

1.2 Australia, Garnaut and climate change clarity demands

All Australia's leading climate scientists recognise the truth of global warming and the urgency with which it must be acted upon. Many of us are urging national political and business leaders to support the Garnaut targets as a minimum requirement for Australia's contribution to a fair and effective global climate agreement. Barry Brook put it well in *Australasian Science* in May 2008 (Brook 2008) when he said, "Active and forthright public communication of science is not only an obligation of scientists, but a critical necessity. This is especially true for climate change and environmental

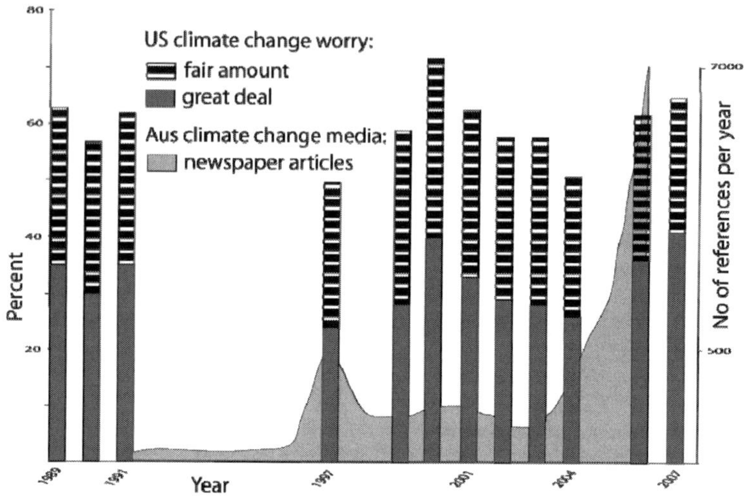

Figure 1 Community awareness of climate change issues shown as a comparison of a set of Gallup polls on environment in the USA from 1989 to 2007 and a count of newspaper articles on climate change in Australia from 1991 to 2006. Gallup asked 'tell me how much you personally worry about the greenhouse effect or global warming' – shown are % replies (left scale) 'a great deal' (dark grey) and 'a fair amount' (stripes) – after Saad (2007) and quoted in Staudt (2008). Power et al. (2008) counted the number of references (right scale – not linear) to 'global warming' or 'climate change' in three of Australia's main daily newspapers – *The Australian, The Age* and *The West Australian.*

sustainability, where we are perilously close to running out of time." The Garnaut Report of climate change costs has dominated 2008.

As highlighted by the Garnaut Report: "... analysis suggests that a global objective of 450 ppm, with discussion of transition to 400 ppm once the 450 ppm goal is being approached with confidence, would better suit Australian interests." The Australian Garnaut Report concluded that an emission reduction target for our nation of 25% below 1990 levels by 2020 would be an equitable contribution to achieving this outcome (Garnaut 2008a). This statement is consistent with the climate science agreed to in the IPCC AR4. *There is no disagreement in the reputable science community regarding these targets.*

The Australian position in the UNFCCC recognises that a number of countries are seeking to define an agreement around a 50% reduction in global emissions by 2050 (Government of Australia 2008) and highlights that:

> a two degree increase in the global average temperature above 1980–1999 levels ... could already see widespread global impacts that would alter in severity from region to region. Australia, which has the driest and most variable climate of the inhabited continents, is projected to be among the first regions to suffer from the severe impacts of climate change.
>
> In considering what constitutes dangerous anthropogenic interference with the climate system, Parties should agree on cooperative approaches that minimise the impacts of climate change at the *lowest achievable stabilisation goal.* [Emphasis added]

In Australia as in most other nations, global warming has risen in the public consciousness. Climate change, now taken to mean global and greenhouse warming (but see Section 3), has been mentioned much more frequently in the media and by politicians over the past couple of years than previously. For example Power et al. (2008) show that there has been an increase in media and political attention given to climate change and that there has also been a corresponding increase in the frequency that people seek out more climate change information (Figure 1). More broadly, Australian scientists carry out more than 12,000 individual research projects in all areas, over 50% of which are funded by taxes. However, in the combined public announcements of the universities and science agencies, only about 1200 outcomes are reported (Cribb 2008). I suspect that the ratio is a little better for climate change but probably not nearly as high as it could usefully be. Here I explore some of the reasons why climate scientists may be diffident about speaking publicly.

People, politicians, policy makers (and breakers) all contend for clear greenhouse communication. It is by no means certain that even now most people are 'sure' about the urgency required for climate change action. For example Figure 1 also shows Gallup poll results for the USA in which concern about greenhouse warming seems to remain roughly constant over time. Climate change experts cannot any longer pretend to be merely passive bystanders. While we continue to hope that good research will prevail in the end, we can no longer avoid considering the claim that, to coin a phrase, 'the end is nigh'! The denialist spin has to be stopped before it does any more damage. In this paper I consider some of the reasons why we are still failing to get across our message with clarity. In Section 2 I explore the difficulties by looking into three cases for which I am personally and closely acquainted with those involved. In Section 3 I take a broader Australian view while the last section contains some conclusions and recommendations for improving climate change communication.

2. Can we all do more to be clearer?

Personally, I have studied climate change for around 35 years (beginning with my PhD in the UK in 1973) and have participated in all four Intergovernmental Panel on Climate Change (IPCC) assessments including as one of the convening lead authors for the *Second Assessment Report* in 1995–96; as an author, a contributing scientist and as a reviewer. I have also worked in university and government laboratories in a number of developed countries on climate change including dissemination of climate change science findings to the public and the mass media (e.g. Henderson-Sellers 1998). Between 2005 and 2008 I led the World Climate Research Programme (WCRP), the international agency organising climate change research in all nations including developing countries' needs and especially focusing on the science underpinning the IPCC Working Group I Reports (Doherty et al. 2009). My time leading the WCRP coincided with the *IPCC Fourth Assessment Report* (AR4) and the award of the 2007 Nobel Peace Prize (WCRP 2008).

However, in common with most of the climate science community, I know how easy it is even for clear communications to be used to argue exactly the opposite case from the one its authors believed they were propounding. One personal example, which I find quite irritating despite the fact that the quotes are not incorrect, is the May 27 2008 blog piece entitled "Science can't prove man's 'carbon footprint' is tipping balance to global warming". This copiously cites my textbook on climate modelling (McGuffie & Henderson-Sellers 2005) in a forlorn effort to disprove global warming.

In this section I consider three specific and intentionally personal examples of what constitutes, in my opinion, rather poor communication of the issues surrounding human-induced climate change. These are:

(i) my own failure to convey my meaning regarding comments on the broadcasting of "The Great Global Warming Swindle" in Australia

(ii) the ABC radio's weighting of airtime on greenhouse issues in the program "Ockham's Razor"

(iii) a media argument between a senior academic at UNSW and climate change 'deniers' on the merits or otherwise of the IPCC process.

These examples are picked because they are very recent (occurring within the past six months), climate change communication topical and because they have either involved me or people whom I know very well. I have known the UNSW Professor, Andrew Pitman, for the whole of his academic life and have participated in his learning and also authored numerous papers and projects with him (e.g. Henderson-Sellers & Pitman 1991 and Henderson-Sellers et al. 2003). I have also worked with both protagonists in the Ockham's Razor 'debate'. I first met Professor Stephen Schneider in 1976 and have collaborated with him frequently since then in, among other ways, all four of the IPCC Assessments. I was a member of the Australian Science and Technology Council (ASTEC) at the same time as Professor Don Aitkin and worked with him on national reports for the then Labor governments of Hawke and Keating. So, in all three cases I know well and respect the integrity of the speakers/writers and honestly believe that all of us were trying very hard to convey our best understanding of the current climate change challenge. Despite this, I believe these illustrative miscommunications have not only been bad in themselves but, more importantly, I think each has negatively affected our ability to get across clear messages to the broader public during this time period and into the future.

2.1 How I managed to confuse a class of students by using excessive emotion in a talk

On 18 August 2008 a second-year student at Macquarie University whom I shall call Chuck emailed me with comments on a lecture I had delivered to his class the previous week. In my lecture I mentioned the British TV program "The Great Global Warming Swindle" and encouraged the class to read Ofcom's response to complaints received in relation to its broadcast in the UK. On this Chuck sent me both good and bad news.

First the good (for me) news (all quotes):

> Your passion on the subject was fantastic and our class certainly was privileged to have a person of such a high reputation presenting one of our lectures;
> I was very impressed with your research and your lecture encouraged me to move forward and change my lifestyle;
> Your lecture helped me to straighten out a few incorrect theories I had in my mind.

So far, so good, a student open and willing to listen, which made hearing the bad news (all quotes) all the more depressing:

> What did surprise me was your reaction to the ABC broadcasting the program;
> I thought it was *scientific* to identify, describe, investigate and explain all aspects of a particular issue or matter;
> Why would you think that it is invalid and 'disgusting' for the ABC to broadcast the program? (I had actually used the word 'disgusting' in relation to the purchasing of this program with Aussie tax-payer funds, when other quality material could have been bought, not about the program itself);
> I thought that university was supposed to expand my thoughts and ideas ... Isn't it the job of scientists to question pre-existing

> truths, not to use 'bully' tactics against scientists (and the public) who disagree with the common point of view?

So in a single lecture I had managed to enrage a previously committed environmentally minded student who felt (sadly) that I was using (or at least was an example of others who teach him using) bully tactics to enforce my view rather than expanding his horizons and allowing scientific 'truth' to emerge in his mind.

2.2 How an excellent science journalist weighed a greenhouse contrarian's view as twice as important as that of an IPCC lead author and established scientist

In the ABC radio program "Ockham's Razor" in May 2008 Robyn Williams oddly gave a double slot to Professor Don Aitkin (27 April and 4 May 2008) to present an 'against greenhouse' view while only allowing Professor Stephen Schneider a single 'slot' to reply (18 May 2008). This was particularly unbalanced since Aitkin's talks were slow and not very dense whereas Schneider's was a fast and furious attempt to try to cover all the points raised by Aitkin in his two-part contrarian discussion. I have no real objection to even equal billing (although the vast majority of views are 'for' greenhouse these days). However, I do question the two-to-one weighting in favour of contrarian discussion. Why did Williams and the ABC believe that this was the most valuable investment of airtime on a public broadcaster?

In introducing Aitkin, Williams himself says,

> Though nine out of 10 Australians are said to be alarmed at climate change, 10% think differently, and Professor Aitkin is one of them. Professor Aitkin gave a lecture on climate to the Planning Institute of Australia, *A Cool Look at Global Warming*. That was a couple of weeks ago, and I thought you might like to hear some of his thoughts, recast for *Ockham's Razor*.

Aitkin's talks can be summarised in these quotations:

> Everyone is entitled to know why such [carbon] taxation is necessary;
>
> [The] 'authority', [of the IPCC] in my view [it] has little place either in science or in a democracy;

> [IPCC] is not, to put it simply, a disinterested body;
> [I am] pretty sure that the evidence to support carbon taxing is not strong;
> I myself would like to see climate change studied more seriously, but without any assumption that human-induced warming was at the heart of it;
> [Australia faces] immediate problems in finding and managing abundant fresh water supplies;
> my strategy is … to unlock the creative impulse in every child and every adult;
> creative society will sit a little more lightly on the planet than an acquisitive, materialist one.

So, it is easy to see that Aitkin is for sustainability and lower environmental footprints i.e. he is in greenhouse terms, a 'good guy'. Also, Aitkin had just given this same talk to a public audience so Ockham's Razor was not his only forum and (likely) Williams had already heard feedback.

For Schneider the task was harder but he rose to Aitkin's challenges in these words:

> the IPCC Fourth Assessment Report said that warming is 'unequivocal';
> the IPCC attempts to be an antidote to this frequent misrepresentative presentation of climate change science as largely speculative.

He responds directly to Aitkin's question 'does it matter if it warms a little?' by saying, "a 2–3 degree warming occurs, the IPCC estimated that there's about a 50–50 chance that approximately 20% to 30% of species known in the world could face extinction."

Schneider replies directly to Aikin's challenge of frequent "use [of] adverbs like 'likely' or 'highly likely'" by saying, "The terms used [by the IPCC] reflect a specified range of certainty, anywhere from extremely unlikely, which is defined as less than 5% chance, to very likely, which is a greater than 90% or extremely likely, which is a greater than 95% chance." Most importantly he replies to Aitkin's challenge about taxation in these words:

> the likelihood of warming beyond a few degrees before the end of this century and its associated impacts, is a better than even bet;
> few security agencies … accept such high odds of potentially dangerous outcomes without implementing hedging strategies;
> risk-averse principles that lead [us] to minimise downside risks of uncertainty;
> macro-economically it's not a very large reduction in the growth rate of the economy to get unhooked from fossil fuels this century, it will be a big blow to those who mine coal or make overweight cars and trucks; but
> we can't hold the sustainability agenda of the planet hostage to maintaining precisely the same jobs that were always there for everybody.

So, Schneider too is arguing for a shift away from our current economy and towards a greatly reduced environmental footprint.

How did two very articulate academics end up debating the need to reduce our environmental footprint when both believe in its virtue? Why did the ABC bring in an authority from the USA to counter Aitkin? And, most pertinently, is there no other view on the climate change challenge that Robyn Williams could perhaps more helpfully have exposed in one of these three (arranged to be wide-ranging) radio programs?

2.3 Why a sensible senior academic ended up on the receiving end of mud hurling in the mass media

On Tuesday 9 September 2008 *The Australian* newspaper ran a piece entitled, "Climate case built on thin foundation" by John McLean.[3] In this opinion piece McLean argued that Professor Ross Garnaut (who had delivered the second part of his Interim Report to government the previous Friday, [Garnaut 2008b]) had made a serious mistake in taking as a starting point the 'claims' (McLean's word) made in the Fourth Assessment Report of the Intergovernmental Panel on Climate Change. McLean suggests that assuming these 'claims' to be correct is foolishness and goes on to suggest

[3] See www.theaustralian.news.com.au/story/0,,24315169-17803,00.html (30/7/09).

that if Garnaut had carefully considered these so-called IPCC claims, "he would have found a very troubling picture". Here is one example of an individual's (a denier's) opinion being raised in opposition to that of many thousands of scientists with claiming equality. How can the media (or anyone reading) believe this to be a sensible 'balance'?

The next day Professor Andrew Pitman the Co-director, Climate Change Research Centre, UNSW had a Letter to the Editor published in *The Australian* entitled, "Most talked about: climate credibility".[4] Pitman's spirited defence of the IPCC and global warming reality ends, "Garnaut had a choice: trust rigorous and independent scientific excellence or a self-proclaimed climate data analyst who is not part of any community contributing to our understanding of the warming trends we observe." He goes on to applaud Garnaut and criticise the Australian Climate Science Coalition that uses " 'science' in the same sense as 'creation-science' and the flat-earth society."

Of course Pitman was immediately attacked. The letters blog for the next day carry a variety of responses that challenge Pitman's credentials (asking about the value of a degree in Geography), his fervour, his tone (described as 'haughty') and the style of his arguments (excessive use of the term 'denier'). The tone of some of the nastier letter/blog comments is quite unpleasant and personal. One blogger asks whether he [Pitman] has ever questioned the role of CO_2 in leading to climate change.

I mention these challenges because I do know a fair amount about Pitman's research and publication record, in part because I have tutored him, supervised his first and second degrees and ultimately published with him. His degrees in "Geography" included some tough atmospheric and climatic courses and a challenging research assignment in the climate modelling branch of the Canadian government's Atmospheric Environment Service (AES) during his PhD. Also, Pitman *has* challenged the climate role of CO_2, or more precisely, he is among a small number of academics who have examined the relative importance of greenhouse gases as compared to land-

---

[4] See blogs.theaustralian.news.com.au/letters/index.php/theaustralian/comments/ipcc_has_the_most_rigorous_assessment_on_earth/ (30/7/09).

use change (e.g. Zhao & Pitman 2002). In the case of the latter he has shown that land changes can (depending on the nature and extent of the change) contribute either a warming or a cooling climatic signal. Of course, if you are a greenhouse denier you almost certainly will think that I am biased in Pitman's favour and perhaps this is so. However, my point is that again the expert (this time Professor Pitman) is criticised for being biased rather than praised for being knowledgeable.

In all three examples described here I believe serious miscommunication occurred. I conclude these were bad in themselves and also that they have negatively affected our ability to get across clear messages to the broader public during this time period and in the future. There would seem to be resonance in Australia with the hesitancy, even fear as described in Section 1, on the part of the international climate change community to state their views strongly and clearly. *This fear of speaking climate change truths is, I believe, very disturbing.*

3. Can national discussions be clearer? Garnaut and CPR in Australia

From these three personal and intentionally selected examples in Section 2 I surmise the following about Australian climate change communication:

(i) Australians wish to hear 'contrarian' greenhouse views and seem to want to evaluate the pros and cons for themselves; and

(ii) expressed vehemence by accepted experts against global warming deniers or in favour of their colleagues seems to be viewed as threatening or bullying and diminishes the authority of that expert and maybe all experts. I explore these two topics in Sections 3 and 4 respectively.

3.1 How Ross Garnaut's important greenhouse message of Friday 5 September was confused by poor media communication

Building on his earlier Interim Report (Garnaut 2008a) released on the 4 July, Professor Garnaut used his 5/9/08 National Press Club presentation to explain his recommendations and why his review had 'very reluctantly' come to the conclusion that tougher reductions in greenhouse gas emissions were not possible either for Australia or for the world at this time.

The Press Club address ('Targets and trajectories') is eloquent, with facts and conclusions correctly argued, and the *Supplementary Draft Report* on targets and trajectories (Garnaut 2008b) is also well explained. In summary Garnaut says that, in his opinion, a global goal of CO_2-e 450 ppm in the atmosphere is not immediately feasible; that therefore Australia's efforts should be directed towards stabilisation of global greenhouse gas concentrations at 550 ppm as a first step towards an ultimate goal of 450 ppm; and that the cost of both is relatively small for Australia – being 1.1% of our Gross Domestic Product (GDP) by 2020 for 550 ppm and 1.6% GDP by 2020 for his 450 ppm scenario. Personally, I do not agree with his conclusion and wish that he had recommended tougher cuts but that is not the point of

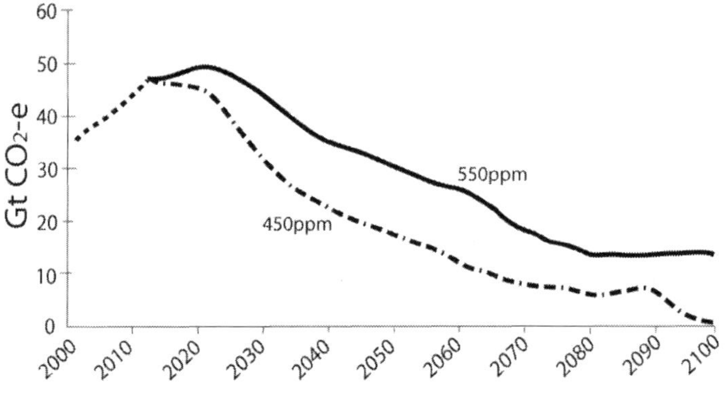

Figure 2 Global emissions trajectories for Garnaut's 550 ppm and 450 ppm atmospheric concentrations. Dotted is observed to date followed by Garnaut's two scenarios for emissions aiming at target atmospheric concentrations stabilising 550 ppm (solid: higher target) and 450 ppm (dashed: lower target) – redrawn from figure 5.1 of Garnaut (2008b, p15, copyright Commonwealth of Australia, reproduced by permission).

this critique. Here I wish only to underline how even a very careful speech and *meticulous report can be betrayed by a sloppy media release.*

The first sentence of the media release (reproduced in Appendix 1) is incorrect. It says (sentence 1), "Australia should put its strongest possible efforts into securing a global agreement to limit emissions to no more than 550 parts per million CO_2-e" (my underlining). But it means, Australia should put its strongest possible efforts into securing a global agreement to limit emissions such that the concentration (or burden) of greenhouse gases in the atmosphere never rises any higher than 550 parts per million CO_2-e. In other words it is NOT the emissions that must be limited to 550 but the resulting total concentration. This is a fundamental mistake and one that should never have been allowed to confuse the opening sentence of a text destined to influence all media coverage of such an important report release.

The media release goes on (in sentence 3) to say, "a global goal of 450 ppm is not immediately feasible, given that emissions are currently around 455 ppm CO_2-e and rising rapidly" (my underlining). This is also wrong, it means, a global goal of 450 ppm is not immediately feasible, given that the atmospheric burden (or concentration) is currently around 455 ppm CO_2-e and rising rapidly.

In view of the fact that Professor Garnaut describes unmitigated greenhouse gas emissions as leading to "a catastrophic outcome for human civilisation and most ecosystems" (media release, last but one paragraph), it is a terrible shame that the release itself does not capture the careful language of the Report and the text of Professor Garnaut's Press Club address.[5] The September Report is not confusing, for example figure 5.1 (reproduced as Figure 2 here) of this report clearly shows the emissions that Garnaut describes as resulting in his two options of ultimate atmospheric burdens of 450 ppm and 550 ppm. I believe that had this graph been included in the media release there would have been a much smaller chance of confusion in subsequent discussion in the mass media and hence the wider Australian public.

---

[5] Garnaut's media release with the *Final Report* (dated 30 September) is much clearer.

My reason for taking issue with these errors is that the main message of the Garnaut Report on 5 September was about (a) strategy (550 ppm first not 450 ppm); (b) its cost (1.1% GDP rather than 1.6% by 2020 respectively) and (c) apportioning cuts in terms of per capita rather than overall national targets (10% total reduction by 2020 being restated as a 30% per capita cut by 2020 and an 80% overall reduction (90% per capita) by 2050 over 2000 levels). This arithmetic is puzzling to work through without an extra distraction of the confusion between emissions and ultimate atmospheric burden (or concentration) of greenhouse gases.

In my view, very sadly, the media release put out by the Review on 5 September contains errors and poor descriptions that render a wide public understanding of a complicated argument more challenging.

Table 1 Global warming described in other terminology might be easier to understand (inspired by Hassol 2008).

Term	Dictionary definition (from Oxford English Dictionary Online 2008)	Problem with term for most people in most communities	Possible improved terminology
Anthropogenic/ generated (increases in greenhouse gases)	Adjective (chiefly of pollution) originating in human activity	Meaning unclear: not a commonly used word. Even when recognised uncertainty exists about whether it means only due to people or due to 'natural' as well as human-created	Human-caused, caused by peoples' actions
Bias (e.g. in data)	Noun inclination or prejudice in favour of a particular person, thing, or viewpoint	Means manipulated or a distortion to most people	Addition (or subtraction) of an amount
Debate (about climate change)	Noun 1 formal discussions in a public meeting or legislature, in which opposing arguments are presented. 2 an argument	Suggests balanced (or aggressive) discussion. Often implies equal weight for and against i.e. in which a regular	Description of well-founded information (facts)

Term	Dictionary definition (from Oxford English Dictionary Online 2008)	Problem with term for most people in most communities	Possible improved terminology
		person cannot yet reasonably draw conclusions	
Enhance / enhancement	Verb increase the quality, value, or extent of	Generally taken to mean improvement	Make larger (perhaps even - make worse)
Positive	Adjective 2 expressing or implying affirmation, agreement, or permission	Good (or nice) thing	Exacerbating, exaggerating
Positive trend	Adjective 5 (of a quantity) greater than zero noun 1 a general direction in which something is developing or changing	Good change which improves a situation	Rising (fast)
Positive feedback	Adjective 2 expressing or implying affirmation, agreement, or permission noun 1 information given in response to a product, performance etc., used as a basis for improvement	Good thing e.g. from one's boss	Furthering or compounding increase (making a bad change worse)
Theory	Noun a supposition or a system of ideas intended to explain something, especially one based on general principles independent of the thing to be explained	A guess or perhaps an hypothesis	Explanation, law, rule, fact

3.2 Why use confusing code about climate change in national documents?

The previous section shows up poor (incorrect) communication, in contrast to Section 2 that suggested that if global warming is described in other terminology it might gather faster and broader community involvement. In this section, the occurrences of the eight words/word groups (Hassol 2008, Table 1) have been reviewed in Garnaut (2008a) together with the Australian government's response to the latter: the Carbon Pollution Reduction (CPR) green paper (2008). The use of these words has been compared with that in the four Summaries for Policy Makers (SPMs) of the IPCC AR4 (i.e. Reports of Working Groups 1–3 and of the Synthesis Report) and two economic reviews commissioned by OECD nations (Stern 2006; Garnaut 2008a). Table 2 lists the number of occurrences of the words in these eight sources. Care must be taken in comparing the totals listed in Table 2 as the texts searched are of very different lengths: the four IPCC AR4 Summaries for Policy Makers are rather dense (~700 words per page) but short (around 20 pages each), the Australian Government CPR green paper has around 500 words per page whereas Stern has over 600 pages and about half the IPCC word density (roughly 450 per page) and finally Garnaut has 548 pages with a still lower word density (around 425 per page).

Generally all these authors take care to use code words clearly. The best example is that the term 'positive' frequently appears in sentences also containing 'negative' and in these cases the meaning is almost invariably clear. However, one wonderfully contradictory example is found in Part V of Stern: "Adaptation will reduce the negative impacts of climate change (and increase the positive impacts), but there will almost always be residual damage, often very large." (Page 405 (3/45) Line 1 in description of figure 18.1, Stern, 2006 – underlining added for emphasis). Even very careful unpicking fails to elicit clarity. What is meant here is presumably that adaptation is a good thing but what is said is that adaptation:

- increases the positive impacts of climate change; and
- reduces the negative impacts of climate change.

Table 2 Number of times selected words/word groups appear in eight very widely read and cited climate change literature.

Document	Garnaut	CPR Greenpaper	Stern Exec	Stern full *	IPCC SPM wg1	IPCC SPM wg2	IPCC SPM wg3	IPCC Syn. SPM	IPCC Syn. Full
Anthropogenic/-generated	25	5	0	10	25	9	2	22	44
Bias	9	1	0	12	0	0	0	0	0
Debate	14	4	0	40	0	0	1	0	1
Enhance	26	31	2	64	0	3	5	1	5
Positive	39	10	2	59	1	1	6	0	4
Positive trend	0	0	1	1	0	0	0	0	0
Positive feedback	4	0	0	15	0	0	0	0	2
Theory	11	7	0	120	0	0	0	0	1

* Footnote to word group searching of Stern Report: the searches were conducted on the web-downloadable version available in August 2008. However, Section VI of Stern (2006) on the web at this time has an oddly renumbered section 23. Pages as far as this are as in the 2006 version. However, page 490 (2006 version) is missing and pages after page 489 are numbered from one to 25 when the original numbering begins again. Herein all these page numbers have had 490 added.

In other words apparently exactly the opposite of the intended meaning. Here analysis focuses on Garnaut (2008a) and CPR (2008).

Garnaut (2008a) in his excellent review of the economic consequences for Australia of global warming and its associated climatic disturbances uses one of these ambiguous words 'enhanced' to mean its reverse in a section heading: "International trade – effects of enhanced impacts in developing nations" (Garnaut 2008a, section 10.2.1, heading 10.2.1, on p261). This is an odd choice of title since this section is precisely about very bad consequences of climatic change impacts on developing countries. Indeed, Garnaut says in this section of these nations that,

> [their] less stable political systems may also mean that societies could fracture in response to extreme climate change impacts … [they are] vulnerable to sea-level rise and disruption of established patterns of rainfall and river flows; [and] Severe shocks can destabilise governance in developing countries and lead to regional conflicts. (Garnaut 2008a, p261)

This is not an enhanced impact in the normal sense of being an improvement or increasing the value of the impact. Garnaut uses the code in line with the widespread phrase usage in climate change circles: "the enhanced greenhouse effect" which means the worsening situation!

Professor Ross Garnaut is the only one of the sources surveyed to have produced tautologies in his text. There are two in the eight-word group search conducted. The first is very forgivable: "The 'sceptics' variously contest the relationship between human-induced anthropogenic emissions and atmospheric concentrations, or the relative importance of the enhanced greenhouse effect and other factors that influence climate." (p48, paragraph 1, line 3 – underlining added to highlight tautology). The second is only a partial tautology around another 'climate change code' word: "Interested stakeholders were encouraged to respond to a series of five issues papers, a discussion paper on the proposed emissions trading scheme and an interim report released in January 2008, all of which stimulated public discussion and debate on some of the most critical issues for climate change mitigation and adaptation in Australia." (pix, preface, paragraph 7, line 5 – underlining added to highlight tautology).

Garnaut also takes a rather catholic view of the best meaning of the word 'debate' with four instances in particular highlighting a wide range of meanings that could cause confusion if these quotes were taken out of context:

> There are nevertheless large uncertainties in the science. While there is a clear majority view that there are high risks, there is debate and honest recognition of limits to knowledge about the times and ways in which the risk will manifest itself. (p1, paragraph 3, line 2 – underlining added to highlight use of word debate)

Here debate is linked to belief (recognition).

> Observation of daily debate and media discussion in Australia and elsewhere suggests that this issue might be too hard for rational policy making. (p2, paragraph 6, line 1 – underlining added to highlight use of word debate)

Here debate seems to mean discussion.

> The Review is not in a position to independently evaluate the considerable body of scientific knowledge, and it is not the intent of this chapter to debate the existence or extent of human-induced climate change. (p47, paragraph 5, line 2 – underlining added to highlight use of word debate)

Here debate means 'argue'.

> In recent public debate and commentary, it has been apparent that industries will seek to influence the design of any such assistance arrangements in ways that maximise their respective returns from the scheme. (p383, paragraph 1, line 1 – underlining added to highlight use of word debate)

Here debate seems to mean discussion or perhaps knowledge sharing.

Overall the Australian government's green paper entitled Carbon Pollution Reduction (CPR 2008) is well written. As it is the least internationally read of the texts analysed here I have spent less time on it in this analysis. However, three of the selected words/word groups give rise to almost classic examples of climate change code usage that could be termed: the good, the bad and the ugly.

The 'Good' a definitional statement for 'anthropogenic':

> Developed countries will be required to take the lead on reducing greenhouse gas emissions, as they have contributed the bulk of the existing stock of human-caused, or so called anthropogenic, greenhouse gases already in the atmosphere. (CPR 2008, p9 [25 of 532], line 1 – underlining added)

The 'Bad', an utterly obscure use of the word 'enhance':

> Being part of the group of countries acting on climate change will enhance our seat at the international negotiating table. (CPR

2008, p9 [25 of 532], paragraph 2, line 1 and again on p70 [86 of 532], paragraph 2, line 1 – underlining added).

And the 'Ugly' – three uses with three meanings of the term 'positive/ positively'':

- to mean 'constructively' – "The third pillar reflects the fact that Australia has the standing and capacity to positively shape the post-2012 outcome." (CPR 2008, p68 [84 of 532], line 1 of pillar 3 line 1– underlining added)
- to have the mathematical meaning (as the text says) 'greater than zero' – "The price will be positive (that is, greater than zero) if permits are scarce – that is, if intended emissions exceed the number of available permits." (CPR 2008, p74 [90 of 532], paragraph 8, line 2 – underlining added)
- in an economic and rather obscure sense probably meaning 'helpfully' – "In the case of energy efficiency, it is likely that the more important market failures relate to … positive spillovers, for example where innovations in one field lead to associated benefits in another unrelated field." (CPR 2008, p285 [301 of 532], paragraph 1, bullet point no. 4 line 1 – underlining added).

There are still less clear meanings of 'positive' which can be deduced from careful reading e.g. "For cropland management, the outcome would only be positive if rainfall conditions from 2008 to 2012 consistently exceeded those in 1990." (CPR 2008, p120 [136 of 532], paragraph 2 – line 94 of box 2.3 line 1 – underlining added). Figure 3 underlines how differently the selected word/word groups are used by the two economists reviewed here (Stern 2006 and Garnaut 2008a) as compared to the CPR green paper and even to the IPCC AR4 even though Working Group 3 of IPCC is composed in part of economists and economic analysts (Table 2).

I must stress that very few of the uses I criticise here are wrong in the fair but broad English sense but many could lead to serious misunderstandings of climate change issues if taken out of context. I have found and shown here that the words 'anthropogenic', 'enhance', and 'positive' are real communication traps. The authors analysed here know this and mostly manage to avoid its pitfalls. The IPCC Synthesis Report is always the last to be written and in many ways the most carefully crafted of these diligent and thorough assessments. In AR4 the full Synthesis Report shares characteristics with the three Summary for Policy Makers (SPMs) in that it uses the term 'anthropogenic/generated' much more than the non-IPCC texts considered here. Garnaut uses this term as much as IPCC WG1 SPM but distributed over 30 times as many pages and eight times as many words.

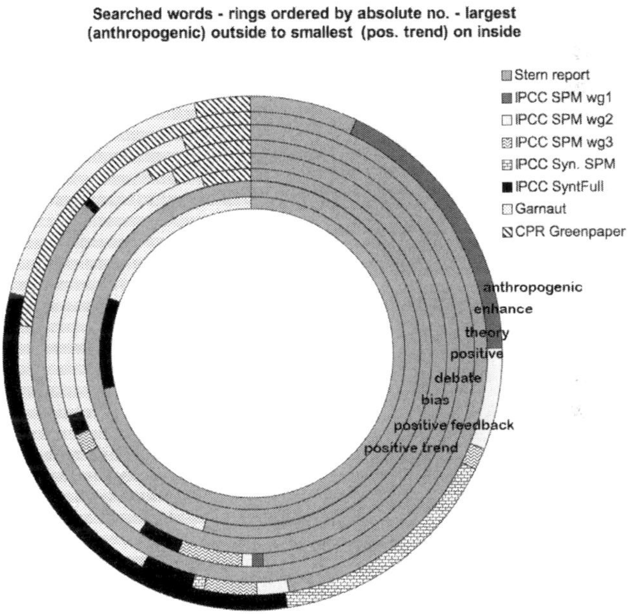

Figure 3 Occurrence of eight words (word pairs) searched for in eight climate change documents (shading). Rings are ordered by absolute number of occurrences from the greatest on the outside (anthropogenic) to the least on the inside (positive trend): anthropogenic, enhance, theory, positive, debate, bias, positive feedback, positive trend.

4. How can we do better in climate change communication?

In this chapter I have illustrated the communication hazards of vehemence, of common climate change terms and of less than careful media dealings. We, the climate change writing community, have, to some extent, created a monster on which Dr Frankenstein might have first smiled and then frowned. Our creature is a public that we have trained to be confused about particular words we like to use and, to some extent, we have even become confused ourselves. Worse still we have helped create a 'them' and 'us' mentality in the mass media which leaves no room for the *'other' greenhouse sceptics.*

Some of the traps and pitfalls of effective climate change communication illustrated in Sections 2 and 3 prompt the question, "What can we do?" Hassol (2008) argues that although many of us do much to communicate climate change knowns and remaining questions we can and should do more and do it more clearly. She particularly exhorts all climate change writers to not overdo "weasel words" nor to pad talks and texts excessively with caveats. In this she has been challenged recently by a respondent who interpreted her recommendations not to 'pad' as a suggestion to oversimplify or 'water down' climate change information (Sedgemore 2008). This incorrect polarisation is, I believe, commonplace in contentious issues that affect the whole community as has been illustrated throughout this chapter. It is this that seems to force climate change communicators, inappropriately, into one of only two groups: 'believers' or 'deniers'.

There is at least one very serious problem in all this climate change code. We, the climate change scientists, are surely to blame – there can be no other word – for 'enhance/enhanced'. The phrase 'enhanced greenhouse warming' has been in the science literature for at least 10 years and probably over 20. There is *no excuse for this terminology which only confuses* 'enhancement' meaning 'getting better' with the actual reality of this enhancement meaning getting worse.

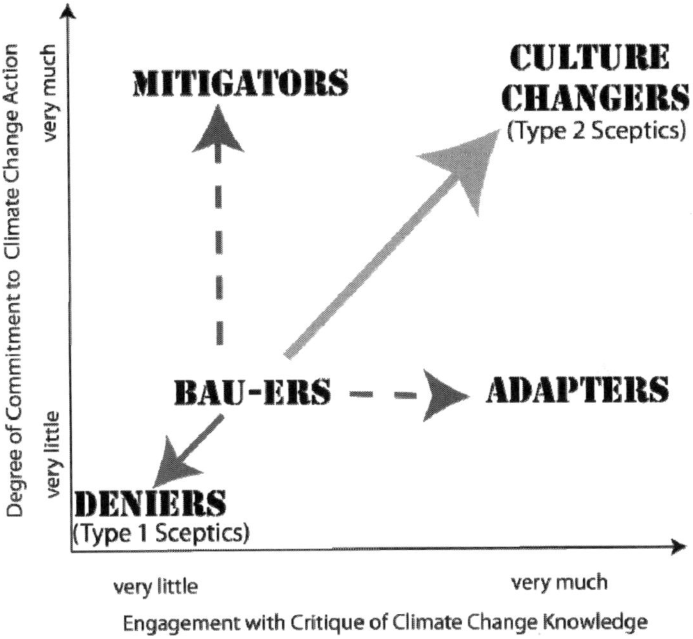

Figure 4 Cartoon of attitudes among climate change communicators including the two types of 'sceptics' described here. Future monitoring of Australian climate change issues could usefully report on the quantity and quality of coverage of the views of such Type 2 Sceptics: the Culture Changers.

4.1 Agree or deny?

It is NOT true to pretend that all climate change experts fully agree with the IPCC AR4. We do not. Many of us are frustrated with some of its processes, angry about some of its statements and boxed-in by its apparent unchallengeable nature. There are at least three types of people who hold strong views on climate change in Australia and probably worldwide: the deniers, the IPCC-agree-ers and the 'other climate change sceptics'. The first group (deniers) get a good hearing in Australia; the second group also as they refer to, and defer to, IPCC AR4. It is the third group of people, amongst whom I count myself, who are seriously short-changed – the people who believe, and with good reasons, that the IPCC understated the threat and the urgency of the climate change challenge. Indeed, I believe there are a range of positions on climate change that are worthy of examination. For example, Sterling and Gray-Donald (2007) propose a four-quadrant model of the characteristics of those involved in Environmentally Sustainable Development (ESD). Here I have modified their categorisation for climate change (Figure 4). The axes are critique of climate change knowledge and commitment to climate change action in terms of which I show five example positions: BAU-ers, Adapters, Mitigators, Cultural Changers and, sadly, still Deniers.

The characteristics of these five positions in Figure 4 are:

Deniers: those too blind to see the 'unequivocal'.

BAU-ers: the base position of business-as-usual involving little or no interest in climate change action and little or no engagement with arguing (or accepting) the science or dominant socio-economic and environmental values. The increasing awareness and acceptance of, e.g. IPCC has forced such people into basic question asking and, with some, interested them in the backward step into Deniers views.

Adapters: the liberal position (e.g. my student, Chuck) is critical of BAU-ers for not taking sufficient notice of climate change adaptation (which is recognised as important). Liberals are also nervous of Mitigators for suggesting that a rush towards heavy-handed emissions reduction in the light of different interpretations, the significant costs and, for some, in view of the Deniers claims and assertions.

Mitigators: the advocacy position, championed by Stern (2006), Garnaut (2008c), developing nations, most NGOs and increasingly by concerned scientists who argue the self-evident urgency of anthropogenic climate change. The slowness of governments' response frustrates these people who also criticise BAU-ers and Adapters for ignorance and weakness respectively.

Cultural Changers: the future position (e.g. as held by Stephen Schneider). These people view unsustainability as arising from deep seated cultural assumptions and articulate the need for urgent cultural change as the only means of achieving both necessary adaptation and urgent mitigation.

It would clearly be a mistake to become too 'locked into' either the Adapters or the Mitigators positions. Ross Garnaut says, "The 'sceptics' variously contest the relationship between human-induced anthropogenic emissions and atmospheric concentrations, or the relative importance of the enhanced greenhouse effect and other factors that influence climate." (p48, paragraph 1, line 3–4, Garnaut 2008a – underlining added to highlight). Garnaut briefly discusses sceptics of two types in his Final Report (Garnaut 2008c, pxvii). Of the 'denier sceptic' he says, "these dissenters hold strongly to the belief that the mainstream science is wrong". He acknowledges a second type of 'sceptic' that he describes as, "climate scientists of professional repute who maintain that the mainstream science embodies misjudgements about quantities".

4.2 Advent of the new climate sceptic

I can agree with Garnaut only this far. I wish now to highlight two ways in which professional and reasonable scientists can differ from the 'mainstream view'. They can either believe that IPCC overstates the degree of change, of impact or of urgency, or they can argue that IPCC *understates* all these features.

The position of Don Aitkin might be said to be illustrative of sceptic type 1 while the other 'sceptics' (type 2) are sceptical of the benefit to people and policy of the IPCC's consensus (Figure 4). In a close analogy with Jeremy Leggett's challenge to the Second World Climate Conference back in 1990 are new assessments of potential interactions between what are termed climate tipping events. The very large-scale impacts and the uncertainty regarding exact thresholds in the climate system have been reviewed by Lenton et al. (2008) who argue that critical thresholds "can be anticipated, because a system that is approaching a tipping point becomes more and more sluggish in its response to perturbations caused by natural variability".

It should be possible to do much more about the quality and coverage of Australia's climate change communication in the future. I ask that we, the type 2 sceptics, be given a voice and as much airtime, mass-media editorials and public support as the deniers (type 1 sceptics) and our colleagues who are sanguine about the speed of the national response to climate change. Indeed, I believe that we could usefully initiate a national *climate change communication monitoring scheme* by means of which we can have a better measure of community concern (cf. Figure 1). This could have a number of components including but not limited to:

- Classification of all climate change communications by ranking them in terms of their 'degree of commitment to climate change action' and their 'level of engagement with critique of climate change knowledge' (cf. Figure 4); and
- Analysis of the use and abuse of confusing climate change terminology (cf. Table 1 and Figure 3).

The first of these should be applied to mass media stories building on and strengthening Power et al. (2008) and separately to research publications to improve our 'less than 10%' public reporting (Cribb 2008). The second would naturally build on the work reported here which highlighted the way in which Garnaut (an economist reviewing the greenhouse literature for government policy decisions regarding emissions reductions) uses the word 'anthropogenic' as compared to the IPCC Synthesis' report writers.

Finally there are immediate actions we can take:

We need to *stop using 'enhance'* in our texts now and we need to *quit talking about the fact of global climate change as a 'theory'*.

We can resist indulging in special pleading i.e. stop media releases like these:

- USA (20 August 2008): Eight scientific organisations urged the next US president to help protect the country from climate change by pushing for increased funding for research and forecasting, saying that more than a quarter of the country's economic output, about $2 trillion, is vulnerable to extreme weather.

- Australia (10 September 2008): the Universities Climate Consortium told the Senate that climate science needs more money to meet demand for climate change professionals noting climate-related research attracted only $129 million in 2004 compared to $198 million for geology and geophysics and $337 million for information technology.[6]

I believe we need to *focus on gaining improved public understanding* of the complexities of climate change knowledge. Since I first made this argument it has been put to me by climate scientists from Australia and the UK that we/they are smart enough to 'walk and chew gum'. In other words that we can and should *both* improve public communication and comprehension *and* invest in relevant new climate science. I agree!

It is clear that climate change will remain a risk-management problem for the foreseeable future. However, the more clearly and quickly we communicate important outcomes and likely damage, the better will be humanity's chances of adaptation. The cleverer we are in the description of relevant and very likely climate change results, the sooner we constrain the potential for some really "dangerous" outcomes that cannot currently be ruled out at less than a 10% chance. One essential ingredient has to be transparency in communication. I strongly urge that Australia establish quantitative and qualitative measures for climate change communication monitoring and improvement.

References

Brook B (2008). Make a stand for good science. *Australasian Science*, 28 May 2008, p42.

[6] The submission was solicited based on statements by [then] Panel Chair of the ARC that Climate Science was getting too fast an increase in funding while Earth Sciences (aka Geology) was only getting a +2%. The changes in question were roughly +$100,000 for Climate Science and +$500,000 for Earth Sciences and correcting this was considered worthwhile. Further, the request was to increase funding to train the next generation of gradudate students – not for research *per se*.

CPR (2008). Carbon Pollution Reduction Scheme. Green Paper, July 2008, Canberra: the Department of Climate Change, Commonwealth of Australia, [quotes from CPR are copyright Commonwealth of Australia and reproduced by permission]. Available at: www.climatechange.gov.au/greenpaper/index.html (30/7/09).

Cribb J (2008). A few words for our sponsor. *The Australian*, 2 April 2008. Available at: www.theaustralian.news.com.au/story/0,,23468035-25192,00.html (30/7/09).

Doherty SJ, Bojinski S, Henderson-Sellers A, Noone K, Goodrich D, Bindoff NL et al. (2009). Lessons learned from IPCC: developments needed to understand and predict climate change for adaptation, 2009. *Bulletin of American Meteorological Society*, doi: 10.1175/2008BAMS2643.1.

Ereaut G & Segnit N (2006). Warm words: how are we telling the climate story and can we tell it better? *Institute for Public Policy Research*. Available at: www.ippr.org/publicationsandreports/publication.asp?id=485 (30/7/09).

Fachon D (2008). Science can't prove man's 'carbon footprint' is tipping balance to global warming. Available at: www.warwickonline.com/pages/full_story/push?article-Op+ed-+Science+can-t+prove+man-s+-carbon+footprint-+is+tipping+balance+to+global+warming%20&id=1483098-Op+ed-+Science+can-t+prove+man-s+-carbon+footprint-+is+tipping+balance+to+global+warming (30/7/09).

Garnaut R (2008a). *The Garnaut climate change review: draft report.* June 2008, [quotes from Garnaut are copyright Commonwealth of Australia and reproduced by permission]. Available at: www.garnautreview.org.au/CA25734E0016A131/pages/all-reports--resources (30/7/09).

Garnaut R (2008b). *The Garnaut climate change review: targets and trajectories. Supplementary draft report.* September 2008. Avaiable at: www.garnautreview.org.au/CA25734E0016A131/pages/all-reports--resources (30/7/09).

Garnaut R (2008c). *The Garnaut climate change review: final report.* Cambridge: Cambridge University Press.
Global Carbon Project (2008). *Carbon budget and trends 2007.* Updated 26 September 2008. Available at: www.globalcarbonproject.org/carbonbudget/index.htm (30/7/09).
Government of Australia (2008). Views on the work program for the ad hoc working group on long term cooperative action. In Views regarding the work program of the ad hoc working group on long-term cooperative action under the convention submissions from parties, Addendum, FCCC/AWGLCA/2008/MISC.1/Add.2, Bonn: UNFCCC.
Hassol SJ (2008). Improving how scientists communicate about climate change. *EOS*, 89(11), 11 March 2008, pp106–07 and Reply to Comments on 'Improving how scientists communicate about climate change'. *EOS*, 89(33), 12 August 2008, pp304.
Henderson-Sellers A (1998). Climate whispers: media communication about climate change. *Climatic Change*, 40(3–4), 421–56.
Henderson-Sellers A (2008). The IPCC report: what the lead authors really think? Talking Point Invited Article. *ERL* (Environmental Research Letters). Available at: environmentalresearchweb.org/cws/article/opinion/35820 (30/7/09).
Henderson-Sellers A & Pitman AJ (Eds) (1991). *Vegetation and climate interactions in semi-arid regions.* Dordrecht: Kluwer.
Henderson-Sellers A, Irannejad P, McGuffie K & Pitman AJ (2003). Predicting land-surface climates – better skill or moving targets? *Geophys Res Lett*, 30(14), 1777–80.
IPCC WG1 SPM (2007). Summary for policymakers. In S Solomon, D Qin, M Manning, Z Chen, M Marquis, KB Averyt, M Tignor & HL Miller (Eds), *Climate change 2007: the physical science basis. Contribution of Working Group I to the fourth assessment report of the Intergovernmental Panel on Climate Change* (pp1–17). Cambridge and New York: Cambridge University Press, www.ipcc.ch/ (30/7/09).
IPCC WG2 SPM (2007). Summary for policymakers. In ML Parry, OF Canziani, JP Palutikof, PJ van der Linden & CE Hanson, (Eds), *Climate*

change 2007: impacts, adaptation and vulnerability. Contribution of Working Group II to the fourth assessment report of the Intergovernmental Panel on Climate Change (pp7–22). Cambridge: Cambridge University Press. Available at: www.ipcc.ch/ (30/7/09).

IPCC WG3 SPM (2007). Summary for policymakers. In B Metz, OR Davidson, PR Bosch, R Dave & LA Meyer (Eds), *Climate change 2007: mitigation. Contribution of Working Group III to the fourth assessment report of the Intergovernmental Panel on Climate Change* (pp3–123). Cambridge and New York: Cambridge University Press. Available at: www.ipcc.ch/ (30/7/09).

IPCC Synthesis Report (2007). *Climate change 2007: synthesis report.* Cambridge and New York: Cambridge University Press. Available at: www.ipcc.ch/ (30/7/09).

IPCC Fourth Assessment Report (2007). The contents of the three complete reports of Working Groups I, II and III are accessible without charge from www.ipcc.ch/ (30/7/09).

Lenton TM, Held H, Kriegler E, Hall JW, Lucht W, Rahmstorf S & Schellnhuber HJ (2008). Tipping elements in the Earth's climate system. *PNAS* (Proceedings of the National Academy of Sciences of the United States of America), 105(6), 1786–93.

McGuffie K & Henderson-Sellers A (2005). *A climate modelling primer.* Third Edition, Chichester: John Wiley and Sons.

Power S, Plummer N, Pearce K, Walland D, Edwards S, Jones D, Gipton S, Holper P &Whitehead R (2008). Changes in Australian attitudes towards global warming. WMO MeteoWorld (April 2008). Available at: www.wmo.int/pages/publications/meteoworld/archive/april08/australia_en.html (30/7/09).

Saad L (2007). Did Hollywood's glare heat up public concern about global warming? Concern about global warming is up slightly over past year. *Gallup News Service.* Available at: www.gallup.com/poll/26932/Did-Hollywoods-Glare-Heat-Public-Concern-About-Global-Warming.aspx (30/7/09).

Schellnhuber HJ, Cramer W, Nakicenovic N, Wigley T & Yoke G (2006). *Avoiding dangerous climate change*. Cambridge: Cambridge University Press.

Sedgemore F (2008). Comment on 'Improving How Scientists Communicate About Climate Change'. *EOS* 89(33), 12 August 2008, pp 304.

Sterling S & Gray-Donald J (2007). Editorial. *Int J Innovation and Sustainable Development*, 2(3/4), 241–48.

Staudt AC (2008). Recent evolution of the climate change dialogue in the United States. *Bull Amer Meteor Soc*, 89, 975–85.

Sterman S (2008). Understanding climate change complacency. Available at: web.mit.edu/newsoffice/2008/climate-sterman-0820.html (30/7/09).

Stern N (2006). *The economics of climate change: the Stern review*. Cambridge: Cambridge University Press.

WCRP (2008). Future Climate Change Research and Observations: GCOS, WCRP and IGBP Learning from the IPCC Fourth Assessment Report, Workshop and Survey Report, GCOS-117, WCRP-127, IGBP Report No. 58, World Meteorological Organization, (WMO/TD No. 1418), January 2008, Geneva.

Zhao M & Pitman AJ (2002). The impact of land cover change and increasing carbon dioxide on the extreme and frequency of maximum temperature and convective precipitation. *Geophys Res Lett*, 29(6), 1078, doi: 10.1029/2001GL013476.

Appendix 1: Garnaut media release (Friday 5 September 2008)

START ON A 550 PPM TRAJECTORY, AIM FOR 450 PPM OR LOWER

Australia should put its strongest possible efforts into securing a global agreement to limit emissions to no more than 550 parts per million CO_2-e and encourage the world onto a lower emissions path as soon as feasible, according to Professor Ross Garnaut's *Targets and trajectories* report released today. Launching the report at the National Press Club today, Professor Garnaut said that a global objective of 450 ppm or lower would suit Australian interests better than a 550 ppm goal.

"However, the Review has reluctantly concluded that international agreement on a global goal of 450 ppm is not immediately feasible, given that emissions are currently around 455 ppm CO_2-e and rising rapidly due to fast global economic growth," said Professor Garnaut. "The path to an ambitious objective of 450 ppm, or lower, is through early progress on 550 ppm," he said. Professor Garnaut said that the emissions trajectories proposed by the Review were based on the 'per capita' allocation of emissions rights, which was a fair global solution and had the best chance of securing international support for sharing a global emissions budget. "Per capita allocations provide the only possible basis for an international agreement that includes developing countries. This is because it takes population growth into account and gradually reduces the weighting of historical starting points over time. "The targets and trajectories that I am proposing for Australia fall out from the per capita division of the global emissions budget as a matter of arithmetic. They are based on a principled approach," he said.

Based on a 550 ppm outcome, Australia's share of the burden would be a 10% reduction (or 30% in per capita terms) by 2020 and an 80% reduction (90% per capita) by 2050 over 2000 levels. If a '450 ppm overshoot' scenario were adopted, emissions would need to be reduced by 25% by 2020 (40% per capita), falling to 90% by 2050 (95% per capita). By comparison, the European Union's per capita target would be a 17% reduction by 2020 and a 69% reduction by 2050. "It's important that any proposals under consideration for the international allocation of emissions reductions 'add-up' to the desired global outcome – there would be no point in global action which was not geared at avoiding the substantial climate change impacts," said Professor Garnaut. During the period from 2010–2012, in which Australia would continue to work within its Kyoto commitment, permits would be sold starting at $20 in 2010, rising each year by 4% plus the percentage increase of the consumer price index.

"This is more or less the price path that the modelling suggests would be followed if there were effective global agreement directed towards stabilisation of global greenhouse gas concentrations at 550 ppm. If a post-Kyoto agreement were struck around 550 ppm, the fixed price is likely to allow relatively seamless transition to a floating price regime," said Professor Garnaut. The report proposes that, in the absence of an international agreement between high-income countries and China, Australia commit to an emissions reduction target of 5% by 2020 (25% per capita), which is in line with the Government's 60% by 2050 emissions target. Any further commitments would be determined in the context of international progress.

The Garnaut-Treasury modelling indicates that the cost to Australia of mitigation would be 1.1% GDP by 2020 under a 550 ppm scenario, and 1.6% GDP by 2020 under a 450 ppm scenario. "The costs of well-designed mitigation, substantial as they are, do not threaten to derail the long-term growth path of Australia, its developing country neighbours or the global economy. Unmitigated climate change probably would," he said. "These increases would not lead to a marginal reduction in human welfare, but a catastrophic outcome for human civilisation and most ecosystems," said Professor Garnaut. "Faced with these risks, it's not a question of whether Australia should reduce emissions, but by how much. The details of the targets and trajectories that the Review is recommending will not be the best for all time. They are the best available to us now," he said. (Also included Table 9.1 of Supplementary Draft Report).

3
Physical monitoring of the Great Barrier Reef to understand ecological responses to climate change

Janice Lough, Scott Bainbridge, Ray Berkelmans and Craig Steinberg

*Abstract**

Significant warming of the Great Barrier Reef (GBR) has already occurred and many components of this complex ecosystem are known to be both sensitive and vulnerable to a changing climate. Assessing biological responses of the GBR requires high-quality, ongoing observations of the physical environment over a range of spatial and temporal scales. The monitoring capabilities of the Australian Institute of Marine Science (AIMS) through automatic weather stations, remote sensing, ocean moorings and an intensive water temperature logger program have already provided critical measurements that link biological responses with the physical environment. This monitoring program is now being enhanced and expanded through the GBR Ocean Observing System (GBROOS). Here, we describe the history and development of some of these monitoring programs. A commitment to such long-term, detailed and regionally specific physical environment monitoring programs is essential to identify how the GBR climate is changing and which parts of this unique ecosystem may be more or less susceptible to the evolving consequences of a rapidly warming world.

---

* Chapter originally submitted in January 2009.

1. The Great Barrier Reef

The Great Barrier Reef (GBR) extends for 2200 km along the Queensland coast between 9°S and 24°S and up to 200 km from shore, encompassing about 35 million hectares. Coral reefs, such as the GBR, are complex ecosystems that are uniquely defined amongst the world's ecosystems in terms of both their biological components and the geological structures they create through the build-up of calcium carbonate (Kleypas et al. 2001; Hopley et al. 2007). At the heart of tropical coral reefs is a mutually beneficial symbiosis between the coral animal and single-celled photosynthetic plants called zooxanthellae. Photosynthetic products of the algae provide the coral host with cheap energy and enable the rapid calcification (building of calcium carbonate skeletons) necessary for corals to form reef structures. In return the algae obtain protection and essential nutrients from the coral host. Together these two organisms produce the amazing variety of coral skeletal growth forms and carbonate structures that can withstand the natural forces of destruction and erosion and provide habitat for the many thousands of reef-associated plants and animals and form the basis of coral reef ecosystems (Veron 2000).

The vast area of the GBR includes nearly 3000 coral reefs, home to 360 species of hard coral, 1500 species of fish, 4000 species of molluscs, 400 species of sponges and the list goes on. Internationally endangered dugongs inhabit extensive sea grass beds. Six of the world's seven marine turtles live on the GBR and hump-backed whales from Antarctica breed there (Hutchings et al. 2008). The GBR is also intimately linked with coastal ecosystems such as mangroves, wetlands and estuaries.

The Outstanding Universal Value of this complex and highly diverse marine ecosystem was recognised through its World Heritage listing in 1981. The GBR is one of a handful of World Heritage areas listed for all four natural criteria: exceptional natural beauty and importance, significant geomorphic or physiographic features, significant ecological and biological processes and significant biological habitats for biological diversity.[1] The GBR is also of

[1] www.gbrmpa.gov.au/corp_site/key_issues/conservation/heritage/world_heritage

economic significance to the Queensland and Australian economies – valued at ~$5.7 billion/year and providing ~66,000 jobs (Access Economics 2007). Coral reefs occupy only ~10% of the GBR area and the rich biodiversity of the inter-reef areas is only just being explored (Picher et al. 2007). We do not even know the full extent of the biodiversity of the GBR ecosystem and new organisms are still being discovered (see CReefs project of the Census of Marine Life).[2]

The GBR, through the efforts of the Great Barrier Reef Marine Park Authority[3] and state and federal initiatives such as the Reef Water Quality Protection Plan,[4] is probably the world's best managed and protected (~30% no-take zones) coral reef ecosystem. It is protected from the local, direct exploitation that has significantly degraded many of the world's coral reefs (see Hughes et al. 2003; Buddemeier et al. 2004; Kleypas & Eakin 2007; Wilkinson 2004, 2007) but it is not, however, immune from the regional consequences of a changing climate due to increasing greenhouse gases.

1.1 A changing climate for coral reefs

The global distribution of coral reefs is constrained to shallow, warm, well-lit, clear and relatively low-sediment waters and is also shaped by the climatological and geological history as well as local bathymetry (Stoddart 1969; Achituv & Dubinsky 1990). Particularly important factors controlling the environmental envelope of coral reefs are water temperatures, salinity, light and the aragonite saturation state of seawater. The latter is a measure of how easily aragonite, the main form of calcium carbonate created by reef-building corals, can form and depends on the concentration of calcium and carbonate ions in seawater, i.e. the oceanic carbonate chemistry (Kleypas et al. 1999, 2006).

The average physical atmospheric and oceanic climate of the GBR controls many biological processes on the reef (Lough 2007; Steinberg 2007). Ocean currents and upwelling affect the connectivity of larval supplies

---

[2] www.creefs.org

[3] www.gbrmpa.au

[4] www.reefplan.qld.gov.au

between reefs and nutrient supplies to reefs. There are significant cross-shelf gradients in community characteristics between terrestrially influenced nearshore reefs and open-ocean-influenced offshore reefs. Latitudinal gradients in average water temperatures affect reef diversity and coral growth rates. Tropical cyclones are an aperiodic source of localised destruction primarily through the large waves they generate (Massel & Done 1983; Puotinen et al. 1997). The two phases of the major source of interannual tropical climate variability, El Niño – Southern Oscillation (ENSO) events, cause distinct and different surface climate anomalies along the GBR (Lough 2007; Steinberg 2007) that can also impact biological processes (Cheal et al. 2007). The health of 47 reefs on the GBR has been assessed annually since 1993 by AIMS' Long Term Monitoring Program – the longest and most extensive continuous record of changes in coral reef communities.[5]

The natural physical environment of coral reefs such as the GBR, at the interface of land, sea and the atmosphere, can be highly dynamic and potentially stressful for reef organisms. Reef organisms have, however, evolved strategies to cope with occasional environmental disturbances and, given sufficient time between such stresses, damage and destruction (as for example, from a tropical cyclone or a low-salinity flood event) would normally be followed by recovery and regrowth. Unfortunately, we have now entered an era of rapid climate change due to human activities altering the composition of the atmosphere that is and will have serious consequences for the maintenance of coral reefs, including the GBR.

1.2 Biological responses to a changing climate

Coral reef ecosystems appear to be particularly sensitive to a changing climate (see review in Lough 2008a) and various components of the GBR have been shown to be both sensitive and vulnerable to climate change (Johnson & Marshall 2007). The following changes in physical environmental parameters are of particular significance for coral reefs and the GBR:

- Warming ocean temperatures

---

[5] www.aims.gov.au/docs/research/monitoring/reef/reef-monitoring.html

- Ocean acidification
- Changing ocean circulation and upwelling patterns
- More intense tropical cyclones causing localised physical destruction
- More extreme rainfall and river flood events affecting the distribution of low salinity waters
- Changes in the frequency and/or intensity of ENSO events
- Rising sea levels drowning some reefs and increasing the potential area for others.

Coral reefs have already provided evidence that they are particularly sensitive to relatively small rises in water temperatures (see various papers in van Oppen & Lough 2009). The symbiotic relationship between coral animals and single-celled algae (zooxanthellae), at the heart of tropical coral reefs, can break down due to a variety of environmental stresses (Baker et al. 2008). When this happens, the coral lose their symbiotic algae and photosynthetic pigments – with the result that the white calcium carbonate skeleton becomes visible through the translucent tissue layer. This process, called "coral bleaching", is not a new phenomenon due to global warming. However, now there are mass coral bleaching events where entire reefs are affected. This is a new phenomenon, and linked to warmer tropical ocean temperatures taking corals closer to their upper thermal limits. Significant coral bleaching affected about half of the GBR in 1998 and again in 2002 and the southern GBR in 2006. Corals appear to be living only 1–2°C below their upper thermal tolerance limit before bleaching occurs (Coles et al. 1976; Jokiel & Coles 1977) and are particularly sensitive during the seasonal water temperature maximum (Berkelmans et al. 2004). The link between unusually warm sea surface temperatures (SSTs) and coral bleaching is now so clearly established that potential coral bleaching conditions can be identified globally in near real-time[6] and summer GBR bleaching potential is now routinely reported.[7] The occurrence and intensity of coral bleaching

---

[6] coralreefwatch.noaa.gov

[7] www.gbrmpa.gov.au/corp_site/key_issues/climate_change/management_responses/coral_bleaching_status_2008-09

can, however, be highly variable at the local scale; within a coral colony, between coral colonies; within a reef and between reefs; and between different coral species (Marshall & Baird 2000). There can be considerable local-scale variations in water temperatures within and between reefs that affect the intensity and occurrence of coral bleaching (Berkelmans 2002; Berkelmans et al. 2004). High irradiance levels in shallow-water reef environments can also enhance the intensity of bleaching (Baker et al. 2008). Local-scale variations that reduce thermal and radiative stress to corals can be linked to water turbidity and water movements, such as mixing, upwelling, tidal range and wave energy (Skirving & Guinotte 2001; Skirving et al. 2006). Identification of bleaching-resistant locations may be critical for recovery of nearby bleached coral populations (Marshall & Schuttenberg 2006).

Warmer ocean water temperatures have also been linked to an increasing frequency of marine disease outbreaks affecting corals and other reef organisms (Harvell et al. 1999; Boyett et al. 2007). Such changes have already been observed on the GBR (Bruno et al. 2007).

A more insidious and alarming effect of increased atmospheric greenhouse gases on marine calcifying organisms, such as reef corals, is progressive 'ocean acidification'. About 30% of the extra carbon dioxide humans have put into the atmosphere has been absorbed by the surface oceans (Feely et al. 2004; Sabine et al. 2004). More carbon dioxide dissolved in the surface ocean lowers the pH, decreases the availability of carbonate ions and thus lowers the saturation state of the major skeletal- and shell-forming carbonate minerals (Kleypas et al. 2006). Changing the ocean chemistry essentially shifts the geochemical equations whereby a range of marine organisms (including corals and the important reef cements, coralline algae) calcify, thus reducing their ability to form their skeletons and shells (Kuffner et al. 2008; Manzello et al. 2008). For coral reefs this results in weaker reef structures more susceptible to erosion and could compromise the habitat for the thousands of reef-associated organisms. Our understanding of the full range of consequences of ocean acidification for the GBR is still limited but they are potentially dire (Hoegh-Guldberg et al. 2007). There is already some field evidence that coral growth is slowing on the GBR (De'ath et al. 2009) and elsewhere (Tanzil et al. 2009). This may be due to increasing thermal stress and changing ocean chemistry.

Warming ocean waters and ocean acidification are two of the main threats to the maintenance of coral reef integrity in a rapidly changing climate. The resultant loss in habitat is just one indirect effect on other reef organisms. The physiological performance of, for example, coral reef fish may also be directly affected by a changing climate (Munday et al. 2008). Understanding ocean circulation patterns and upwelling on the GBR is also of critical importance as water movement and connectivity between reefs affects many processes within the GBR (Munday et al. 2009). Understanding how the physical environment is changing locally requires ongoing and high-quality observations.

1.3 Large-scale warming of the GBR

Global land and sea temperatures are already warming in response to increasing amounts of greenhouse gases in the atmosphere and are projected to warm further unless drastic greenhouse gas mitigation strategies are rapidly and globally instigated (Meehl et al. 2007; Trenberth et al. 2007). Average global temperatures for the most recent 30-year period (1979–2008) are +0.6°C warmer than for the last 30 years of the 19th century (1871–1900) (Figure 1a). Over the period 1870–2008 the rate of warming was 0.05°C/decade but this rate has accelerated to 0.12°C/decade for the period 1950–2008. SST s averaged over the GBR region have also warmed (Figure 1b) and are now +0.4°C warmer for the most recent 30-year period compared to the end of the 19th century. As for global average temperatures, the rate of warming of GBR waters has accelerated from 0.04°C/decade over the whole period to 0.10°C/decade over the recent 1950–2008 period.

Spatially, the rate of recent warming increases from north to south along the GBR (Figure 2) and the rate tends to be greater for annual minimum than annual maximum SSTs. Recent, 1950–2007, warming along the GBR has already resulted in significant southward shifts of average climate zones (Lough 2008b). Such climate changes will result in significant changes in Australia's marine ecosystems (Poloczanska et al. 2007).

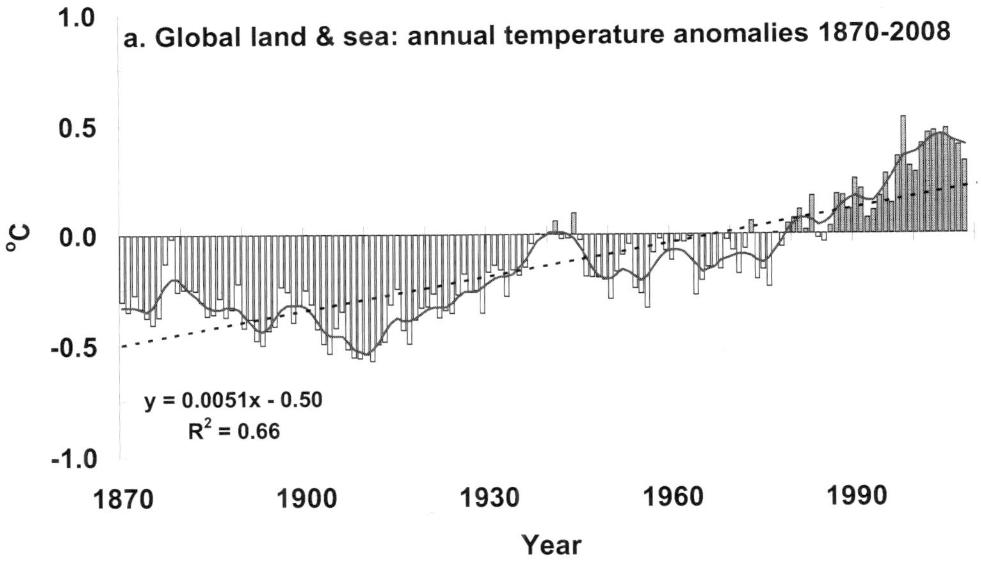

Figure 1a Annual temperature anomalies from 1871–2008 average for global land and sea temperatures, HadCRUT3V (data sources: Brohan et al. 2006; Rayner et al. 2003).

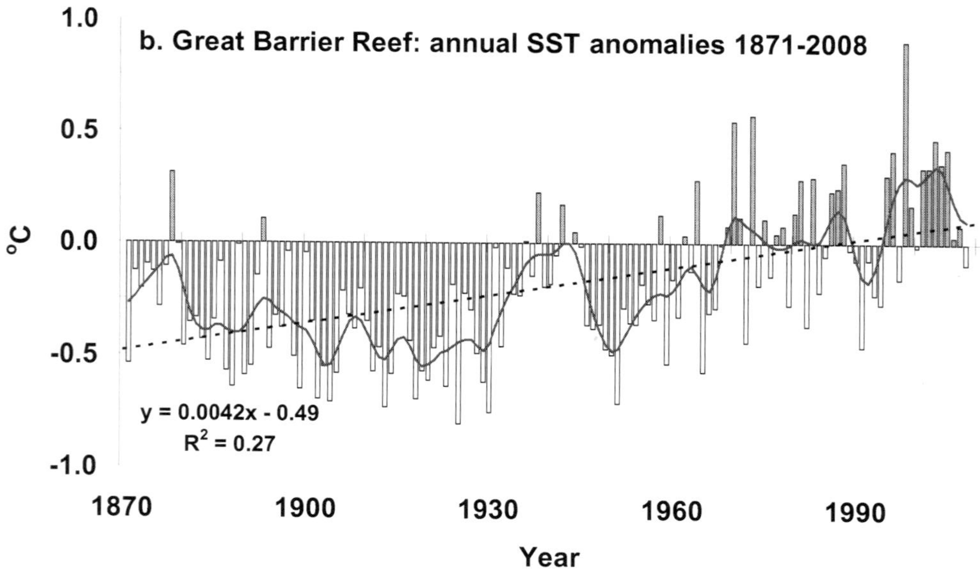

Figure 1b Annual temperature anomalies from 1871–2008 average for Great Barrier Reef sea surface temperatures (HadlSST) (data sources: Brohan et al. 2006; Rayner et al. 2003).

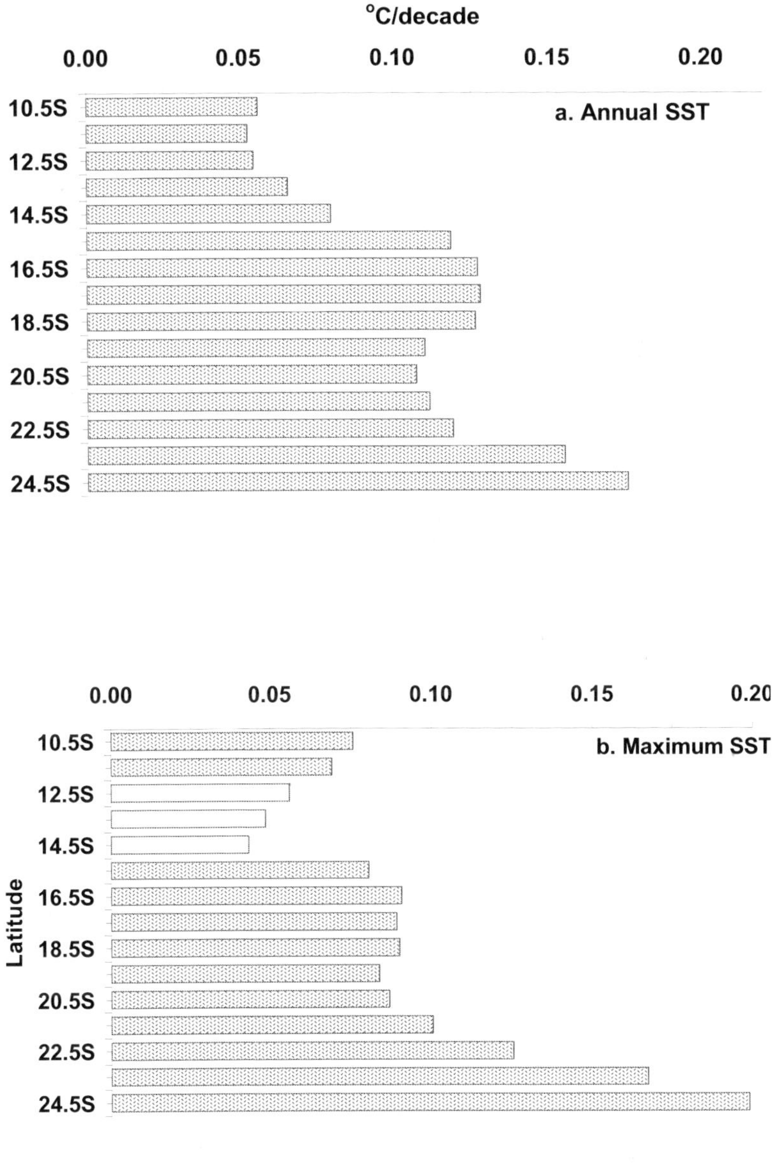

°C/decade
0.00
0.05
0.10
0.15
0.20
a. Annual SST
10.5S
12.5S
14.5S
16.5S
18.5S
20.5S
22.5S
24.5S
0.00
0.05
0.10
0.15
0.20
b. Maximum SST
Latitude
10.5S
12.5S
14.5S
16.5S
18.5S
20.5S
22.5S
24.5S

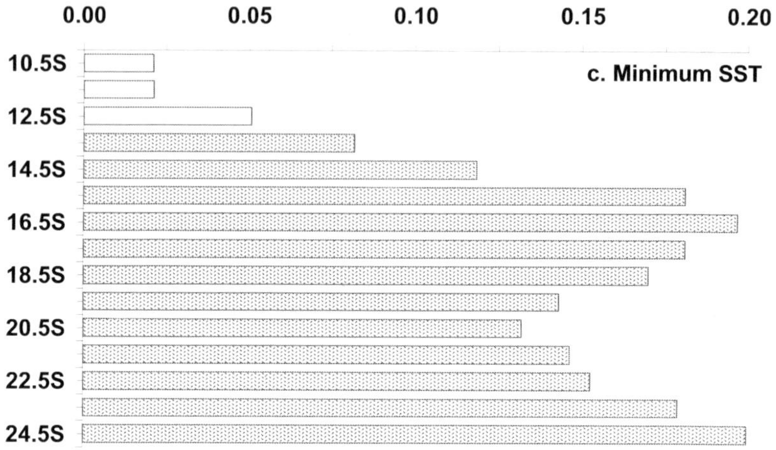

Figure 2 Linear trend (°C/decade), 1950–2007 for 1° latitude bands along the GBR for a) annual, b) annual maximum and c) annual minimum SSTs. Shaded bars indicate linear trend significant at 5% level (HadlSST, Rayner et al; 2003).

Average, large-scale data, as presented in Figure 1, provide solid evidence that the climate of the GBR is already significantly changing as a regional response to global climate change (Lough 2007). However, to clearly identify how a changing climate will affect the various biological components of the GBR requires high-resolution, detailed physical environmental observations at the scale of reef processes (Wolanski 2004). In the following sections we describe a number of ocean climate monitoring systems that provide the basis for linking changes in the physical environment of the GBR to biological responses.

2. Monitoring the physical environment of the GBR: Automatic Weather Stations (AWS)

Prior to 1980, most instrument-based environmental data on the GBR consisted of collections of instantaneous measurements recorded by on-site observers. Advances in electronics and instrumentation in the late 1970s and early 1980s, in particular the low-power complementary metal-oxide semiconductor (CMOS) microprocessor, allowed meteorological and oceanographic variables to be recorded unattended, at regular intervals and either stored on board or telemetered automatically to a central collection point. This was a major innovation for marine research where access to remote sites is severely limited. As a result it opened up many new areas of research in oceanography and ecosystem processes.

2.1 AWS – history and development

AIMS first implemented weather monitoring stations in 1980 at Hinchinbrook Island and Rib Reef, using a tape-recorder system requiring regular exchanging of data cassettes. These locations were, at that time, important for research into mangroves, water circulation and reef chemistry. In 1982 these stations were converted to real-time monitoring stations using high-frequency radio telemetry. Real-time data provided the opportunity to respond quickly to faults and minimise data gaps. The primary driver for more and better weather monitoring stations came from advances in computing in the early to mid-1980s when numerical modelling of water flows on the GBR became a major research focus. Modelling larval dispersal, reef connectivity and Crown-of-Thorns starfish outbreaks were important new applications. A new land-based station was commissioned in 1983 at Cape Bowling Green and reef-based stations were added on a floating tourist pontoon at John Brewer Reef in 1986 and at Myrmidon Reef in 1987. When the John Brewer pontoon was removed, this station was re-located to a similar tourist pontoon at Hardy Reef in the Whitsunday area in 1989 where it still operates. Additional stations came online at Davies Reef and Agincourt Reef in 1990. Climate change and coral bleaching became a new research focus after the major coral bleaching event in 1998 (Berkelmans & Oliver 1999). By this time the value of having a 15+ year dataset of weather and sea temperatures on the reef was clearly evident. Adequate resourcing of the weather stations, particularly for maintenance, had been an issue for some years before this event. This was, however, resolved soon after bleaching was demonstrably linked to water temperature, a parameter predicted to rise with climate change.

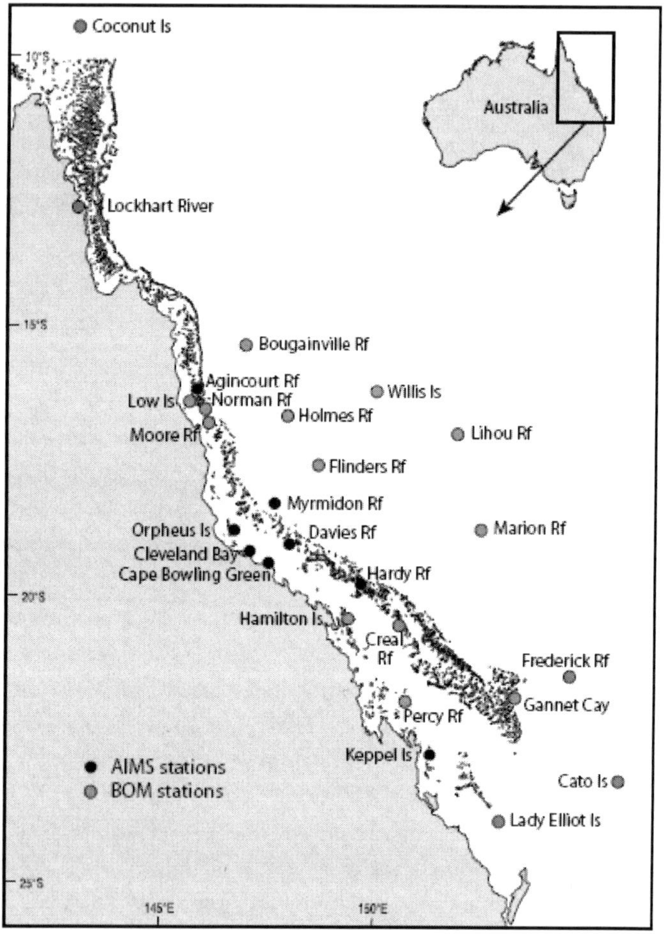

Figure 3 Current automatic weather stations on the GBR owned or managed by AIMS (black) and BOM (grey).

With the clearly increased susceptibility of inshore reefs during this event, additional stations were commissioned in partnership with the Great Barrier Reef Marine Park Authority (GBRMPA) at Cleveland Bay (1999), Halftide Rocks (Keppel Islands, 2000) and Orpheus Island (2002) in a three-way partnership with James Cook University (JCU; Figure 3). Over the years, a number of other stations have operated but have since been decommissioned either for logistical reasons or because research programs ceased.

2.2 AWS – current status

AIMS currently operates eight AWSs on the GBR (Figure 3) and one in Western Australia at Ningaloo Reef. For historical and practical reasons, the range of sensors on the AWS are limited to a number of basic meteorological variables (wind speed and direction, air temperature, above-water photosynthetically active radiation [PAR] and barometric pressure) and one oceanographic variable (sea temperature). These variables are measured at half-hourly intervals. Current and decommissioned stations and the type of data collected by each station are listed in Table 1.

All data (current and historical) are archived and served through the AIMS weather station portal.[8] There has been considerable effort to quality control the available data to correct for sensor drift and malfunction. Data are provided at three levels of quality assurance and quality control (QA/QC): Raw (uncontrolled, unprocessed), Level 1 (automatic rules applied to remove points) and Level 2 (manual visualisation of all data and correction where possible; suspect data points removed in Level 1 corrected where possible). Level 2 processing is generally undertaken bi-monthly in a batch process.

2.3 AWS – future opportunities and challenges

The real value of AIMS' AWS network data lies in the length of the records (25+ years for some stations) with relatively few gaps. As such, these data can be applied to address more complex research problems such as the influence of intra-seasonal, medium and longer-term climate variations (El Niño – Southern Oscillation and Pacific Decadal Oscillation) on marine ecosystems (e.g. Anthony et al. 2004). Capturing detailed environmental data during multiple disturbance events, such as coral bleaching, also allows a more comprehensive understanding of the physiological and ecological responses

---

[8] www.aims.gov.au/docs/data-centre/weatherstations.html

Table 1 List of past and current automatic weather stations operated by AIMS and the parameters measured.

Station	Start date	End date	Air temp	Water temp1 (~1m)	Water temp2 (5 – 9m)	Wind sp & dir	PAR	Barom pressure	Rain
Rib Reef	1980	1985	✓	×	×	✓	✓	✓	×
Hinchinbrook Island	1980	1985	✓	×	×	✓	✓	✓	✓
Myrmidon	1987	current	✓	✓	✓	✓	✓	✓	×
John Brewer	1986	1988	✓	×	✓	✓	✓	✓	×
Cape Bowling Green	1983	current	✓	×	×	✓	✓	✓	×
Daintree	1997	1998	✓	×	×	✓	✓	✓	✓
Agincourt	1991	current	✓	×	✓	✓	✓	✓	×
Davies	1991	current	✓	✓	✓	✓	✓	✓	×
Hardy	1989	current	✓	✓	✓	✓	✓	✓	×
Ningaloo	1997	current	✓	×	×	✓	✓	✓	×
Cleveland Bay1	1990	1992	✓	×	✓	✓	✓	✓	×
Cleveland Bay2	1999	current	✓	✓	✓	✓	✓	✓	×
Cape Cleveland	1993	1996	✓	×	×	✓	✓	✓	×
Halftide Rocks (Keppels)	2000	current	✓	✓	✓	✓	✓	✓	×
Orpheus Island	2002	current	✓	✓	✓	✓	✓	×	✓

and, ultimately, better predictions of future responses with continued global warming.

Although operating a network of real-time monitoring stations in the marine environment is expensive, the resultant benefit of better and more complete datasets is hard to overestimate. Aside from driving numerical models, monitoring environmental conditions in real-time enables nowcasting and forecasting of disturbances such as bleaching. This enhances science opportunities and assists reef managers in responding to these events (Hendee & Berkelmans 2002; Marshall & Schuttenberg 2006; Hendee et al. 2007, 2008). Through innovative management approaches, it also allows for reactive monitoring and management of activities which have the potential to impact on the reef such as dredging and coastal developments (Berkelmans et al. 2002). With the evolution of the weather station network into a real-time environmental monitoring network involving many more stations and parameters (see GBROOS), the challenge for the scientific community is to explore these datasets, develop better models of ecological and physical processes and deliver products which can be used to further advance research and management of the reef and provide better information to the wider community.

Significant opportunities exist to improve integration of monitoring data and value-adding to existing infrastructure. For example, the Bureau of Meteorology (BOM) operates up to 20 AWS on or near coral reefs in Queensland, none of which measure water temperature or any other oceanographic parameter. Fitting BOM weather stations with water temperature sensors would greatly increase the value of these stations to marine science and would be relatively inexpensive. Given the increased mandate for collaboration and data sharing among Commonwealth-funded agencies (see GBROOS), this would seem a logical next step.

Better integration of in situ and remotely-sensed monitoring data provides another opportunity to add value to our existing environmental monitoring infrastructure. In situ measurements have the advantage that data are locally accurate, although they lack spatial coverage. In contrast, remotely sensed data often lack local accuracy but offer wide spatial coverage. Combining the two to generate regionally specific algorithms and real-time calibration/validation of remotely sensed data using in situ data is an area of active research, particularly in chlorophyll, turbidity, SST and ocean current monitoring.

3. *Monitoring the physical environment of the GBR: temperature loggers*

3.1 Temperature loggers – history and development

As with the AWSs, development of a water temperature logger network on the GBR was largely driven by the availability of cheaper technology. However, while the first AWSs were deployed in the early 1980s the first temperature loggers were not deployed till the early 1990s. The reason for this delay is primarily to do with cost. Temperature loggers were available and indeed used in the 1980s but they were expensive and their use was generally limited to well-funded research areas such as oceanography in non-reef environments (see Oceanographic Moorings). Early temperature loggers were large and even in the early 1980s cost >$20,000 each, precluding widespread deployment. Significant reductions in costs (<$500 each) and size in the early 1990s allowed more widespread and multi-depth deployments. This was the first time that loggers could be deployed at sites representative of what reef corals actually experienced. The Seatemps temperature logger program was initiated by GBRMPA with the initial deployments at Magnetic Island in 1991 and at Orpheus Island in 1993. This expanded to a GBR-wide program in 1995 with deployments at many sites visited annually by the AIMS Long-Term Monitoring Program.[9] The temperature logger program was transferred to AIMS in 1999.

The present logger program relies on many volunteers who exchange the loggers with preparation, data handling and program coordination undertaken centrally. As such the choice of sites in the program are, to a large extent, determined by researchers, tourist operators and government agencies with fidelity and commitment to particular sites. After the initial establishment of the GBR-wide program, additional funding helped expand the program to Queensland ports in 1998 (Ports Corporation of Queensland), inshore GBR reefs in 2005 (GBRMPA) and Commonwealth Marine Protected Areas in the Coral Sea and Western Australia in 2006 (Department of Environment, Water, Heritage and the Arts).

---

[9] www.aims.gov.au/docs/data-centre/reef-monitoring-surveys.html

The temperature logger program is now well known and has made a valuable contribution to marine science across many disciplines including the development of bleaching thresholds (Berkelmans 2002), understanding fish migration (Speare & Steinberg 2001), seabird foraging success (Erwin & Congdon 2007), coral disease outbreaks (Bruno et al. 2007), upwelling monitoring, turtle nesting, physical oceanography, validation/calibration of paleo- and remotely sensed temperature data (Lough 1999; Alibert et al. 2003), risk assessments for introduced pests, marine stinger outbreaks, coral spawning, seagrass and coral and fish monitoring. An important application of these data has been the development of time-integrated thermal bleaching and mortality thresholds for reefs of the GBR (Berkelmans 2002, 2009). These demonstrate how the thermal thresholds for bleaching vary with location on the GBR and also provide the basis, combined with the AWS water temperature data, for near-real-time assessment of bleaching risk.[10] The

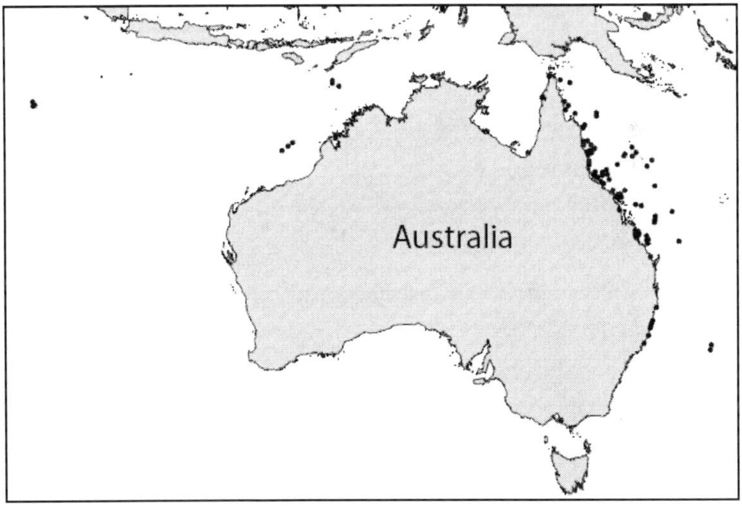

Figure 4 Map of current temperature logger locations in tropical and subtropical reefs of Australia.

---

[10] www.aims.gov.au/docs/data-centre/weatherstations.html

benefit of a shared data resource was self-evident and partnerships were formed with other agencies running their own logger programs with AIMS hosting data collected by partners. These partnerships have expanded the logger program to include sites in northern New South Wales and Lord Howe Island (New South Wales Marine Parks Authority), Australian Indian Ocean territories (Department of the Environment, Water, Heritage and the Arts, Parks Australia North) and Western Australia (Department of Environment and Conservation) (Figure 4).

3.2 Temperature loggers – current status

The Seatemps logger program currently consists of approximately 120 sites with a deployment of over 240 loggers. At most sites, loggers are deployed at multiple depths generally representative of shallow reef flat and deeper reef slope communities. Loggers usually record at half-hourly intervals and are exchanged approximately annually.

Retrieved loggers are sent to AIMS for downloading, data cleaning and archiving. As for the AWSs, data are manually inspected for outliers and systematic errors corrected where possible by comparing data from other loggers at the same site or nearby sites. Loggers are double or triple calibrated prior to each deployment to achieve a desired accuracy of ±0.1°C or better with minimal sensor drift over the 12-month deployment period. As of the end of 2008, the Seatemps database contained >18 million records, all of which are made available through the AIMS Data Centre Web portal.[11]

3.3 Temperature loggers – future opportunities and challenges

Expansion of the Seatemps logger program has been facilitated by a growing demand by researchers and reef managers for basic water temperature data. This trend is likely to continue with more partners and more activity within partnerships. As a result the logger program is likely to continue to expand. It will be a challenge to adequately resource the coordination and administration of the program, particularly in the areas of data handling, logger calibration and quality control of data.

---

[11] www.aims.gov.au/docs/data-centre/seatemperatures.html

The cost of loggers has continued to decline, while data quality and reliability (particularly at the low-cost end) has improved. Advances in circuitry and memory capacity now provide the opportunity to deploy loggers at sites visited infrequently (>2 years), at high temporal resolution (every 10 minutes), high precision and high accuracy. Achieving an accuracy of <±0.1°C, however, still requires manual calibration and post-processing of data since factory calibration remains coarse (up to ±0.3°C or more). Continued advances in technology, reduction in sensor cost and the ability to manage marine fouling will allow the range of sensors used on loggers for medium to long-term deployment to be increased. Potential parameters include pressure (depth/tide), turbidity, underwater light, dissolved oxygen, chlorophyll and conductivity sensors. Together these will enable considerably more detailed monitoring of the environments of marine communities in the future.

4. Monitoring the physical environment of the GBR: Great Barrier Reef Ocean Observing System (GBROOS)

4.1 GBROOS – history and development

Systematic and cost-effective observing systems throughout Australia's extensive ocean territory are a key requirement for assessing the nature and consequences of ongoing climate change. GBROOS is a geographic node of the Australian Integrated Marine Observing System (IMOS – see Chapter 9) and is funded by the Australian federal and Queensland state governments under the National Collaborative Research Infrastructure Strategy (NCRIS). Launched in 2008, this observing system is designed to measure and monitor the impact of the Coral Sea on the GBR, how the flow of oceanic water into the reef matrix affects corals and associated ecosystems, as well as how changes in water movement and water characteristics (e.g. temperature) affect the reef. The ability to link and model processes at a range of spatial and temporal scales is fundamental to understanding change in the GBR.

In a number of cases, GBROOS builds on and extends earlier long-term monitoring projects. A brief summary of the development of historical GBR observations is provided for each data type in the following sections.

4.2 GBROOS – current status

The experimental design of GBROOS supports the need to link processes occurring at a range of scales from regional (thousands of kilometres) down to the flow around individual coral heads (centimetres). This layered approach includes the ability to validate higher-level synoptic instruments with smaller-scale and higher temporal resolution instruments. Data platforms include satellites, underway sampling, fixed oceanographic moorings, sensor networks and ocean radar. A basic array of sensors has been progressively deployed since 2007 and the capability and extent of this network will be increased over the next few years.

4.2.1 Satellite observations

High-resolution (1km) synoptic satellite observations of the GBR from Townsville commenced in 1988 with the installation of an L band satellite reception facility at JCU and an image-processing workstation at AIMS. The facility was funded from the Northeast Australian Satellite Imagery System (NASIS) consortium members. Initially National Oceanic and Atmospheric Administration Advanced Very High Resolution Radiometer (NOAA-AVHRR) imagery was collected allowing mapping of SSTs of north-eastern Australian waters, the Coral Sea, and beyond, at 1 km resolution (Skirving et al. 2001). Prior to this satellite reception installation, imagery for the northern GBR was not captured by the Commonwealth Scientific and Industrial Research Organisation (CSIRO) station located in Melbourne. The facility, therefore, filled a gap in the national network which now provides full coverage of Australia's seas. Data are distributed nationally to CSIRO and the Australian Bureau of Meteorology (BOM) to produce a national composite SST product and are also used operationally to improve weather forecasts. The data are archived through the IMOS Australian Oceans Distributed Active Archive Centre (AO-DAAC) (King et al. 2009).

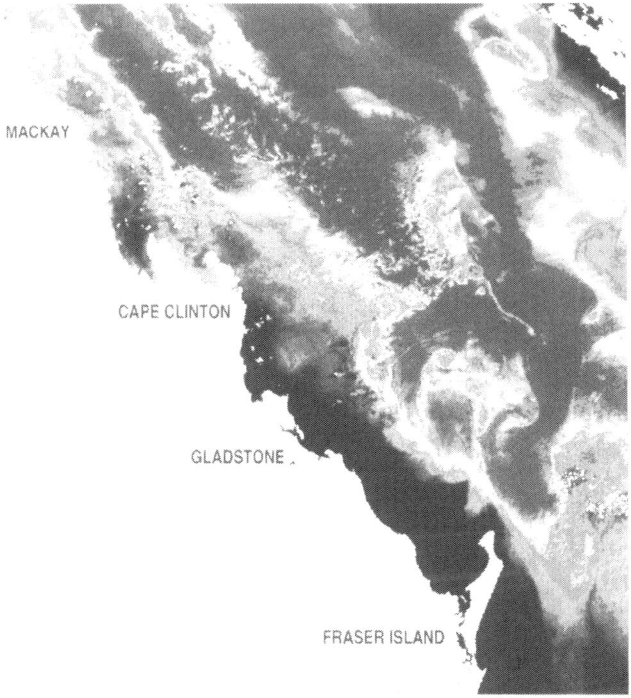

Figure 5 SST image of the Southern GBR showing a range of complex surface circulation features: speckled temperature in the Pompey and Swains reef matrix; well-defined East Australian Current (EAC) east of the shelf edge that recirculated in the vicinity of the Capricorn Bunker Group of reefs between Fraser Island and Gladstone; strong shelf edge front between the cool shelf waters and warm EAC.

SST imagery has been useful in mapping the major ocean currents of the GBR and Coral Sea and in revealing their complex and dynamic structures such as meandering eddies and frontal regions (Kleypas & Burrage 1994; Burrage et al. 1996). Figure 5 shows an example of a feature-rich SST image of the southern GBR. The SST data has also been used to calculate heat stress and operational tools are being developed to map and predict areas at risk of coral bleaching (Liu et al. 2003; Maynard et al. 2008).

In 1997 an additional L band facility was installed at AIMS allowing reception of the Sea Wide Field of view Sensor (SeaWiFS) aboard the Seastar satellite. The complete archive is held by the SeaWiFS project at the National Aeronautics and Space Administration's Goddard Space Flight Center[12] and an atlas is viewable on the AIMS' Data Centre web portal[13] (see Mahoney et al. 2006).

In 2007 the remote sensing capacity for the GBR was further enhanced with a dual X and L band satellite receiving station to supplement the existing L band receiver. This gives access to newer higher spectral and spatial resolution Moderate Resolution Imaging Spectroradiometer (MODIS) satellites as well as the lower L bandwidth satellites such as NOAA AVHRR and SeaWiFS. The two L band systems allow better optimisation of passes to reduce data loss. The satellite data are processed to give SST and ocean colour products such as chlorophyll-a data using existing processing algorithms.

Two sample products for the GBR from the MODIS imagery are shown in Figure 6. This image shows a major flooding event from the Burdekin River, south of Cape Bowling Green (Figure 3), and the extent to which the plume follows the coast in the GBR lagoon and eventually disgorges into the Coral Sea through the Grafton Passage off Cairns. The chlorophyll-a algorithm (Figure 6b) attempts to measure the concentration of this substance in oceanic waters. Successive imagery can also be used qualitatively in coastal regions to trace the extent to which low chlorophyll-a oceanic waters intrude into the higher chlorophyll-a and turbid GBR lagoon.

---

[12] oceancolor.gsfc.nasa.gov/SeaWiFS/

[13] www.aims.gov.au/docs/data-centre/satellite-observing.html

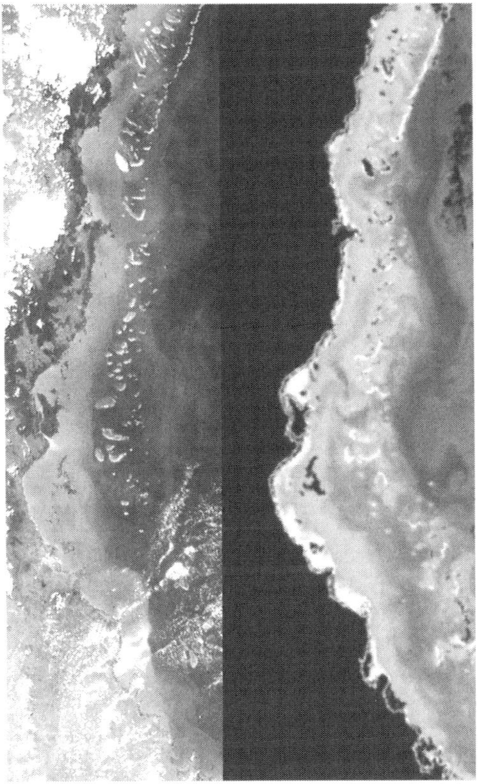

Figure 6 MODIS imagery of the Central GBR a) black and white image of the Burdekin River plume and other rivers along the wet tropics, 10 January 2007 and b) Chlorophyll-a image from 22 April 2004 on the right. The EAC can be seen here as low chlorophyll-a waters (darker grey).

In coastal waters the algorithm breaks down due to interference from viewing the seafloor, dissolved organic matter and suspended sediments. Research programs within the Marine Tropical Science Research Facility (MTSRF)[14] are attempting to differentiate the different contributions of these

[14] www.rrrc.org.au/mtsrf/index.html

factors by developing regionally specific algorithms. Ocean colour has also proved a useful alternative to SSTs in tropical and equatorial waters for identifying complex circulation features. SSTs tend to show fewer significant gradients than ocean colour and high levels of atmospheric water vapour can often mask the temperature signal.

Satellite data is validated by a range of real-time underway and static sampling stations. A radiometer, mounted on a commercial ferry in the Whitsundays, collects incoming irradiance data and surface reflectance data daily as the ferry travels to and from the reef. Data are downloaded daily and used to validate and correct the NOAA AVHRR SST data product. A number of buoy-mounted radiometers are also being deployed to give static validation locations. The SST validation data are used to develop more reliable algorithms for tropical waters and to facilitate the development of data products specifically suited to the complex tropical atmosphere (Beggs et al. 2009).

An ocean colour reference station is being installed at the end of the Lucinda jetty in north Queensland to measure incoming, outgoing and in-water radiance. A proposed set of reef-based stations will be matched to in-water sensing stations to correlate incoming radiation with in-water radiation and the levels of photosynthetically active radiation (PAR).

4.2.2 Underway sampling

At the next level below remote sensing data and validation stations is a range of under-way sampling equipment installed on several research and commercial vessels. Commercial vessels are useful as they transit a known path and can provide long-term consistent measurements of some basic water parameters such as temperature, salinity, turbidity and chlorophyll. This data can be used to validate the larger remote sensing data as well as forming a useful data set in its own right.

AIMS' two research vessels (RV Cape Ferguson and RV Solander) have been fitted with instrumentation and regularly transit the northeast and north-west tropical seas of Australia. They thus provide a large geographic spread of data and it is possible for the vessels to sample particular areas in response to unusual events (e.g. coral bleaching, river flood plumes), which is not possible with commercial ferries.

4.2.3 Oceanographic moorings

Direct ocean current measurements are sparse mainly due to the harsh environment and cost of mounting regular ship-based deployment and recovery programs. An early review of the physical oceanography of the GBR (Pickard et al. 1977) relied on ship-based observations recorded as notes on navigational charts and limited instrumental observations close to ports. The then inadequacy of observations and understanding of the physical oceanography of the GBR was highlighted as a result of a meeting held by GBRMPA and the Australian Academy of Sciences (Middleton 1983). The resulting report recommended targeting a number of key oceanographic processes across a range of spatial and temporal scales. Research over the next decade by AIMS, CSIRO, JCU and the University of New South Wales produced several process-based studies and supported early numerical modelling, summarised in Wolanski (1994).

A number of observational experiments in the early 1980s by JC Andrews and E Wolanski from AIMS attempted to establish permanent measurements but these ended after only a few years, due to lack of funding. In 1987 the Transports of the East Australian Current System (TEACS) heralded the first sustained measurement program aiming to capture interannual to decadal variability of the major current systems of the GBR (Burrage et al. 1995). This program was supported by AIMS and the CRC Reef and has been continued and expanded with the start of GBROOS.

The GBROOS mooring array consists of four pairs of sites. Each pair has a deep-water mooring located on the continental slope typically in 70–200m of water and a shallow outer shelf mooring typically in 35–70m (Figure 7). These moorings measure current profiles using acoustic Doppler techniques supplemented with surface and bottom Wetlabs Water Quality Monitors that measure temperature, pressure, conductivity (salinity), dissolved oxygen, fluorescence and turbidity.

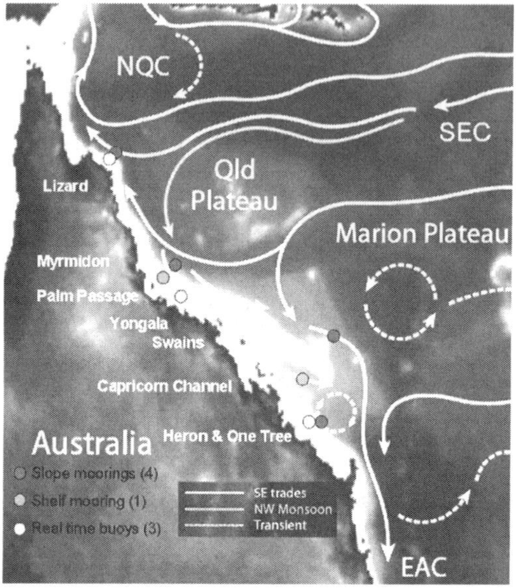

Figure 7 Location of the GBROOS Moorings array. Arrows show persistent ocean surface currents.

The most northerly moorings, near Lizard Island, measure ocean flows and water properties of the northern arm of the South Equatorial Current as it moves north to form the Hiri Current as well as the on-shelf movement of water through a convoluted reef passage into the GBR lagoon. The central pair of moorings is located off Townsville where the outer reef matrix is quite sparse and Coral Sea water easily intrudes into the main GBR lagoon through the deep and wide channels of the Palm and Magnetic Passages (Brinkman et al. 2002). This area provides a significant inflow of ocean water and much of the southward movement of water though the GBR lagoon. These flows are assisted by the change in orientation of the shelf edge from north-south to northwest-southeast. This south-eastward movement of water occurs both along the outside of the reef matrix as the East Australian Current (EAC) due to the barrier nature of the outer reefs and, via inflow of water in the Townsville region, through the lagoon of the GBR.

The southern pair of moorings consists of one at the mouth of the Capricorn Channel measuring the water exchange at the notional southern exit point of the GBR lagoon and one mooring on the south-eastern tip of the Swains measuring water travelling past the outside extremity of the Swains reef matrix. At this point the reef forms a dense barrier between the ocean and lagoon waters. Finally, there are three moorings near Heron and One Tree Islands in the Capricorn Bunker region of the GBR. Here the EAC meanders significantly and regularly forms a recirculation that impinges onto the Capricorn Bunker shelf and occasionally intrudes north-westward up the Capricorn Channel (Figure 5).

Figure 8 Network repeater sensor-pole on Heron Island.

4.2.4 Sensor networks

The final component of GBROOS is a series of sensor-network deployments at seven sites; four at island research stations (Heron and One Tree Islands in the south, Orpheus Island in the central GBR and Lizard Island in the north) and three reef sites (Davies, Rib and Myrmidon reefs in the central GBR). At each site a high-speed back-haul data link is installed, using traditional line-of-sight microwave satellite communications or 3G mobile telephone where the site is within range.

Once the communications link is installed, an on-water wireless data network is created using a series of poles (Figure 8) located around the reef using spread-spectrum radios. A number of buoys, ranging from 1.4 m to 2.5 m in size, equipped with sensors are located in this network (Figure 9). The initial suite of sensors deployed includes SeaBird Electronics SBE39 (for temperature and depth) and SBE37 (for temperature, depth and salinity) along with a series of thermistor strings.

Figure 9 Sensor buoy at Heron Island.

The deployment at One Tree Island is shown in Figure 10. This coral reef system is characterised by a shallow sand interior and a coral pavement rim that drops down into the main lagoon with several coral micro-atolls, typically 50–150m in diameter. A sensor pole is installed in the centre of the micro-atoll on the sandy bottom with a sensor string running across the atoll to the coral rim and up and over the rim down into the main lagoon. The design has a temperature and pressure sensor (SBE39) at the base of the pole (Sensor 'A' in Figure 10) and a thermistor string running into the lagoon (sensor 1–6 in Figure 10) with a second SBE39 at the end of the thermistor string (Sensor 'B' in Figure 10). The Sea-Bird sensors provide a reference for the cheaper thermistor string while the pressure sensor gives information about tidal movement and subsequent flushing.

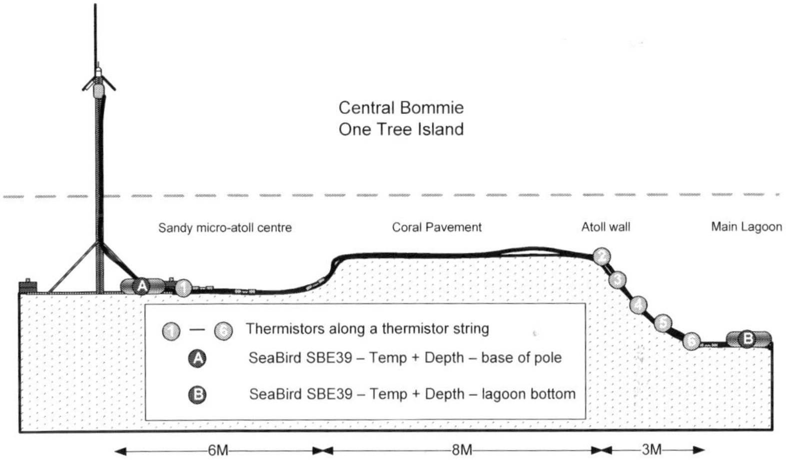

Figure 10 Deployment diagram for Central Bommie – One Tree Island.

Initial deployments include mostly traditional oceanographic instruments matching those used on the oceanographic moorings; this gives a link between the mooring instrumentation and that of the sensor networks. The design of the equipment allows any standards-based sensor to be attached including video and image sensors. The intention is to provide the platform and infrastructure for future deployments as new sensors become available. Areas currently under investigation include pCO_2 (for ocean acidification research) as well as biosensors. An example of a new biosensor is the CellSense® sensor that uses *Escherichia coli* bacterial cells for rapid eco-toxicity analysis. The sensor uses ferricyanine to divert electrons from the respiratory system of the immobilised bacteria to a suitable carbon electrode. The resulting current is a measure of bacterial respiratory activity and any changes due to pollutants or other processes can be detected as a change in the magnitude of the current (see Rodriguez-Mozaz et al. 2004). While these novel sensors are not ready for long-term deployments they open a new range of techniques and opportunities for fine scale biochemical sensing within the marine environment. The main problems with sensor deployments are fouling of the sensor head, sensor drift and maintaining sensor reliability in the harsh marine environment.

4.2.5 Ocean radar

GBROOS is also facilitating the installation of a high frequency ocean radar system in the Capricorn Bunker area of the GBR. The installation has one transceiver on Tannum Sands on the Queensland coast and one on Lady Elliot Island. This gives an area of coverage over most of the Capricorn Bunker reefs including Heron and One Tree Islands. The radar provides instantaneous surface currents and waves for both vector and velocity (Heron et al. 2006).

4.3 GBROOS – future opportunities and challenges

The GBROOS deployments are designed to provide data to support understanding of how processes at differing scales interact. This will provide information about how ocean processes are reflected in within-reef water dynamics and how this in turn links to the responses observed in the biological components of the GBR. The present observational focus has been on the temperatures and movements of ocean waters. There is a clear need to expand the range of routinely observed variables to include, for example, light (PAR, UV and total solar radiation) given its significance in enhancing and suppressing coral bleaching occurrence and intensity.

In the Heron Island/One Tree Island area the remotely sensed satellite imagery and ocean surface radar together with the mooring array is focused on measuring the complex dynamics of the EAC recirculation and its impact on the outer shelf. This wealth of data will provide improved insights into how physical environmental changes impact coral reef ecology in this sector of the GBR. Sustained, long-term measurements allow observation of important processes and detection of long-term trends. The data are currently supporting a range of hydrodynamic modelling studies which aim to produce real-time and high-resolution predictive models of this area (Webster et al. 2007; Schiller et al. 2008, 2009).

A fundamental challenge to ocean observing is dealing with the real-time data provided by the new sensor networks. Such real-time data can guide sampling regimes and scheduling field work to effectively monitor biological responses. One example is taking samples for bio-markers of stress, such as stress proteins, to monitor the impact of a warm-water event on corals. Another example is using the entire environmental history of a coral colony to better understand the set of conditions likely to lead to adverse biological responses rather than just having a number of snapshots or data from other areas and extrapolating to the area of interest.

Real-time data makes data capture and management more complex. Reliable communications systems are needed between reefs often 50 or more kilometres from any existing infrastructure to the main data processing centre. Reliable communications are also needed for buoys and other structures that move and are prone to fouling, corrosion and physical wear and tear. Finally, as the number of sensors increases the load on processing and storage of real-time data also increases.

It is common for logged data to go through extensive quality control processes, often involving comparisons with other data and manual checking. This level of validation is not possible with real-time data; such data have only cursory quality control applied mostly via an automated set of checks. Users of the data need to understand the condition of the data and use it accordingly. This is common in the meteorological community where real-time data is a fundamental part of the modelling and predictive work but relatively new to the marine and environmental communities.

The data are currently checked by a series of database checks that test for out of bounds data (values too high or low to be valid), values that change suddenly (rate of change) and values that lie outside of three standard deviations from the long term mean. The system does not delete data but flags it as bad using the quality control flags used by the Argo project (UNESCO/IOC/IODE & MAST, 1993. Manual of Quality Control Procedures for validation of Oceanographic Data – UNESCO Manuals and Guides 26). The quality control system is linked to an alert system so that if large amounts of data are found to be bad then an email is sent to the operator to alert them to the potential issue. The output of the quality control systems is a 'Level-1' data product that has automated quality control applied and all values flagged. All data products that come from the Level-1 data are colour coded with the quality control flag to ensure users understand the quality of the data they are using.

Higher level products are also produced using manual quality control measures to deliver Level-2 products where points that are known to be bad are removed and other more complex corrections are applied (such as manual de-spiking and de-trending to account for sensor drift). It is anticipated that the Level-2 data product is what most end users will use; only people requiring real-time or raw data will use the Level-1 product. Note that the Level-1 data is made available within 10 to 20 minutes of collection; the Level-2 data is only made available every few months or so.

The management of real-time sensor data is being developed by a number of groups. One of these, the Open Geographic Information Systems Consortium (OGC), has developed a technology 'stack' called the Sensor Web Enablement (SWE). This provides standards and reference software to facilitate the integration and discovery of sensors and sensor data (OGC 2008. Sensor Web Enablement Working Group).[15] The goal is for software and human systems to be able to discover sensors and then access and process the resulting sensor information into integrated data products. The standards also cover issues such as monitoring sensors, controlling and changing the sensing parameters and activating sensors based on other events or parameters.

---

[15] www.opengeospatial.org/projects/groups/sensorweb

The current GBROOS sensor deployments use traditional technologies with non-intelligent sensors and most of the processing and control is undertaken in the data centre. New 'smart' sensors have a rules-based system by which the sensor can alter its sampling and sensing characteristics, monitor its own health and recalibrate or re-set as required, communicate with the network and determine the best form of communication based on its set of rules, and can process the data providing quality controlled synthesised data as well as raw measurements. In the future, sensors on a reef buoy, for example, will be able to 'talk' to sensors on other platforms, access a range of other data such as tidal, remote sensing imagery and weather forecasts, and run all this through an internal analytical engine to determine what to measure, what sampling regime to implement and what data to send back.

These new flexible designs moves us towards a system that can model events as they unfold using data the sensor collects itself and data it obtains from other sensors. While these scenarios are some way off the underlying technologies are being developed now and there are a number of terrestrial examples that showcase much of this functionality.

The ultimate goal of the GBROOS infrastructure is to build improved understanding and hence better models of the interactions between large-scale oceanic processes and smaller-scale reef processes. The ability to use real-time data in data-assimilating models will result in a new generation of predictive models for the tropical marine environment of the GBR. The link between changes in our complex ocean systems and the observed biological responses is vital to understanding the impact of a changing climate on the GBR. GBROOS aims to provide some of the observations that, through modelling and visualisation tools, will increase our understanding and open up new ways of managing and sustaining the GBR.

5. The need for ongoing monitoring of the GBR in a rapidly changing climate

The GBR is a huge, complex and dynamic ecosystem that has aesthetic, social and economic value both nationally and internationally. Its degradation or even loss as a consequence of human-induced climate change would be a tragedy (Veron 2008). Drastic reductions in human greenhouse gas emissions are necessary to avert such a worst-case scenario. Even with such globally-driven changes in our relationship with energy production, climate of the GBR will be changing for the foreseeable future. Successful management and protection of the GBR in a changing climate requires the best possible knowledge of how the three-dimensional physical environment is changing that can be linked to ecological responses. This requires ongoing commitment to GBR ocean climate monitoring and research investment linking physical to biological outcomes. The GBR, through the various monitoring programs described here, has a significant advantage over many other coral reef environments in this regard and such programs may increase the chances of preserving parts of the GBR in a rapidly changing physical environment.

Acknowledgements

Long-term and ongoing monitoring of the physical environment of the GBR would not be possible without the skilled contributions of AIMS technical workshops, GBROOS staff, crews of AIMS research vessels and AIMS' field operations staff (past, present and future). GBROOS is funded by the Integrated Marine Observing System (IMOS) – an initiative of the Australian Government being conducted as part of the National Collaborative Research Infrastructure Strategy, and the Queensland State Government. The support of the Tropical Marine Network (University of Sydney, Australian Museum, University of Queensland and James Cook University) who run research stations at One Tree, Lizard, Heron and Orpheus Islands, respectively, on the GBR is also acknowledged. Financial contributors to the temperature logger program over the years include AIMS, GBRMPA, CRC Reef, Ports Corporation of Queensland and the Department of Environment, Water, Heritage and the Arts. The temperature logger program would not be possible without an army of volunteers from various government agencies, tourist operators, academia, conservations groups and private individuals who have helped to exchange temperature loggers. AIMS also thanks

Quicksilver Cruises (Agincourt Reef) and Fantasea Cruises (Hardy Reef) who have hosted and helped maintain Automatic Weather stations on their tourist pontoons as well as assisting with the temperature logger program. The authors also thank the reviewers for their detailed and helpful comments on the original version of this chapter.

References

Access Economics (2007). Measuring the economic and financial values of the Great Barrier Reef Marine Park 2005–6. Report by Access Economics Pty Ltd for Great Barrier Reef Marine Park Authority, February 2007. Available from www.gbrmpa.gov.au.

Achituv Y & Dubinsky Z (1990). Evolution and zoogeography of coral reefs. In Z Dubinsky (Ed), *Ecosystems of the world* 25: *coral reefs.* Amsterdam: Elsevier.

Alibert C, Fallon SJ, Kinsley L, McCulloch MT, Berkelmans R & McAllister F (2003). Source of trace element variability in Great Barrier Reef corals affected by the Burdekin flood plumes. *Geochimica Cosmochimica Acta,* 67, 231–46.

Anthony KRN, Ridd PV, Orpin AR, Larcombe P & Lough JM (2004). Temporal variation of light availability in coastal benthic habitats: effects of clouds, turbidity and tides. *Limnology and Oceanography,* 49, 2201–11.

Baker AC, Glynn PW & Riegl B (2008). Climate change and coral reef bleaching: an ecological assessment of long-term impacts, recovery trends and future outlook. *Estuarine, Coastal and Shelf Science,* 80, 435–71.

Beggs H, Verein R, Kippo H, Underwood M, Barton I, Steinberg C et al. (2009). Enhancing ship of opportunity sea surface temperature observations in the Australian region. 9th International Conference on Southern Hemisphere Meteorology and Oceanography. Melbourne, Australia, 9–13 February 2009. Available at www.bom.gov.au/events/9icshmo/manuscripts/PT_A2_Beggs.pdf.

Berkelmans R (2002). Time-integrated thermal bleaching thresholds of reefs and their variation on the Great Barrier Reef. *Marine Ecology Progress Series,* 229, 73–82.

Berkelmans R (2009). Bleaching and mortality thresholds: how much is too much? In MJH van Oppen & JM Lough (Eds), *Coral bleaching: patterns, processes, causes and consequences,* (pp103–19). Ecological Studies, Vol. 205, Berlin, Heidelberg: Springer-Verlag.

Berkelmans R & Oliver JK (1999). Large scale bleaching of corals on the Great Barrier Reef. *Coral Reefs,* 18, 55–60.

Berkelmans R, De'ath G, Kininmonth S & Skirving WJ (2004). A comparison of the 1998 and 2002 coral bleaching events on the Great Barrier Reef: spatial correlation, patterns and predictions. *Coral Reefs,* 23, 74–83.

Berkelmans R, Hendee JC, Marshall PA, Ridd PV, Orpin AR & Irvine D (2002). Automatic weather stations: tools for managing and monitoring potential impacts to coral reefs. *Marine Technology Society Journal,* 36, 29–38.

Boyett HV, Bourne DG, Willis BL (2007). Elevated temperature and light enhance progression and spread of black band disease on staghorn corals of the Great Barrier Reef. *Marine Biology,* 151, 1711–20.

Brinkman R, Wolanski E, Deleersnijder E, McAllister F & Skirving W (2002). Oceanic inflow from the Coral Sea into the Great Barrier Reef. *Estuarine, Coastal and Shelf Science,* 54, 655–68.

Brohan P, Kennedy JJ, Harris I, Tett SFB & Jones PD (2006). Uncertainty estimates in regional and global observed temperature changes: a new dataset from 1850. *Journal of Geophysical Research* 111, D12106, doi: 10.1029/2005JD006548.

Bruno JF, Selig ER, Casey KS, Page CA, Willis BL, Harvell CD, Sweatman H & Melendy AM (2007). Thermal stress and coral cover as drivers of coral disease outbreaks. *PLoS Biology,* 5(6), e124, doi:10.1371/journal.pbio.0050124.

Buddemeier RW, Kleypas JA & Aronson RB (2004). *Coral reefs and global climate change.* Arlington, VA: Pew Center on Global Climate Change.

Burrage DM, Steinberg CR, Skirving WJ & Kleypas JA (1996). Mesoscale circulation features of the Great Barrier Reef region inferred from NOAA satellite imagery. *Remote Sensing of the Environment,* 56, 21–41.

Burrage D, Steinberg C, Bode L, Black K (1995). *Long-term current observations in the Great Barrier Reef.* State of the Great Barrier Reef World Heritage Area Report, Technical Workshop, Townsville, pp27–29. November 1995.

Cheal AJ, Delean S, Sweatman H & Thompson AA (2007). Spatial synchrony in coral reef fish populations and the influence of climate. *Ecology,* 88, 158–69.

Coles SL, Jokiel PL & Lewis CR (1976). Thermal tolerance in tropical versus subtropical Pacific reef corals. *Pacific Science,* 30, 159–66.

De'ath G, Lough JM & Fabricius KE (2009). Declining coral calcification on the Great Barrier Reef. *Science,* 323, 116–19.

Erwin CA & Congdon BC (2007). Day-to-day variation in sea-surface temperature reduces sooty tern *Sterna fuscata* foraging success on the Great Barrier Reef, Australia. *Marine Ecology Progress Series,* 331, 255–66.

Feely RA, Sabine CL, Lee K, Berelson W, Kleypas J, Fabry VJ & Millero FJ (2004) Impact of anthropogenic CO_2 on the $CaCO_3$ system in the oceans. *Science,* 305, 362–66.

Harvell CD, Kim K, Burkholder JM, Colwell RR, Epstein PR, Grimes DJ et al. (1999). Emerging marine diseases – climate links and anthropogenic factors. *Science* 285: 1505–10.

Hendee JC & Berkelmans R (2002). Expert system generated coral bleaching alerts for Myrmidon and Agincourt Reefs, Great Barrier Reef, Australia. In *Proceedings 9th International Coral Reef Symposium, Bali, 2000,* Ministry of Environment, Indonesian Institute of Sciences and International Society for Reef Studies, 2, 1099–104.

Heron ML, Willis B, Prytz A, Cetina-Heredia P, Mao Y, Hoegh-Guldberg O et al. (2006). *HF ocean surface radar monitoring for coral bleaching in the Great Barrier Reef. OCEANS* 2006, September, 1–5.

Hoegh-Guldberg O, Mumby PJ, Hooten AJ, Steneck RS, Greenfield P, Gomez E et al. (2007), Coral reefs under rapid climate change and ocean acidification. *Science,* 318, 1737–42.

Hendee JC, Gramer L, Kleypas JA, Manzello D, Jankulak M & Langdon C (2007). The integrated coral observing network: sensor solutions for sensitive sites. In *Proceedings of 3rd International Conference on Intelligent Sensors, Sensor Networks and Information Processing.* 2–7 December 2007, Melbourne, Australia.

Hendee J, Gramer L, Manzello D & Jankulak M (2008). Ecological forecasting for coral reef ecosystems. In Proceedings 11th International Coral Reef Symposium 7–11 July 2008, Ft. Lauderdale, Florida.

Hopley D, Smithers SG & Parnell KE (2007). *The geomorphology of the Great Barrier Reef: development, diversity and change.* Cambridge: Cambridge University Press.

Hughes TP, Baird AH, Bellwood DR, Card M, Connolly SE, Folke C et al. (2003). Climate change, human impacts and the resilience of coral reefs. *Science,* 301, 929–33.

Hutchings P, Kingsford M & Hoegh-Guldberg O (Eds) (2008). *The Great Barrier Reef: biology, environment and management.* Collingwood, Vic.: CSIRO Publishing.

Johnson JE & Marshall PA (2007). *Climate change and the Great Barrier Reef: a vulnerability assessment.* Great Barrier Reef Marine Park Authority and Australian Greenhouse Office, Australia.

Jokiel PL & Coles SL (1977). Effects of temperature on the mortality and growth of Hawaiian reef corals. *Marine Biology,* 43, 210–18.

King EA, Mak PPY, Turner PJ, Smith G, Suber K, Paget MJ et al. (2008). A distributed data system for marine research. eResearch Australasia 2008, 29 September – 1 October 2008, Melbourne, Australia.

Kleypas JA & Burrage DM (1994). Satellite observations of surface circulation patterns in the southern Great Barrier Reef, Australia. *International Journal of Remote Sensing,* 15, 2051–63.

Kleypas JA & Eakin CM (2007). Scientists' perceptions of threats to coral reefs: results of a survey of coral reef researchers. *Bulletin of Marine Science,* 80, 419–36.

Kleypas JA, McManus JW & Menez LAB (1999). Environmental limits to coral reef development: where do we draw the line? *American Zoologist,* 39, 146–59.

Kleypas JA, Buddemeier RW & Gattuso J-P (2001). The future of coral reefs in an age of global change. *International Journal of Earth Science,* 90, 426–37.

Kleypas JA, Feely RA, Fabry VJ, Langdon C, Sabine CL & Robbins LL (2006). *Impacts of ocean acidification on coral reefs and other marine calcifiers: a guide for future research.* Report of a workshop held in St Petersburg FL 18–20 April 2005, sponsored by NSF, NOAA and the US Geological Survey.

Kuffner IB, Andersson AJ, Jokiel PL, Rodgers KS & Mackenzie FT (2008). Decreased abundance of crustose coralline algae due to ocean acidification. *Nature Geoscience,* 1, 114–17.

Liu G, Skirving W & Strong AE (2003). Remote sensing of sea surface temperatures during 2002 Barrier Reef coral bleaching. *EOS,* 84(15), 137–44.

Lough JM (1999). *Sea surface temperatures on the Great Barrier Reef: a contribution to the study of coral bleaching.* Great Barrier Reef Marine Park Authority Research Publication No. 57.

Lough JM (2007). Climate and climate change on the Great Barrier Reef. In JE Johnson & PA Marshall (Eds), *Climate change and the Great Barrier Reef: a vulnerability assessment,* (pp15–50). Great Barrier Reef Marine Park Authority and Australian Greenhouse Office, Australia.

Lough JM (2008a). A changing climate for coral reefs. *Journal of Environmental Monitoring,* 10, 21–29.

Lough JM (2008b). Shifting climate zones for Australia's tropical marine ecosystems. *Geophysical Research Letters,* 35, L14708, doi: 10.1029/2008GL034634.

Lynch TP, Mclaughlan D, Hughes D, Cherry D, Critchley G, Allen S et al. (2008). A National Reference Station infrastructure for Australia – using telemetry and central processing to report multi-disciplinary data streams for monitoring marine ecosystem response to climate change. Oceans 2008 MTS IEE Oceans, Poles and Climate: Technological Challenges. 15–18 September 2008, Quebec City, Canada.

Mahoney MV, Slivkoff MM, Rehbein MA, Crossman DJ & Steinberg CR (2006). AIMS-CRC Reef remote sensing web atlas. Australian Institute of Marine Science. Available at: www.aims.gov.au/docs/data-centre/satellite-observing.html.

Manzello DP, Kleypas JA, Budd DA, Eakin CM, Glynn PW & Langdon C (2008). Poorly cemented coral reefs of the eastern tropical Pacific: possible insights into reef development in a high-CO_2 world. *Proceedings of the National Academy of Sciences USA,* 105, 10450–55.

Marshall PA & Baird AH (2000). Bleaching of corals on the Great Barrier Reef: differential susceptibilities among taxa. *Coral Reefs,* 19, 155–63.

Marshall P & Schuttenberg H (2006). Adapting coral reef management in the face of climate change. In JT Phinney, O Hoegh-Guldberg, J Kleypas, W Skirving & A Strong (Eds), *Coral reefs and climate change; science and management,* (pp 223–41). Coastal and Estuarine Studies 61, American Geophysical Union.

Massel SR &Done TJ (1993). Effects of cyclone waves on massive coral assemblages on the Great Barrier Reef: meteorology, hydrodynamics and demography. *Coral Reefs,* 12, 153–66.

Maynard JA, Turner PJ, Anthony KRN, Baird AH, Berkelmans R, Eakin CM, Johnson J, Marshall PA, Packer GR, Rea A & Willis BL (2008). ReefTemp: an interactive monitoring system for coral bleaching using high-resolution SST and improved stress predictors, *Geophysical Research Letters,* 35, L05603, doi:10.1029/2007GL032175.

Meehl GA, Stocker TF, Collins WD, Friedlingstein P, Gaye AT, Gregory JM et al. (2007). Global climate projections. In Solomon S, Qin D, Manning M, Chen Z, Marquis M, Averyt KB et al. (Eds), *Climate change 2007: the physical science basis. Contribution of Working Group*

I to the fourth assessment report of the Intergovernmental Panel on Climate Change, Cambridge, UK and New York, NY, USA: Cambridge University Press.

Middleton JH (1983). Report of the scientific discussion meeting on the physical oceanography of the Great Barrier Reef region: held at the University of New South Wales 5–6 July 1982 with support from the Great Barrier Reef Marine Park Authority and the Australian Academy of Science. Great Barrier Reef Marine Park Authority Technical Memorandum 5, GBRMPA, Townsville, Australia 42pp. Available at: www.gbrmpa.gov.au/corp_site/info_services/publications/tech_memorandums/tm005/technical_memorandum_5.

Munday PL, Jones GP, Pratchett MS & Williams AJ (2008). Climate change and the future for coral reef fishes. *Fish and Fisheries,* 9, 261–85.

Munday PL, Leis JM, Lough JM, Paris CB, Kingsford MJ, Berumen ML & Lambrechts J (2009). Climate change and coral reef connectivity. *Coral Reefs,* doi 10.1007/s00338-008-0461-9.

Pickard GL, Donguy JR, Henin C & Rougherie F (1977). *A review of the physical oceanography of the Great Barrier Reef, Western Coral Sea.* Canberra: Australian Government Publishing Service.

Pitcher CR, Doherty P, Arnold P, Hooper J, Gribble N et al. (2007). *Seabed biodiversity on the continental shelf of the Great Barrier Reef World Heritage area.* AIMS/CSIRO/QM/QDPI, CRC Reef Research Task Final Report.

Poloczanska ES, Babcock RC, Butler A, Hobday AJ, Hoegh-Guldberg O, Kunz TJ et al. (2007). Climate change and Australian marine life. *Oceanography and Marine Biology: An Annual Review,* 45: 407–78.

Puotinen ML, Done TJ & Skelly WC (1997). *An atlas of tropical cyclones in the Great Barrier Reef region, 1969–97.* CRC Reef Research Centre Technical Report No. 19, Townsville: CRC Reef Research Centre.

Rayner NA, Parker DE, Horton EB, Folland CK, Alexander LV, Rowell DP et al. (2003). Globally complete analyses of sea surface temperature,

sea ice and night marine air temperatures, 1871–2000. *Journal of Geophysical Research,* 108: 4407, doi: 10.1029/2002JD002670.
Rodriguez-Mozaz S, Marco M, Lopez de Alda M & Barceló D (2004). Biosensors for environmental applications: future development trends. *Journal of Pure and Applied Chemistry,* 76: 723–52.
Sabine CL, Feely RA, Gruber N, Key RM, Lee K, Bullister JL et al. (2004). The oceanic sink for anthropogenic CO_2. *Science,* 305: 367–71.
Schiller A, Ridgway KR, Steinberg CR & Oke PR (2009). Dynamics of three anomalous SST events in the Coral Sea. *Geophysical Research Letters,* 36, L06606, doi: 10.1029/2008GL036997.
Schiller A, Oke PR, Brassington G, Entel M, Fiedler R, Griffin DA & Mansbridge JV (2008). Eddy-resolving ocean circulation in the Asian-Australian region inferred from an ocean reanalysis effort. *Progress in Oceanography,* 76, 334–65.
Skirving W & Guinotte J (2001). The sea surface temperature story on the Great Barrier Reef during the coral bleaching event of 1998. In Wolanski E (Ed), *Oceanographic processes of coral reefs. physical and biological links in the Great Barrier Reef,* (pp301–13). Boca Raton, Florida: CRC Press.
Skirving WJ, Mahoney MV & Steinberg CR (2001). Sea surface temperature atlas of the Great Barrier Reef, 1990–2000. Version 1. Australian Institute of Marine Science & CRC Reef Research Centre. CD-ROM.
Skirving W, Heron M & Heron S (2006). The hydrodynamics of a bleaching event: implications for management and monitoring. In JT Phinney, O Hoegh-Guldberg, J Kleypas, W Skirving & A Strong (Eds), *Coral reefs and climate change science and management,* (pp145–61). Coastal and Estuarine Studies 61, Washington, DC: American Geophysical Union.
Speare P & Steinberg CR (2001). Environmental factors influencing the activity of black marlin. In E Wolanski (Ed), *Oceanographic processes of coral reefs: physical and biological links in the Great Barrier Reef,* (pp237–53). Boca Raton, Florida: CRC Press,

Steinberg C (2007). Impacts of climate change on the physical oceanography of the Great Barrier Reef. In JE Johnson & PA Marshall (Eds), *Climate change and the Great Barrier Reef: a vulnerability assessment,* (pp51–74). Great Barrier Reef Marine Park Authority and Australian Greenhouse Office, Australia.

Stoddart DR (1969). Ecology and morphology of recent coral reefs. *Biological Reviews,* 44, 433–98.

Tanzil JTI, Brown BE, Tudhope AW & Dunne RP (2009). Decline in skeletal growth of the coral *Porites lutea* from the Andaman Sea, South Thailand between 1984 and 2005. *Coral Reefs,* doi: 10.1007/s00338-008-0457-5.

Trenberth KE, Jones PD, Ambenje P, Bojariu R, Easterling D, Tank AK et al. (2007). Observations: surface and atmospheric climate change. In S Solomon, D Qin, M Manning, Z Chen, M Marquis, KB Averyt, M Tignor & HL Miller (Eds), *Climate change 2007: the physical science basis. Contribution of Working Group I to the fourth assessment report of the Intergovernmental Panel on Climate Change,* (pp235–336). Cambridge, UK and New York, NY, USA: Cambridge University Press.

van Oppen MJH & Lough JM (Eds) (2009). *Coral bleaching: patterns, processes, causes and consequences.* Ecological Studies Volume 205, Berlin, Heidelberg: Springer-Verlag.

Veron JEN (2000). *Corals of the world.* Townsville: Australian Institute of Marine Science.

Veron JEN (2008). *A reef in time: the Great Barrier Reef from beginning to end.* Cambridge, Massachusetts: Harvard University Press.

Webster IT, Brinkman R, Parslow J, Prange J, Stevens ADL & Waterhouse J (2007). *Review and gap analysis of receiving-water water quality modelling in the Great Barrier Reef.* CSIRO Water for a Healthy Country Flagship.

Wolanski E (1994). *Physical oceanographic processes of the Great Barrier Reef.* Boca Raton, Florida: CRC Press.

Wolanski E (Ed) (2001). *Oceanographic processes of coral reefs: physical and biological links in the Great Barrier Reef.* Boca Raton, Florida: CRC Press.
Wilkinson C (Ed) (2004). *Status of coral reefs of the world: 2004.* Townsville, Australia: Global Coral Reef Monitoring Network, Australian Institute of Marine Science.
Wilkinson C (Ed) (2008). *Status of coral reefs of the world: 2008.* Townsville, Australia: Global Coral Reef Monitoring Network, Australian Institute of Marine Science.

4
Are the deep waters of the Weddell Sea still warming?

Robin Robertson

*Abstract**

The Warm Deep Water (WDW) in the Weddell Sea warmed from the 1970s to the 1990s, and then cooled starting in the mid-1990s. The question is what has happened with the WDW since then, and what is happening now?

In a previous study, temporal temperature changes in the Weddell Sea deep water masses were evaluated using a historical data set encompassing Conductivity/Temperature/Depth (CTD) and bottle data from 1912 to 2000. The deep water masses investigated were WDW and Weddell Sea Deep Water (WSDW). This data set has now been extended to 2005 and reanalysed for temporal trends in the WDW and WSDW. A general warming trend was present in the WDW from the 1960s to the present. Two significant cooling episodes occurred during this period: one associated with the Weddell Polynyas of the mid-1970s and the other in the mid- to late 1990s. From the 1970s to the 1990s, the warming trend was ~0.012 ±0.007°C yr^{-1}. But the low values during the Weddell Polynyas bias this trend, and a more representative value is ~0.007±0.004°C yr^{-1} from 1960 to 2005. The warming trend is comparable to the global average surface water warming observed by Levitus et al., to the warming of the WSBW in the central Weddell Sea observed by Fahrbach et al., and to the surface ice temperature warming from 1970 to 1998 in the Weddell Sea observed by Comiso. The higher temperatures, along with a slight

* Chapter originally submitted in December 2008.

freshening, resulted in less dense WDW. The core of the WDW also moved higher in the water column by 150–200 m. There appeared to be a warming trend in the Weddell Sea Deep Water (WSDW) below 1500m; however, high data uncertainties prevented identification of a well-defined temporal trend. Warming of the deep water in the Weddell Sea can have important implications for Antarctic Bottom Water (AABW) formation, melting of ice shelves and pack ice, and the regional ocean-atmosphere heat transfer; however, no definitive impacts have yet been identified.

1. Introduction

Changes in the deep waters of the Weddell Sea have been investigated as part of several programs, including Deep Ocean Ventilation Through Antarctic Intermediate Layers (DOVETAIL) (Muench & Hellmer 2002; Gordon et al. 2001), the Southern Ocean component of Consortium for Ocean's Role in Climate (CORC) (Visbeck et al. 2001), and the set of repeated transects across the Weddell Sea collected by Alfred-Wegener-Institut (Schröder & Fahrbach 1999; Fahrbach et al. 1994, 2001). Observational data from these programs indicated warming of the Weddell Sea Warm Deep Water (WDW) by ~0.012±0.007 °C yr^{-1} from the 1970s to the early 1990s (Robertson et al. 2002) (hereafter referred to as RVGF2002). In the mid-1990s, the warming trend ceased and the WDW cooled (Fahrbach et al. 2004). So what has happened since the mid-1990s and why does it matter?

Warming of the Weddell Sea deep waters (Figure 1) has significant climate implications, potentially affecting several climate triggers. These triggers include melting of the ice shelves and the associated sea level rise, modifications of sea ice coverage and the resultant effects on the ocean-atmosphere heat flux, and deep water production with its effects on the global thermohaline circulation. Many of the Antarctic ice shelves are melting, particularly those in the Amundsen and Bellingshausen Seas and along the Antarctic Peninsula (Rignot et al. 2008). Melting can lead to collapse, such as that of the Larsen B Ice Shelf in the Weddell Sea in 2002 and

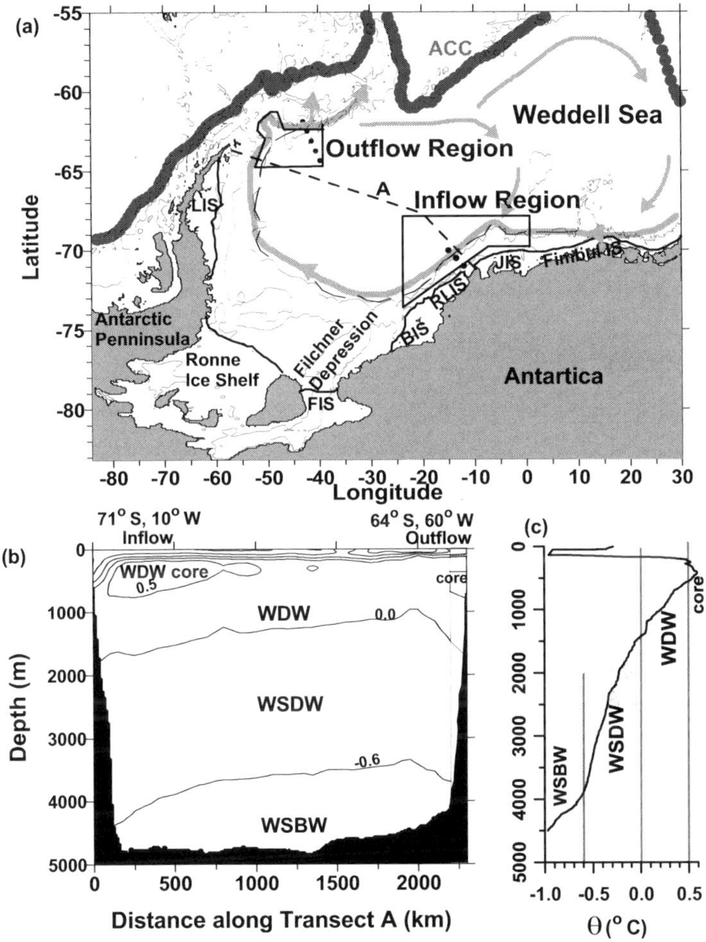

Figure 1 a) The bathymetry for the Weddell Sea with contours shown at 500 and 3000 m. Transect A is indicated by the dashed line and labelled. The location and direction of the core of the Weddell Gyre current are indicated by the light gray line with arrows. The southern extent of the Antarctic Circumpolar Current is shown by a series of large dark gray dots and labeled ACC (coordinates were kindly provided by A Orsi [personal communication]). The edges of

the Larsen, Ronne, Filchner, Brunt, Riiser-Larsen, Jelbert, and Fimbul Ice shelves are indicated by solid lines and the ice shelves are labelled (LIS for the Larsen Ice Shelf; FIS for the Filchner Ice Shelf; BIS for Brunt Ice Shelf, RLIS for Riiser-Larsen Ice Shelf, and JIS for Jelbert Ice Shelf). The regions used to determine the inflow and outflow characteristics are indicated by boxes and labelled accordingly. b) Typical hydrography along transect A from east to west. The potential temperature θ, is shown contoured at 0.5°C, 0.35°C, 0.0°C, –0.6°C and –1.3°C. The deep water masses, including Warm Deep Water (WDW), Weddell Sea Deep Water (WSDW), and Weddell Sea Bottom Water (WSBW), are identified. Surface waters are not labeled. The warm core of WDW near the inflow and outflow regions with θ values greater than 0.5°C and 0.35°C, respectively, are denoted "core". c) A sample potential temperature, θ, profile with depth for the Weddell Sea (at 66°S 48°W). WDW, WSDW and WSBW portions of the profiles are identified.

of the Wilkens Ice Shelf on the west side of the Antarctic Peninsula in 2008. Although melting of the ice shelves does not contribute to sea level rise since the ice shelf is floating, collapse of the ice shelves allows glacier flow to accelerate and dump more glacial ice into the ocean, which does increase sea level (Scambos et al. 2004; Rignot et al. 2008). In the Antarctic, the ocean is warmer than the air through much of the year and heat is lost from the ocean to the atmosphere . The source of heat in this region is the WDW), a subsurface flow coming in from the Antarctic Circumpolar Current (ACC). Mixing processes bring some of the WDW heat to the surface (RVGF2002) where it contributes to melting of sea ice. In some regions of the Weddell Sea, sensible heat polynyas form when heat from the WDW is brought to the surface (Gordon 1978; Gordon et al. 2007). The ocean has a much lower albedo than sea ice. Consequently, there is a much greater absorption of solar radiation and less reflection from the ocean surface compared to the sea ice. This absorbed solar radiation warms the ocean surface, impeding sea ice formation and inducing further warming of the surface. Sea ice also acts as an insulator, preventing heat loss from the ocean. Warmer deep waters will increase the available heat for melting of the sea ice and contribute to warming of the upper ocean and increased heat transfer from the ocean to the atmosphere (Gordon & Huber 1990; McPhee & Martinson 1994; McPhee et al. 1996; Gordon 1998; Muench et al. 2002). As much as 60% of the Antarctic Bottom Water (AABW) for the global thermohaline circulation is produced in the Weddell Sea (Figure 1) (Deacon 1937; Orsi et al. 1999). AABW is formed through mixing in the Weddell Sea. WDW or Modified Warm Deep Water (MWDW) mixes with shelf or ice-shelf waters resulting in Weddell Sea Deep Water (WSDW), which is known as AABW when it exits the Weddell Sea, sometimes undergoing further transformation. Changes in the deep waters, WDW or WSDW, in the Weddell Sea impact AABW production and thus the global thermohaline circulation.

This paper updates the earlier RGVF2002 paper, extending the earlier data set to 2005. The observational data set, described in Section 2, was utilised to evaluate changes in the temperature of the WDW and WSDW masses in the Weddell Sea using historical Conductivity/Temperature/Depth (CTD) data collected after 1950 in the region west of 10°E. The circulation of the Weddell Sea is briefly outlined in Section 3. Section 4 describes the observed temporal trends in the data set and Section 5 discusses the mechanisms for change of the deep waters and implications of warming deep waters on the ice shelves, sea ice, and deep water formation. A brief conclusion is given in Section 6.

2. Historical hydrographic data set

The observational data set was assembled from the available CTD and bottle data for the Weddell Sea from the National Oceanographic Data Center (NODC) data set (Conkright et al. 1998) and assorted sources (see RVGF2002 for details). This data set was updated with data from CORC/Arches, kindly provided by A Gordon and B Huber, and repeat transect data collected by the Alfred-Wegener-Institut (Fahrbach et al. 1994, 2001) as kindly provided by E Fahrbach. In situ temperature, T, conductivity, C, and pressure, P, from these CTD profiles were converted to potential temperature, θ, salinity, S, and depth, D, using the UNESCO routines (Fofonoff & Millard 1983). The data and editing process are fully described in RVGF2002. The editing primarily consisted of removing profiles with constant temperatures and/or salinities at all depths, although occasional obvious spikes were also deleted. After editing, 3078 profiles remained, ranging in time from 1912 to 2005. The temporal and spatial distributions of these profiles are shown in Figure 2. The data were quite sparse before 1960 (Figure 2).

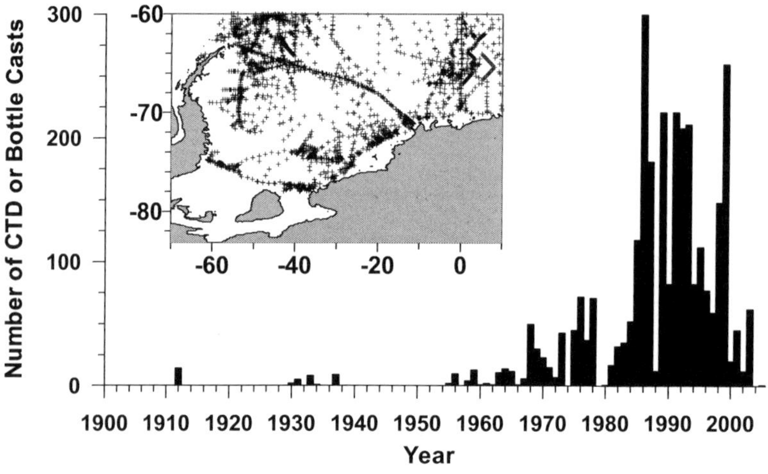

Figure 2 The distribution of the CTD and bottle data profiles used in the study with time, with the inset showing the locations of the CTD and bottle data profiles as crosses.

The data uncertainties varied for each cruise, but generally decreased for the more recent data. As a general guideline, the uncertainties for the data as described in the literature for the respective cruises are ± 0.003°C for θ and ±0.002 psu for S after 1990, ±0.003°C for θ and ±0.003 psu for S from the 1980s, ± 0.006°C for θ and ± 0.005 psu for S from the 1970s, and ± 0.02°C for θ and ± 0.01 psu for S from the 1960s. Data prior to 1960 have uncertainties equivalent to or greater than those of the 1960s. A few profiles have higher uncertainties than those typical for the decade. For example, uncertainties for a cruise from 1987 to 1988 were estimated at ±0.01°C for θ and ± 0.01 psu for S (Gouretski & Danilov 1993).

To aid the later temporal comparisons, an effort was made to reduce seasonal variability. In the Weddell Sea, the strongest seasonal variability occurs in the water column above the seasonal thermocline, which reaches to about 200 m. As a result, the upper 200 m were removed from all profiles. In the waters below the seasonal thermocline, the seasonal cycle, was believed to be smaller than the changes reported here [J Toole, personal communication]. Furthermore, since most of the data were collected during the austral summer, little seasonal aliasing was expected. In order to keep the austral summer months grouped together, the year was defined as starting in October and running through September.

3. *Water mass definitions and the general circulation*

3.1 Water mass definitions

Three primary deep water masses are present in the Weddell Sea (Figure 1a): WDW, WSDW, and Weddell Sea Bottom Water (WSBW) (Figure 1b). As in RVGF2002, these water masses are defined by potential temperature, θ(°C), and salinity, S (psu), ranges as given in Table 1. The basic hydrography across the Weddell Sea can be seen in Figure 1b, which shows the smoothed temperature with depth along transect A (World Ocean Circulation Experiment (WOCE) line SR4) (dashed line in Figure 1a). Since transect A passes through both the inflow and outflow regions of the Weddell Gyre (Figure 1a), the warmer core of inflowing WDW in the eastern Weddell Sea with potential temperatures greater than 0.5°C can be seen on the left side of Figure 1b and the cooler core of outflowing WDW on the right side of Figure 1b.

As seen in a sample profile for the Weddell Sea collected in the outflow region near 64.25°S, 48.25°W, the potential temperature decreases gradually with depth from ~0.4°C to ~−0.6°C (Figure 1c). Due to this gradual decrease, there is no distinct division between WDW and WSDW. The most common upper boundary for the WSDW is the 0.0°C isotherm (Foster & Carmack 1976; Weppernig et al. 1996; Meredith et al. 2000). As in RVGF2002, this was adopted as the upper limit of WSDW. Near the bottom of the potential temperature profile (Figure 1c), a strong temperature gradient exists between –0.6 °C and –0.8 °C and marking the transition zone between WSDW and WSBW. Different investigators have defined the lower potential temperature boundary for the WSDW as either –0.6°C, –0.7°C, or –0.8°C (Table 1). Here, the warmer value (–0.6°C), which is more appropriate for the deeper water depths of the utilised data, is used to ensure that all cold contributions are counted as WSBW.

3.2 General circulation

The major feature of the circulation in the Weddell Sea is the clockwise flow of the Weddell Gyre (grey arrows in Figure 1a). Circumpolar Deep Water from the ACC (thick black line in Figure 1a) enters the Weddell Sea at ~30° E and is incorporated into the Weddell Gyre becoming WDW (Deacon 1979; Orsi et al. 1993; Gordon & Huber 1995). Although a majority of the CDW enters the region east of the prime meridian, there are some injections of WDW west of the prime meridian (Fahrbach et al. 2004). The WDW transits the gyre, cooling as it progresses, primarily through mixing with cooler water types (Gordon 1998; Schröder & Fahrbach 1999). The WDW mixes with shelf waters, forming WSDW and/or WSBW. Additionally, WSDW and WSBW formed over the continental shelf and slope flow down the continental slope and enter the deep basin along the transect, particularly west of the Filchner Depression. Due to the density and depth of the WSBW, it is confined to the basin. WSDW lies at a shallower depth (Figure 1b) and passes over the ridges, exiting the basin. The WSDW leaves the Weddell Sea, primarily through the Scotia Sea, where it is then transformed by further mixing into AABW) (Locarnini et al. 1993; Whitworth et al. 1994). This AABW fills the global deep ocean as part of the thermohaline circulation, which is also known as the global conveyor belt. Gordon (1998) and Fahrbach et al. (1998b) give comprehensive descriptions of the hydrography and circulation in the Weddell Sea.

The spatial structure of the thermal stratification along the Weddell Gyre is evaluated in depth in RVGF2002 and only key points are covered here. The maximum potential temperature, θ_{MAX}, was used as a metric to evaluate the

WDW. θ_{MAX} was much warmer in the water column in the south-eastern Weddell Sea (left side of Figure 1b), exceeding 0.5°C, compared to the north-western Weddell Sea (right side of Figure 1b), with values less than 0.5°C. θ_{MAX} was also much deeper in the water column in the south-eastern Weddell Sea, with an average depth of θ_{MAX} of ~330 m compared to ~580 m in the western Weddell Sea (RVGF2002). The warmest portion of the WDW section of the profile generally occurs just below the permanent pycnocline. Consequently, erosion of the upper portion of the WDW will remove the warmest portion of the WDW and both deepen and decrease θ_{MAX}. This erosion results from mixing events which can be caused by storms, polynyas, passage of an iceberg, or mid-water column mixing events from plumes of Ice Shelf Water or Shelf Water.

Although the data set encompassed most of the Weddell Sea (Figure 2), both RVGF2002 and this study focus on the deep waters flowing around the Weddell Gyre. The most extensive data coverage in time occurred in two regions, an inflow and an outflow region (boxed areas in Figure 1a). These regions were selected to evaluate temporal trends and variability. The inflow region in the south-eastern Weddell Sea was ~30° west of the region where the CDW enters the Weddell Gyre (Orsi et al. 1993, 1999). The outflow region covers the area just upstream from where WSDW exits the Weddell Gyre (Gordon 1998). Only areas within the inflow and outflow with water depths greater than 1000 m were included in order to avoid shelf process effects. The travel time between the inflow and outflow regions has been estimated to be ~2–4 years, using mean speeds of ~0.04–0.08 m s^{-1} and following Fahrbach et al. (1994, 2001). Spatial variability within these two regions was primarily a function of the water depth. As a result the inflow and outflow profiles in Figures 3 and 4 were sorted by water depth, respectively. The water depth determined which portion of the Weddell Gyre was sampled with the heart of the core sampled in water depths of 4000 and 5000 m (leftmost two sets of profiles in each panel of Figures 3 and 4) and the edge of the WDW core for the shallower profiles (rightmost set of profiles in each panel of Figures 3 and 4). Consequently, the spatial decorrelation scales are dependent on the slope of the continental slope in the cross-shelf/cross-gyre direction and range from >100 km within the core in the off-slope direction to <20 km over the slope. This latter distance is roughly equivalent to the Rossby radius in this region. Along the gyre or along-slope, the spatial correlation scale is much greater and the spatial trend along the gyre is fully discussed in RGVF2002.

During three consecutive years, 1974–76, satellite imagery of the Weddell Sea revealed a large open ocean area surrounded by sea ice, a polynya. The Weddell Polynyas are open ocean or sensible heat polynyas, forming deep convective cells. These convective cells can reach 3000 m depth at times (not shown) and cool the water column. The mechanism for formation of the Weddell Polynya is still unknown. Gordon et al. (2007) suggest that it is connected with salinity anomalies and occurs during a prolonged negative Southern Annular Mode (SAM). Grigg and Holbrook (2001) note that open ocean polynya are less stable than the brine rejection process typical of coastal polynyas, which is consistent with both the absence of polynyas and the weak polynyas since the 1970s. These polynyas persisted during the entire winter during these years. Although, some small polynyas with lifetimes of a week or two have occurred since the 1970s, no extensive, winter-long polynyas like those of 1974–76 have recurred (Gordon et al. 2007).

4. *Temporal temperature trends*

4.1 Warm Deep Water

To evaluate the temporal trends, θ profiles in both regions were grouped according to year and water depth (Figures 3 and 4 for the inflow and outflow regions, respectively). The data prior to 1950 were not included as a separate panel, but were included in the background data set (gray profiles). In both Figures 3 and 4, the profiles shift from the left portion of the group to the right portion in the panels from the 1950s to 2005 indicating a general warming in the inflow and outflow regions, respectively. The warmest θs occurred during the 1990s and 2000s in both the inflow and outflow regions, with the highlighted profiles appearing on the left side of the group (Figures 3a-b and 4a-b, respectively). Most of the coldest WDW, as evidenced by the low θs and the highlighted profiles appearing on the right side of the group, occurred during the 1970s, the decade of the Weddell Polynyas (Figures 3d and 4d) (Gordon 1978). The coldest θ values occurred during the decade when the Weddell Polynyas occurred (Figures 3d and 4d), despite only a few of the inflow region profiles being collected near the area of polynyas. In the deep basin inflow profiles during the polynya time, (the rightmost group in Figure 3), a profile with a much thinner WDW layer and cooler water from 200 m to below 2000 m can be seen. The profiles prior to 1970 had θ values similar to those of the 1980s (Figures 3e, 3f, 4e, and 4f), indicating a return to pre-polynya conditions.

To further evaluate the apparent warming, annual average profiles of θ were determined for the inflow and outflow regions from 1960 to 2005. The average annual profiles were calculated for three different water depth ranges capturing the core of the gyre flow (3000 to 3500 m, 3500 to 4000 m, and 4000 to 4500 m). θ_{MAX} was determined for each of these average profiles. The potential temperatures of the inflow region were adjusted for distance along the transect using the spatial trend (taken from Figure 5 in RGVF2002) and using 12.5° W as a reference point. The average of these θ_{MAX} values for three depths representing the core are shown in Figure 5a with the values from the inflow region as triangles and those from the outflow region as dots.

The years 1973 through 1978 and 1984 had the lowest θ_{MAX} values for the inflow and outflow regions (Figure 5a). One of the two 1984 θ_{MAX} inflow temperatures was ~0.3°C lower due to the presence of a ~200 m thick cold intrusion in the WDW. The 1984 outflow profile had no evidence of large intrusions, and there was no apparent reason for the anomalously low value. The other years with low θ_{MAX} values occurred near the time of the Weddell Polynya and their low values are believed to be associated with it.

RVGF2002 found a warming trend of ~0.012 ±0.007 °C yr^{-1} (dashed lines in Figure 5a) subsequent to the Weddell Polynya years (1970s to 1990s). But the anomalously low temperatures of the 1970s biases this warming trend to a higher rate than when the entire temporal period is used. The observed warming trend was ~0.007±0.004°C yr^{-1} from 1960 to 2005 (black lines in

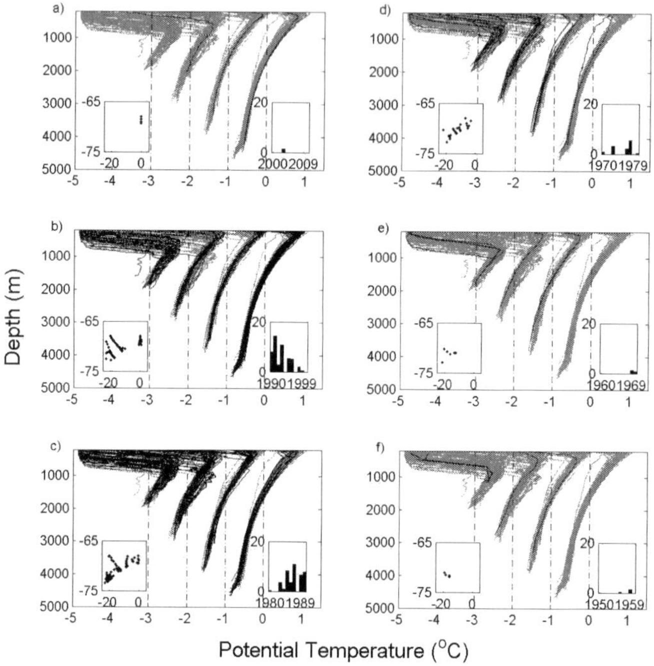

Figure 3 Profiles of potential temperature, θ, for the inflow region for the water depth ranges of 1000 to 2000 m, 2000 to 3000 m, 3000 to 4000 m, and 4000 to 5000 m. The profiles are offset −1°C for each depth range, with the deepest range having no offset. The profiles are highlighted for each decade in black for a) 2000 to 2009, b) 1990 to 1999, c) 1980 to 1989, d) 1970 to 1979, e) 1960 to 1969, and f) 1950 to 1959. The pre-1950 data are included in all panels in gray. Each panel includes a plot of the CTD and bottle sites and a histogram of the percentage of occurrence of profiles for each year of the panel. Dashed lines indicate the zero line for each set of profiles.

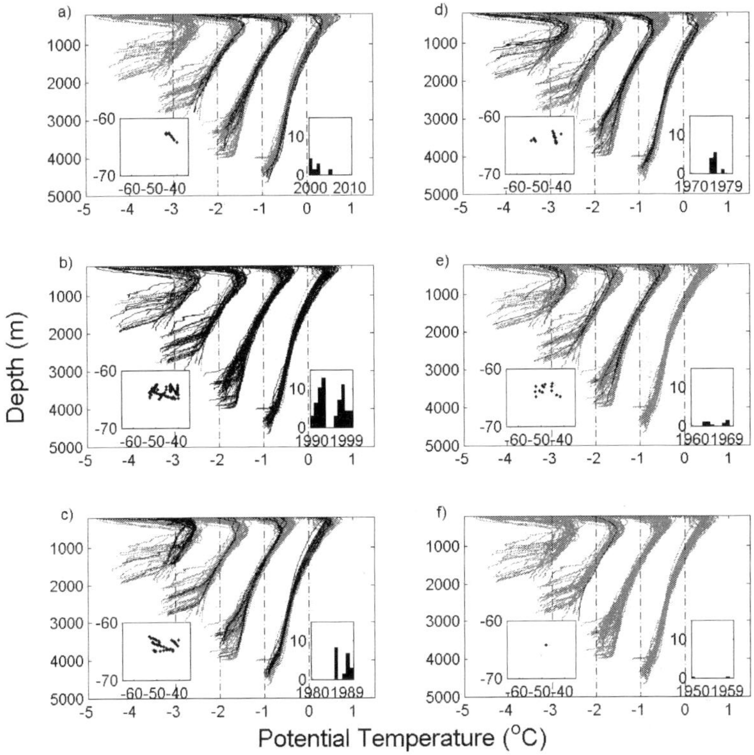

Figure 4 Profiles of potential temperature, θ, for the outflow region for the water depth ranges of 1000 to 2000 m, 2000 to 3000 m, 3000 to 4000 m, and 4000 to 5000 m. The profiles are offset -1°C for each depth range, with the deepest range having no offset. The profiles are highlighted for each decade in black for a) 2000 to 2009, b) 1990 to 1999, c) 1980 to 1989, d) 1970 to 1979, e) 1960 to 1969, and f) 1950 to 1959. The pre-1950 data are included in all panels in gray. Each panel includes a plot of the CTD and bottle sites and a histogram of the percentage of occurrence of profiles for each year of the panel. Dashed lines indicate the zero line for each set of profiles.

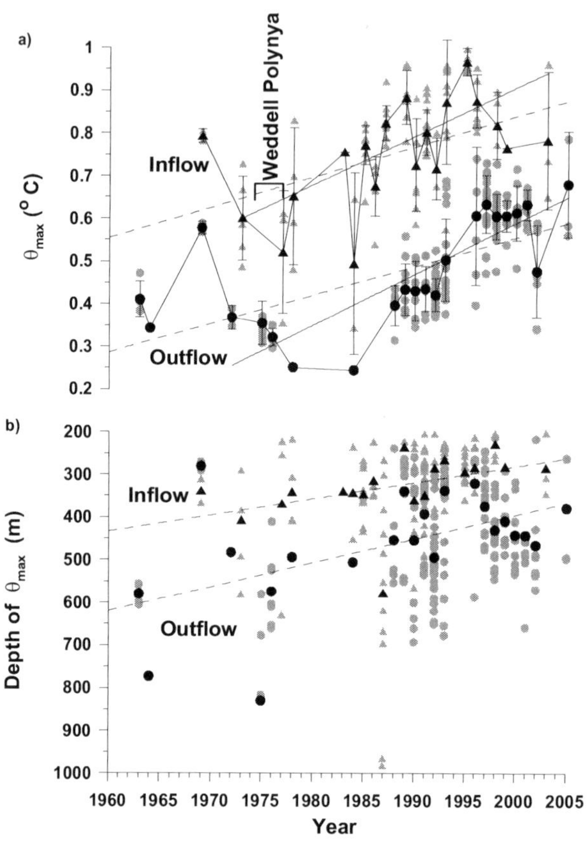

Figure 5 The average a) θ_{MAX} and b) depth of θ_{MAX} over time for the inflow region (black triangles) and the outflow region (black dots) from profiles where the water depth is between 3000 and 4500 m. The light triangles and light dots are the values for the individual profiles in the inflow and outflow regions, respectively. The error bars represent one standard deviation of the values. The dashed black lines are fitted trends for the inflow and outflow regions from 1950 to 2005, respectively. The solid lines are the fitted trends for the inflow and outflow regions from 1970 to 2005, respectively.

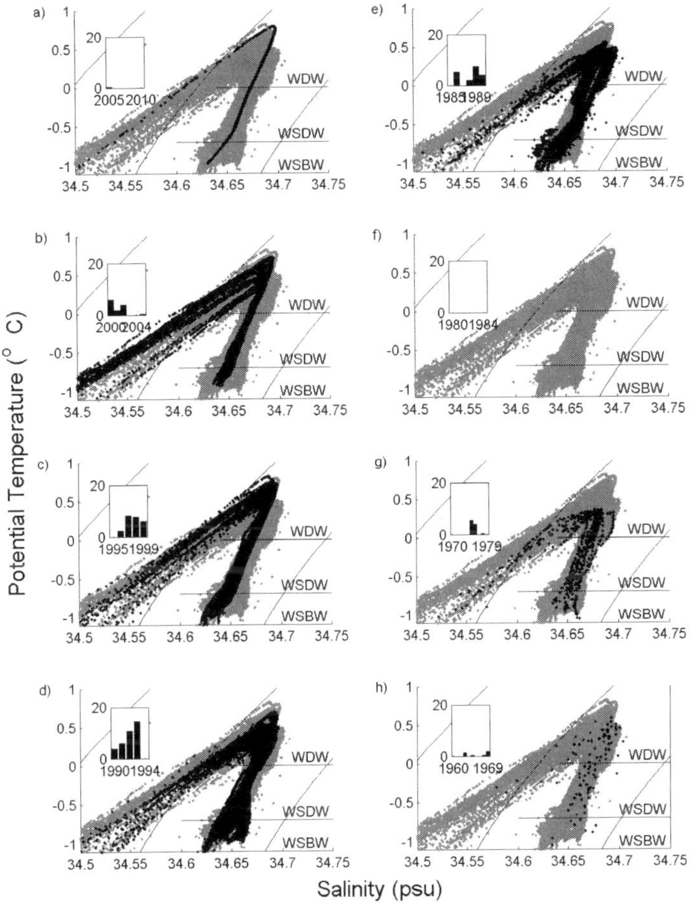

Figure 6 *θ-S* diagrams for the outflow region with the values highlighted for the time periods in black a) 2005 to 2010, b) 2000 to 2004, c) 1995 to 1999, d) 1990 to 1994, e) 1985 to 1989, f) 1980 to 1984 (no points), g) 1970 to 1979, and h) 1960 to 1969. Isopycnal lines are indicated for $\sigma_\theta(S,\theta,0)$. The ranges of the WDW, WSDW, and WSBW are shown. The pre-1960 data are included in all panels in gray. Each panel includes a histogram of the percentage of occurrence of profiles for each year of the panel.

Figure 5a) as determined using a least squares regression. Data before 1960 were too sparse to include in the calculation. It should be noted, however, that each region had only one set of profiles after 2001. Both of these trends are robust to the spatial variability resulting from the wide spatial area used for the inflow region, since similar trends were found without spatial adjustment.

A shallower WDW layer is more vulnerable to surface mixing processes, such as entrainment of surface waters and erosion of the upper portion of the WDW. And a shallower WDW layer is more likely to affect the surface ice cover. During the years of the Weddell Polynya, θ_{MAX} was located deeper in the water column, roughly 200 to 300 m, than after 1980 (Figures 3, 4 and 5b). The unusually deep locations of θ_{MAX} for 1987 for the inflow region were associated with intrusions. Other than 1987, there was a trend for θ_{MAX} to shoal with time in both the inflow and outflow regions (Figure 5b). The shoaling is stronger for the outflow region than the inflow region.

How does the temporal warming pattern manifest itself in θ-S diagrams for the outflow region? From the θ-S diagrams, the WDW core density decreased with time (Figure 6). The least dense WDW occurred in 2005 (Figure 6a). The densest WDW was during 1976, which is consistent with the time required for deep water from the first Weddell Polynya to transit to the outflow region (Figure 6g). It should be noted that due to calibration errors in some of the data, salinity corrections were made. Nonetheless, the density change is robust to these adjustments.

To further illuminate density changes in the WDW with time, the mean potential density, σ_θ, was determined for the WDW and the core WDW, defined as θ >0.0°C and θ >0.5°C, respectively. The mean WDW σ_θ was relatively constant from the 1960s to 1990 when it suddenly decreased in density (diamonds in Figure 7). Subsequent to this shift, the mean WDW density increased with time, but did not return to the density prior to 1990. The density of the WDW core also appears stable from the 1960s to 1990, although there was no core WDW during the time of the Weddell Polynya, since the WDW did not exceed 0.5°C (circles in Figure 7). Like the mean σ_θ, for the WDW, the mean σ_θ, for the WDW core also decreased subsequent to

1990. It appeared have peaked between 2000 and 2004 and decreased thereafter. However, since only one profile is available after 2005, this apparent decrease is not conclusive. The mean $\sigma_\theta$seems to shift around 1990. The Weddell Polynya shows up primarily as the lack of WDW core during the 1970s with a negligible increase in WDW σ_θ. Although the polynya will generate denser water, the temperature range restriction for the definition of WDW limits the mean σ_θ. When evaluating these changes, it is important to consider that there are limitations on the mean $\sigma_\theta$of WDW. If the water cools below 0.0°C or 0.5°C for the core, the water is no longer WDW and will no longer be part of the average. This effectively puts a lower limit on $\sigma_\theta$and this lower limit is dependent on the salinity and pressure of the WDW. With the uncertainties in the data, it is dangerous to determine a temporal trend from this WDW σ_θ data. However, it appears that the WDW and WDW core underwent a shift in density around 1990.

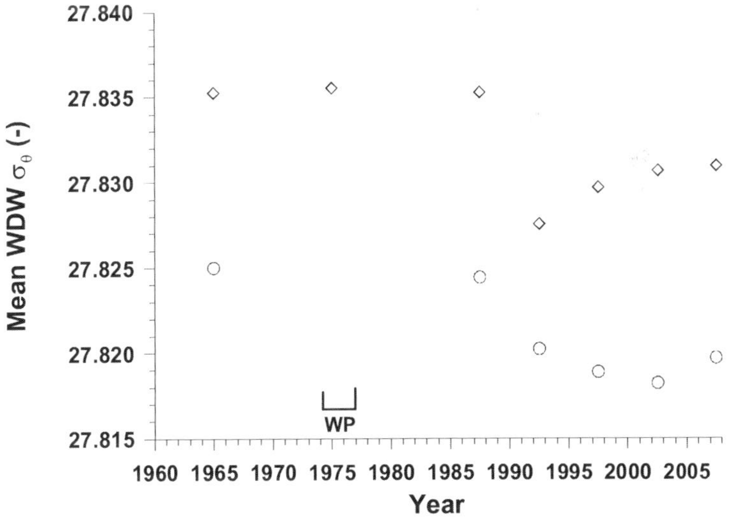

Figure 7 The mean density for the WDW (diamonds) and the WDW core (circles) for the corresponding time periods as Figure 6. The time of the Weddell Polynya is indicated by WP.

4.2 Weddell Sea Deep Water

The temporal behavior of the WSDW was investigated for the inflow and outflow regions in water depths greater than 4000 m using two 1000 m depth bands from 1500 to 3500 m (Figure 8). Similar to the WDW, the coldest average WSDW potential temperatures for these depth bands occurred from 1975 to 1977, during or just after the Weddell Polynya (Figure 8). Although the warmest average potential temperatures occurred after 2000 and there does seem to be a warming trend, the large error bars prevent evaluation of a definitive trend. Fahrbach et al. (1998a) used a shorter record (1989–95) from an array of moorings in the central Weddell Sea and determined a warming trend of ~0.01°C yr^{-1} for the WSBW. Agreeing with his observation, the data do show a strong warming during this short time period, and it persists after this period, but it does not generate a trend which exceeds the uncertainties (Figure 8).

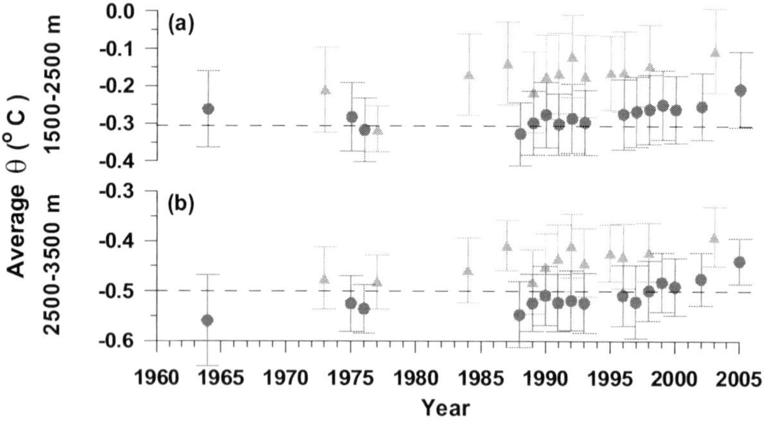

Figure 8 The average potential temperature, *θ*, for the inflow (triangles) and outflow (dots) regions for the depth ranges of a) 1500 to 2500 m and b) 2500 to 3500 m over time. The error bars represents one standard deviation of the values. Dashed reference lines indicate –0.3°C and –0.5°C in a) and b), respectively.

5. *Discussion*

5.1 Other observed warming trends

The observed WDW warming trend of 0.007±0.004 °C yr^{-1} is comparable to the average warming of the surface of the global ocean since the 1950s, 0.31°C or ~0.007°C yr^{-1} (Levitus et al. 2000) and the warming trend of ~0.009°C yr^{-1} for the WSBW from 1989 to 1995 observed by Fahrbach et al. (1998a, 2004) in the central Weddell Sea. The observed WDW warming trend also agrees with the warming on the poleward flank of the ACC of 0.006±0.001°C yr^{-1} from 1960 to 2000 (Böning et al. 2008). Although there is no direct connection between the surface ice temperature and the WDW), and changes in the atmosphere are likely to play a stronger role, the trend was also comparable to warming of the surface ice from 1970 to 1998 in the Weddell Sea, ~0.01–0.02°C yr^{-1}, observed in satellite data by Comiso (2000). Events such as the Weddell Polynya, 1974–76, lowered the WDW and WSDW potential temperatures by a few tenths of a degree and were associated with most of the lowest values in our record (Figures 3, 4 and 5). This agrees with the observed average and a maximum cooling in the deep basin during the Weddell Polynya of 0.2°C and 0.4°C, respectively, reported by Gordon (1982). Cool WDW occurred in 1973 and may indicate either that the path of the circulation was altered and the warm Circumpolar Deep Water flowing into the region missed our locations or cooler Circumpolar Deep Water entered the region at this time. Other than the Weddell Polynya of the 1970s and a period of low sea ice in the 1990s, a warming trend existed with temperatures from the 1960s to the present (Figure 5a). Variability was present in the warming trend, which was not uniform and showed strong fluctuations (Figure 5a).

5.2 Potential mechanisms for changes

There are three basic mechanisms for warming of the upper layers: changes in the inflow waters, changes in the speed of the gyre, or changes in the cooling processes within the Weddell Sea. If the warming is due only to the inflow of warmer water or the increased inflow of warm water, both the inflow and outflow waters should change with a ~2- to 4-year lag due to transit around the gyre. Changes in the speed of the gyre will more strongly impact the outflow region than the inflow region. For example, if the velocity of the gyre increases due to increased wind stress, the transit time around the gyre is reduced. If the mixing processes remain the same, the cooling during transit will also be reduced. The result will be an apparent warming of the outflow waters relative to the inflow waters, although increased wind stress could also modify the inflow waters. As a rough estimate using a two-dimensional model with the along gyre spatial trend and typical diffusion rates, an increase (reduction) in gyre velocity of 20% was estimated to increase (reduce) the outflow θ_{MAX} by ~0.07°C. Differences due to changes in the cooling processes occurring within the gyre should affect the outflow waters more than the inflow waters. For example, cooling by a large polynya will affect the outflow waters much more than the inflow waters, although the conditions for development of a large polynya might be linked to changes in the inflow waters. Smedsrud (2005) suggested lower sea ice coverage near Maud Rise, indicating a possible polynya, in the 1990s is responsible for the cooling trend in the late 1990s; however, Fahrbach et al. (2006) disagreed and attributed the changes to differences in the inflow.

Since the trends for the changes in θ_{MAX} are roughly equivalent for the inflow and outflow regions (Figure 5a), the primary cause of warming is believed to be due to a change in the inflow waters rather than a change in the gyre spin or cooling processes within the Weddell basin. There appears to have been less inflow of CDW or a cooler version of CDW during the early 1970s and inflow of warmer or a higher volume of CDW in the 1990s, particularly the late 1990s. The difference in the shoaling of WDW is an example of the inflow and outflow regions behaving differently. Higher wind speeds would induce shoaling, which would be evident in both regions. Higher wind speeds would also speed the flow of the gyre, increasing the warming in the outflow region. Although this mechanism cannot be ruled out, there is no evidence of change in wind speed affecting the temperatures, primarily due to a data set with gaps. It is likely that fluctuations in θ are due to processes occurring along the traverse of the Weddell Gyre. These processes should both dampen some fluctuations and induce others dependent on the occurrence of their forcing processes, particularly for the outflow region. Some examples of these processes are mixing of waters flowing out from the ice shelves and iceberg effects. The grounding of several large icebergs in 1987–88 is believed to have decreased Ice Shelf Water, decreased WSDW production, and increased WSDW salinity (Nøst & Østerhus 1995).

Changes in the inflow, either in volume or in temperature, can result from shifts in the location of the ACC modifying the inflow of Circumpolar Deep Water into the Weddell Gyre. The Antarctic Oscillation/Southern Annular Model (AAO/SAM) is associated with a shift in the position of the Weddell Front and the ACC and changes in sea ice extent (Hall & Visbeck 2002).

The AAO/SAM is an annular mode similar to the Arctic Oscillation, where the atmospheric pressure conditions fluctuate between positive and negative indices reflecting large-scale, meridional pressure differences (Thompson & Wallace 2000). Although it occurs at irregular intervals, it has had a strong positive trend in recent years (Dickson et al. 2001) (solid black line in Figure 9). A connection was found between the SAM and El Niño/La Niña by Fogt and Bromwich (2006). El Niño often occurs during the negative SAM and La Niña during the positive SAM. The positive SAM has been found to be associated with a more southerly position of the ACC, which would allow warmer water to enter the Weddell Sea, and with stronger winds, which would spin up the Weddell Gyre (Hall & Visbeck 2002). Since 1950, the SAM has trended towards its positive phase (Figure 9). The intensification of westerly winds during the past decade is consistent with this positive SAM trend (Böning et al. 2008). And the positive SAM trend coincides with the observed poleward migration of the ACC, the warming of the upper 1000 m between 50–60°S, and the southward shift of the polar fronts (Gille 2008). This poleward displacement of the ACC explains 87% of the temperature trend for the waters of the Southern Ocean (north of 60°S) below 200 m (Gille 2008). Furthermore, a link was proposed between El Niño – Southern Oscillation (ENSO) and SAM and the Weddell Polynya by Gordon et al. (2007), with a faster, smaller radius gyre during a negative SAM/El Niño and a slower larger radius Weddell Gyre during a positive SAM/La Niña. The negative SAM, which often accompanies an El Niño, is associated with a colder, drier atmosphere. This induces more evaporation increasing sea ice formation and brine rejection. The result is an increase in dense water formation. If a sufficient amount of this dense water breaks through the permanent pycnocline, it triggers deep convection, bringing heat from the WDW to the surface and initiating a sensible heat polynya. This results in significant cooling of the WDW and loss of heat to the atmosphere, as occurred in the 1970s (Gordon et al. 2007).

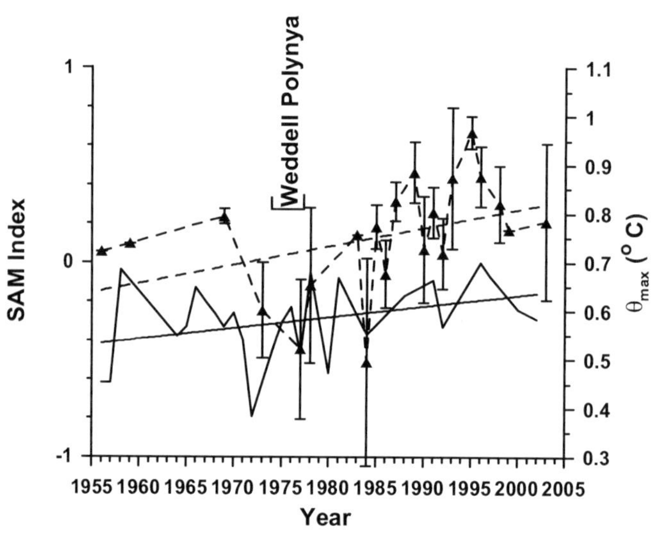

Figure 9 The Southern Annular Mode (SAM) index as kindly provided by M Visbeck. The solid line is the fitted trend from 1950 to 2005. The SAM is overlaid on the average θ_{MAX} over time for the inflow region (triangles) from 3000 to 4500 m water depth. The error bars represent one standard deviation of the values. The dashed line is the fitted trend for θ_{MAX} from 1950 to 2005.

The SAM shows an increasing trend from 1955 to the present (solid black line in Figure 9) which corresponds to the warming trend in the inflow region (dashed line in Figure 9). The cooling of the late 1990s also coincides with a decreasing SAM. Although it is not definitive, it appears the WDW temperature may be influenced by the SAM. The SAM correlated with the inflow region θ_{MAX} values with a coefficient of 0.44±0.29 (95% confidence). Although the uncertainty is large, the correlation met the criteria for significance. Correlations of 0.4–0.6 are typically considered significant for climate parameters (i.e. Hall & Visbeck 2002). Changes within the Weddell Gyre could also be affecting the deep water; however, again there is insufficient data to conclusively identify them.

5.3 Implications

Changes in the deep waters of the Weddell Sea have the potential to affect ice shelves, sea ice, and deep water production. Ice shelves melt from below with the heat flowing into the cavity as WDW. Warming of WDW has the potential to increase basal ice shelf melting, reduce sea ice coverage and thickness, and decrease deep water production. But what evidence exists that this is happening in the Weddell Sea?

5.3.1 Ice shelves

Warmer deep waters will increase melting of the ice shelves, potentially leading to collapse, like that of the Larsen B Ice Shelf in 2002. This has been observed to be occurring broadly in the Antarctic (Rignot & Jacobs 2002) and specifically for the Pine Island and Thwaites ice shelves in the Amundsen Sea (Rignot et al. 2008). This melting of ice shelves has accelerated, since the 1960s (Rignot et al. 2008). However, in the eastern Weddell Sea where the WDW enters the gyre, the Jelbert, Fimbul, Brunt and Larsen-Riiser ice shelves (Figure 1a) are showing only a slight mass loss (Rignot et al. 2008). Further along the gyre, the Filchner-Ronne ice shelves are presently increasing in mass or are in balance (Rignot et al. 2008). By the time WDW reaches the Filchner-Ronne Ice Shelf region, it has cooled by more than 0.6°C from when it enters the Weddell Sea (RGVF2002). Although melting of the ice shelves along the west side of the Antarctic Peninsula has increased by a factor of 140%, ice shelves on the east side do not appear to be melting faster (Rignot et al. 2008). By the time the WDW reaches the Larsen ice shelves, it has transited the entire Weddell Gyre. Although warmer WDW occurs in this region compared to several decades ago, it is much cooler than the warm water entering the Pine Island or Thwaites ice shelf cavities in the Amundsen Sea and does not induce as much melting. Potentially, a small amount of WDW may sneak through the Scotia Sea and be injected onto the continental shelf, bypassing the Weddell Gyre. Shifts of the ACC further south would increase these injections, leading to increased warming. This could have contributed to the demise of the Larsen B Ice Shelf. But this mechanism has not been confirmed (Rignot et al. 2008) and it is believed that a majority of the flow in this region follows the Weddell Gyre undergoing cooling along the way (Robertson et al. 1995). So presently there is no evidence that warmer WDW is affecting the ice shelves of the Weddell Sea.

5.3.2 Sea ice

Warming WDW potentially can affect sea ice. Mixing events, which penetrate the permanent pycnocline, can transport heat from the WDW core to the surface (Robertson et al. 1995; RGVF 2002). This has the potential for the WDW to melt more surface ice. In portions of the Antarctic, including the western side of the Antarctic Peninsula, sea ice concentrations have decreased from 1979 to 2004 (Stammerjohn et al. 2008). And Comiso (2000) found a warming of the sea ice from 1970 to 1998. Furthermore, there is a general trend for a shorter sea ice season in the Antarctic, particularly on the west side of the Antarctic Peninsula (Stammerjohn 2008). However, there are no clear trends for the sea ice concentrations or thickness in the areas of the inflow or outflow regions (Stammerjohn 2008). There is a link between the SAM and sea ice extents on the western side of the Antarctic Peninsula with the northerly winds associated with a positive SAM resulting in less sea ice coverage than average (Stammerjohn 2008). However, changes in sea ice extents have been attributed to air temperatures, which warmed by $2.9^{o}Cyr^{-1}$ from 1951 to 2005 inducing more atmospheric melting (Stammerjohn 2008). As a result, effects of WDW warming appears to be a minor factor on sea ice coverage when compared to the stronger effects of atmospheric changes.

5.3.3 Antarctic Bottom Water Formation

A majority of the Antarctic Bottom Water (AABW) is formed in the Weddell Sea (Rintoul 2007). AABW formation will be affected by changes in the WDW, since it is a primary source of water. Generally deep waters in the Antarctic have been observed to be warming and freshening since the 1960s (Jacobs 2006; Rintoul 2007), with the freshening being driven by ice shelf melting (Jacobs 2006). In the Weddell Sea, there is the aforementioned warming trend. Changes in salinity are less clear due to data uncertainties, but a slight freshening trend is apparent in Figure 6. Both warming and freshening lead to less dense deep waters. The WDW of the Weddell Sea has been observed to become less dense over the past few decades with a shift in σ_θ occurring ~1990 (Figure 7). Less dense and fresher AABW has been observed in the outflow downstream of the dense outflow (Jacobs 2006) and fresher, less dense AABW has recently been observed in some parts of the Antarctic. However, decreasing AABW production has not definitively been identified in observations (Rintoul 2007). Looking at model studies for further insight, models predict a decrease in AABW production. Sen Gupta et al. (2009) evaluated the output of a set of global climate models used for the International Panel on Climate Change (IPCC) Fourth Assessment Report for changes in the Southern Hemisphere. Evaluating the results of 15 models for the last 20 years of the 20th century, the mean of the model results indicated a southern shift of the ACC coinciding with a positive trend of the SAM. Changes to the first order were attributed to modifications in atmospheric forcing. The models also indicated an increase in WDW being upwelled and drawn onto the shelf and a decrease in AABW formation with a mean of −11.1 Sv. However, an estimate of the temperature change of the WDW was not provided. Although this is not a definitive verification of either the observations or the model results, it is good sign that they support each other during the period from 1980 to 2000.

6. Conclusions

In conclusion, the WDW of the Weddell Sea appears to be warming since the 1960s, with an anomalously cool period during and subsequent to the

Weddell Polynya of the mid-1970s. The warming trend appears to be associated with the SAM. WDW is also becoming less dense, which has implications for bottom water formation and the global thermohaline circulation. However, as yet, the warmer WDW has had no discernable effects on the ice shelves or sea ice, although weak evidence exists supporting AABW production of the Weddell Sea. This may change with either more observational data or if the WDW and/or WSDW warm further. It should be noted, that this analysis is based on a limited observational data set with sparse sampling both temporally and spatially and the resulting potential for biases and large uncertainties in the results.

Acknowledgements

I am grateful to E Fahrbach, K Nicholls, A Gordon and B Huber for providing me with observational data for this work. Also thanks are due to M Visbeck for the SAM data. The careful editing efforts of Neil Holbrook and an anonymous reviewer are greatly appreciated.

References

Böning CW, Dispert A, Visbeck M, Rintoul SR & Schwarzkopf FU (2008). The response of the Antarctic Circumpolar Current to recent climate change, *Nature Geoscience,* 1, doi:10.1038/ngeo362.

Carmack EC & Foster T (1977). Water masses and circulation in the Weddell Sea. In *Polar oceans: proceedings of the Polar Oceans Conference*, Montreal, 1974, (pp151–65). Calgary, Canada: Arctic Institute of North America.

Comiso JC (2000). Variability and trends in Antarctic surface temperatures from in situ and satellite infrared measurements. *J of Climate*, 13, 1674–96.

Conkright ME, Levitus S, O'Brien T, Boyer TP, Antonov JI & Stephens C (1998). World Ocean Atlas 1998 *CD-ROM Data Set Documentation*, NODC Internal Rep. 15, Silver Springs, MD, 16 pp.

Deacon GER (1937). The hydrology of the Southern Ocean. *Discovery Reports,* 15, 1–124.

Deacon GER (1979). The Weddell gyre. *Deep-Sea Res,* 26A, 981–95.

Dickson RR, Hurrell JW, Bindoff NL, Wong APS, Arbic B, Owens B et al. (2001). The world during WOCE. In G Siedler & J Church (Eds), *Ocean circulation and climate,* (pp557–83). WOCE IPO, 77, International Geophysics Series, Academic Press.

Fahrbach E, Harms S, Rohardt G, Schröder M & Woodgate RA (2001). Flow of bottom water in the northwest Weddell Sea. *J Geophys Res,* 106, 2761–78.

Fahrbach E, Hoppema M, Rohardt G, Schröder M & Wisotzki A (2006). Causes of deep-water variation: comment on the paper by LH Smedsrud "Warming of the deep water in the Weddell Sea along the Greenwich meridian: 1977–2001". *Deep-Sea Res I,* 53, 574–77.

Fahrbach E, Hoppema M, Rohardt G, Schröder M & Wisotzki A (2004). Decadal-scale variations of water mass properties in the deep Weddell Sea. *Ocean Dynamics,* 54, 77–91.

Fahrbach E, Meyer R, Rohardt G, Schröder M & Woodgate RA (1998a). Gradual warming of the Weddell Sea Deep and Bottom Water. In *Filchner-Ronne Ice Shelf Program, Report No. 12,* (pp24–34). Bremerhaven, Germany: Alfred-Wegener-Institut.

Fahrbach E, Rohardt G, Schröder M & Strass V (1994). Transport and structure of the Weddell Gyre. *Deep-Sea Res I,* 41, 389–411.

Fahrbach E, Schröder M & Klepikov A (1998b). Circulation and water masses in the Weddell Sea, In *Physics of Ice-Covered Seas,* (pp569–603). Vol 2. Helsinki: Helsinki University Press.

Fofonoff NP & Millard RC (1983). Algorithms for computation of fundamental properties of seawater, *UNESCO Tech Rep Mar Sci,* 44.

Fogt RL & Bromwich DH (2006). Decadal variability of the ENSO teleconnection to the high latitude South Pacific governed by coupling with the Southern Annular Mode. *J Climate,* 19, 9790–97.

Foster TD & Carmack EC (1976). Temperature and salinity structure in the Weddell Sea, *J Phys Oceanogr,* 6, 36–44.

Gille S (2008). Decadal-scale temperature trends in the Southern Hemisphere Ocean, *J of Climate*, 21, 4749–65.
Gordon AL (1978). Deep Antarctic convection west of Maud Rise. *J Phys Ocean*, 8, 600–12.
Gordon AL (1982). Weddell deep water variability. *J Mar Res*, 40, 199–217.
Gordon AL (1998). Western Weddell Sea thermohaline stratification. In SS Jacobs & RF Weiss (Eds), *Ocean, ice, and atmosphere: interactions at the Antarctic continental margin*, (pp215–40). Antarctic Res Series, Vol. 75, Washington, DC: American Geophysical Union.
Gordon AL & Huber BL (1990). Southern Ocean winter mixed layer. *J Geophys Res,* 95, 11,655–72.
Gordon AL & Huber BL (1995). Warm Weddell deep water west of Maud Rise. *J Geophys Res*, 100, 13,747–53.
Gordon AL, Visbeck M & Huber B (2001). Export of Weddell Sea deep and bottom water. *J Geophys Res*, 106, 9005–17.
Gordon AL & Visbeck M (2007). A possible link between the Weddell Polynya and the Southern Annular mode. *J Climate*, 20, doi: 10.1175/JCLI4046.1.
Gouretski VV & Danilov AI (1993). Weddell gyre: structure of the eastern boundary. *Deep-Sea Res*, 40, 561–82.
Grigg SB & Holbrook NJ (2001). The impact of polynyas on stability of the thermohaline circulation as simulated in a coupled ocean-atmosphere-sea ice box model. *Geophys Res Lett*, 28, 767–70.
Hall A & Visbeck M (2002). Synchronous variability in the southern hemisphere atmosphere, sea ice, and ocean resulting from the annular mode. *J of Climate*, 15, 2043–3057.
Levitus S, Antonov JI, Boyer TB & Stephens C (2000). Warming of the world ocean. *Science,* 287, 2225–29.
Jacobs SS (2006). Observations of change in the Southern Ocean. *Phi Trans R Soc*, 364, doi:10.1098/rsta.2006.1794.

Locarnini RA, Whitworth III T & Nowlin Jr WD (1993). The importance of the Scotia Sea on the outflow of Weddell Sea Deep Water. *J Mar Res*, 51, 135–53.
McPhee M & Martinson DG (1994). Turbulent mixing under drifting pack ice in the Weddell Sea. *Science*, 263, 219–20.
McPhee MG, Ackley SF, Guest P, Huber BA, Martinson DG, Morison JH et al. (1996). The Antarctic Flux Zone Experiment. *Bull American Met Soc*, 77, 1221–32.
Meredith MP, Locarnini RA, Van Scoy KA, Watson AJ, Heywood KJ & King BA (2000). On the sources of Weddell Gyre Antarctic bottom water. *J Geophys Res*, 105, 1093–104.
Muench RD & Hellmer HH (2002). The International DOVETAIL Program: an overview. *Deep-Sea Res*, 49, 4711–14.
Muench RD, Padman L, Howard SL & Fahrbach E (2001). Upper ocean diapycnal mixing in the northwestern Weddell Sea. *Deep-Sea Res*, 49, 4711–14.
Nøst OA & Østerhus S (1995). Impact of grounded icebergs on the hydrographic conditions near the Filchner Ice Shelf, Antarctica, In SS Jacobs & RF Weiss (Eds), *Ocean, ice, and atmosphere: interactions at the Antarctic continental margin*, (pp267–84). Antarctic Res Series, Vol. 75, Washington, DC: American Geophysical Union.
Orsi AH, Johnson GC & Bullister IL (1999). Circulation, mixing, and production of Antarctic Bottom Water. *Progress in Oceanography*, 43, 55–109.
Orsi AH, Nowlin Jr WD & Whitworth III T (1993). On the circulation and stratification of the Weddell Gyre. *Deep-Sea Res*, 40, 169–203.
Rignot E, Bamber JL, van den Broeke MR, Davis C, Li Y, Jan van de Berg W & van Meijgaard E (2008). Recent Antarctic ice mass loss from radar interferometry and regional climate modelling. *Nature Geoscience*, 1, 106–10, doi:10.1038/ngeo102.
Rignot E & Jacobs SS (2002). Rapid bottom melting widespread near Antarctic ice sheet grounding lines. *Science*, 296, 2020–23.

Rintoul SR (2007). Rapid freshening of Antarctic bottom water formed in the Indian and Pacific Oceans. *Geophys Res Lett*, 34, doi:10.1029/2006GL028550.
Robertson R, Padman L & Levine MD (1995). Fine structure, microstructure and vertical mixing processes in the upper ocean in the western Weddell Sea. *J Geophys Res*, 100, 18,517–35.
Robertson R, Visbeck M, Gordon AL & Fahrbach E (2002). Long-term temperature trends in the deep waters of the Weddell Sea. *Deep-Sea Res*, 49, 4791–806.
Scambos TA, Bohlander JA, Shuman CA & Skvarca P (2004). Glacier acceleration and thinning after ice shelf collapse in the Larsen B embayment, Antarctica. *Geo Res Lett*, 31, doi: 10.1029/2004GL020670.
Schröder M & Fahrbach E (1999). On the structure and the transport of the eastern Weddell Gyre, *Deep-Sea Res II*, 46, 501–27.
Sen Gupta A, Santoso A, Taschetto AS, Ummenhofer CC, England MH & Trevena J (2009). Projected changes to the Southern Hemisphere ocean and sea-ice in the IPCC AR4 climate models. *J Climate*, doi: 10.1175/2008JCL12827.1
Smedsrud LH (2005). Warming of the deep water in the Weddell Sea along the Greenwich meridian: 1997–2001. *Deep Sea Res I*, 52: 241–58.
Stammerjohn SE, Martinson DG, Smith RC, Yuan X & Rind D (2008). Trends in Antarctic annual sea ice retreat and advance and their relation to El Niño – Southern Oscillation and southern annular mode variability. *J Geophys Res*, 113, doi:10.1029/2007JC004269.
Thompson DWJ & Wallace JM (2000). Annular modes in the extratropical circulation. Part I: Month-to-month variability. *J of Climate*, 13, 1000–16.
Visbeck M, Gordon A, Smethie B, Schlosser P, Toole J, Huber B & Krahmann G (2001). The CORC/ARCHES observing system for Weddell Sea Deep and Bottom Water variability. *Clivar Exchanges*, 6, 23–25.

Weppernig RP, Schlosser P, Khatiwala S & Fairbanks RG (1996). Isotope data from Ice Station Weddell: implications for deep water formation in the Weddell Sea. *J Geophys Res*, 101, 25,723–39.

Whitworth III T, Nowlin Jr WD, Locarnini RA & Smith SG (1994). Weddell Sea Shelf Water in the Bransfield Strait and Weddell-Scotia Confluence. *Deep-Sea Res*, 41, 629–41.

5
Monitoring the atmosphere and hydrology

David Walland

*Abstract**

Monitoring the earth's climate is crucial. Australia, largely through the Bureau of Meteorology, has been collecting observations of the atmosphere and hydrological cycle for large parts of the past century. The resulting data underpins our knowledge about climate, climate variability and change in Australia. Climate data is collected largely for domestic purposes but is also required under international obligations. The data is collected to very high standards and every effort is made to improve the quality of our climate data where possible, for instance dealing with issues of data inhomogeneity. The climate record incorporates data from a number of networks including surface, upper air, atmospheric chemistry, surface radiation and of course hydrology networks. The resulting climate record gives us a wonderful resource to examine trends in our climate and assist in the process of attribution of these trends to climate change.

Introduction

The primary indications of a changing climate over the land-surface can be found in observations of the atmosphere and hydrosphere. Common indicators of climate change include globally averaged temperature increases, regional rainfall changes, changes in the position of the subtropical ridge, and overland run-off into streams. All rely on high quality, reliable

---

* Chapter originally submitted in January 2009.

measurement of atmospheric and hydrological variables undertaken by National Meteorological and Hydrological Services around the world.

Such agencies primarily measure the atmosphere and hydrosphere to understand and predict short timescale variability or 'weather', but over long time periods the data provide the basis for the climate records. The difference between 'weather' and 'climate' is important but so is the close relationship. Weather is what we experience, for example, at any point in time or throughout a short period such as a day or over a week. Climate is derived from the statistics of weather such as the daily averages, trends, variability and extremes over an extended period. It is important to recognise that there is a continuum of change on all timescales from daily weather, through monthly, seasonal and interannual cycles to long-term climate variability and change variability. Climate, like weather, is not static in time or space.

The principal database for describing the climate of Australia, its variability and its long-term trends, is provided by the national meteorological network, established, operated and maintained by the Bureau of Meteorology under the authority of the *Meteorology Act 1955*. More recently, the *Water Act 2007* has extended the mandate of the bureau to hold hydrological and related meteorological data collected by a wide variety of water agencies throughout Australia. Long and homogeneous records are critical for examining trends as well as the detection and attribution of climate change, and there are sufficient, reliable records to provide a reasonably comprehensive description of Australia's climate over the past 100 years.

The international context

The data collected by the Bureau of Meteorology on the weather and climate of Australia serves more than the domestic market. The data also contribute to the World Weather Watch program of the World Meteorological Organization (WMO), which underpins the basic global scale meteorological and hydrological observing and data processing systems. Australia has international obligations under the Convention of the World Meteorological Organization to supply for international exchange meteorological and related data collected under its jurisdictions. Under WMO Congress Resolution 40 and through participation in a number of international scientific programs, Australia has agreed to make data to defined specifications available to other countries and to those international programs.

The Manual on the Global Observing System (WMO 1995) is the primary reference document that defines what, how and when WMO Member countries and states should observe meteorological and related data in order to meet their international obligations as WMO Members. A related publication, the Guide on the Global Observing System (WMO 2007), provides detailed guidance on how to establish, operate and manage networks of stations to make these observations. Adherence to such guidance material is critical to achieve consistency across countries and to meet minimum requirements for a wide variety of applications. Other documents provide guidance for specialised networks, for instance the WMO Guide to Hydrological Practices (WMO 1994).

This international commitment to maintain a comprehensive observations program and make it freely available provides a strong supplementary rationale for Australia's monitoring effort. It is important to note, however, that Australia benefits far more than it contributes to this vast international exchange of data. In particular, as a non-satellite operator, Australia benefits greatly from remote sensing from space. Data collected from both polar orbiting and geostationary satellites are largely responsible, for example, for the marked improvement in weather forecasts generated by Numerical Weather Prediction models. This is especially true for Australia, which is surrounded by vast areas of oceans from which it is difficult to gather detailed in situ data. Satellites also aid the monitoring of broad scale climate oscillations such as the El Niño – Southern Oscillation.

Data quality

It is important that all observations meet certain standards and standardisation principles so that measurements at all sites are comparable. There are international guidelines that govern the operation of observing stations, instruments and networks to ensure that, when followed, records will achieve the required standards and will be internally consistent.

To help guarantee that observations are adequate for climate and climate change assessment, data must be as homogenous as possible. That is, the observations must be comparable over time. Any variations in a climate record that could have resulted from changes over time in the location, instrument type or the surrounding environment compromise the integrity of the record and lessen its utility for monitoring climate variability and change. Such changes can introduce artificial biases and fluctuations equal to or greater than the true climatic signal. Any unavoidable or inadvertent changes introduced to a network should be managed in accordance with the Global Climate Observing System (GCOS) Climate Monitoring Principles. Clearly, not all data holdings will be of the same quality. It is important, however, that at least a sub-set of observing stations be maintained at a minimum standard to ensure the highest possible quality. These stations are referred to as the reference climate stations (RCS).

Over the past decade, increasing human activity has placed pressure on maintaining the pristine environment of some climate reference stations. Developments such as buildings and asphalt car parks can affect the nature and structure of the environment surrounding observing sites. Even natural changes, such as the growth of a tree or tree stand can bias the data towards the recording of the local microclimate at the expense of the background macroclimate, which the site is intended to represent. Recently, developments at one RCS resulted in 22% of the natural vegetation within 100 m radius of the site being changed to asphalt. In another case the construction of a 10-m aircraft hanger obstructed the wind flow within a short distance of the site anemometer.

Figure 1 provides an example of how a change in site can affect a climate record. The old and new sites for the observing station at Yunta in South Australia have very different characteristics. Historically the Stevenson screen was in a very poor location with significant built infrastructure crowding the screen. This microclimatic environment would have retained heat during the day and reduced airflow. The new site located in an open area meets required

Figure 1 Old (left) and new (right) sites for the meteorological equipment at Yunta, South Australia. As the new site is more representative of the background climate, a significant warm bias resulting from the local microclimatic influences (buildings, gas tank, fence, etc.) must now be removed from data recorded at the old site. Such corrections are best carried out when there is an extended period of overlapping records.

standards but a systematic reduction in temperature at Yunta has resulted,something that must be accounted for in the record if the data are to be used in a high-quality dataset.

The first meteorological satellite, TIROS,[1] was launched nearly 50 years ago and since then the capacity to observe a wide range of characteristics of the earth's atmosphere and surface has grown considerably. An increasing reliance, therefore, is being placed upon remote sensing to derive climate information. The expansion in the number of satellite programs together with active research on algorithms to derive useful information from the data collected means that remote sensing data are being used to supplement and in some cases supersede in situ measurement of various meteorological and related fields. Examples include upper air temperature, soil moisture and vegetative cover. Just as with data directly observed at the earth's surface, homogeneity remains a key issue with remotely sensed data from satellites. Many satellite missions, especially those associated with research programs, have a limited lifespan, and when the instrument fails the data stream is lost. A replacement instrument may not be flown in a later satellite or when it is

[1] Television InfraRed Observation Satellite.

the instruments are calibrated differently. An exception has been the US National Oceanic and Atmospheric Administration (NOAA) series, which has now established a 30-year record of reasonably comparable remote sensing observations. Continuity of these observations into the future will be maintained through the European MetOp series which features a number of instruments whose heritage is sourced from the NOAA satellites.

Recently a great deal of thought has been given to ensuring the 'climate quality' of remote sensing data by the Committee on Earth Observation Satellites (CEOS). There has been much cooperation between CEOS and the Global Climate Observing System (GCOS) community (UNFCCC 2006). The increasing role that satellite information will play in the formulation of the climate record has been recognised, and it is clear that the value of remote sensing to operational systems will depend on sustained data streams of homogenous data. Protocols have been established to help this happen including an agreement between GCOS and CEOS to monitor a set of Essential Climate Variables.

While data sourced from satellites will play a larger role in establishing a global climate record, it is important to note that in situ data will always be essential to a well-balanced observing system. The measurement accuracy of surface-based observations, for instance the vertical temperature profiles by radiosonde, are considerably more accurate than currently available by satellite; vertical satellite profiles over land are still limited by their inability to resolve boundary layer effects. There are some parameters such as atmospheric pressure that satellites cannot measure. Even allowing for the increasing reliance on remote sensing data, in situ data are crucial for calibrating the remotely sensed data, which for most part consist of measurements of radiance or reflectivity, which in turn require complex processing to derive the desired outputs of variables like temperature and humidity. To get these calculations right and to understand the associated errors, remotely sensed data need to be calibrated against data observed at the same time and location using in situ measurements, either on the ground or from instruments flown on weather balloons. The relative shortness of the satellite record also means that century-scale analyses will depend on in situ observations for many decades to come.

National meteorological network

There are several components to Australia's meteorological network. The largest and oldest component is the land or surface-based measurements. Upper air measurements are expensive to collect and thus not collected over the same network density. Measuring atmospheric chemistry is also essential, as the composition of the atmosphere has been changing due to human influences. It is important to try and measure background levels as far away as possible from the influence of local emission sources. For instance spikes in emissions over cities or continuous emissions from power generation sites will strongly modulate the baseline atmospheric chemistry. The make-up of the various networks is considered in more detail in the sections below.

Surface data

The basic surface component of Australia's national meteorological network includes some 868 manual and automatic meteorological observing stations (Figure 2a), more than 7000 daily rainfall stations, including around 800 that measure rainfall intensity, located on the Australian mainland, on remote islands and in Antarctica. The Bureau of Meteorology staffs 58 of the manual observing stations and approximately 420 are operated by volunteers and paid cooperative observers, with equipment and overall coordination provided by the bureau. There are currently 593 automatic weather stations (AWS) in the network, many co-located with manual observation sites. The vast majority of the daily rainfall observations are taken by volunteers.

The key variables observed depend on the type of station but typically include air temperature, air humidity, barometric pressure, wind direction and speed, and rainfall. At stations where there is a human observer, there are also measurements of weather, visibility, cloud type/amount, perhaps evaporation as well as any other notable phenomena.

Weather observation was historically an inherently manual activity, requiring the interpretation and description of prevailing conditions as well as reading values from measuring devices. The bureau's data collection operations are now more automated but still require manual involvement for more complex observations, for at-source quality management, for system operation and for general station management, plus the supporting network and systems planning and management. There is a need to take further advantage of opportunities for automation as technology improves and the expense for manual observation becomes higher than the bureau can bear. A balance must be found however, between this level of automation and the need for the highest quality stations that can be best assured through manual observation.

The land surface network underpins Australia's climate monitoring capability. It provides an enormous quantity of high-quality information from which a basic understanding of the spatial and temporal variability of the climate of this country derives. It is also the key to the ability to detect any changes arising from human activities or from natural climate drivers. Australia supports climate change detection through the maintenance of 103 Reference Climate Stations (RCS) (Figure 2b). These stations have been selected to provide continuous homogeneous records of high quality, for use in detecting and monitoring long-term climate trends. Most RCSs have at least 30 years of record and are relatively free from changes due to increased urbanisation. Currently, 82 of these RCSs have an AWS at the site. Electronic fieldbooks are used throughout the RCS network at the remaining non-AWS sites to improve data quality and to achieve increased speed in reporting.

Data from the RCS network form the backbone of a number of high-quality research datasets that have been developed to monitor changes in the climate, as measured in daily and annual mean temperature, daily and monthly rainfall totals and evaporation over Australia.[2] The high quality data have been corrected for inhomogeneities caused by changes in, for example, site location and instrumentation or observation practice. In addition to providing information on regional climate trends, Australia's RCS network also makes an important contribution to monitoring global climate trends, with 68 stations having been designated as part of the GCOS Surface Network (GSN) (Figure 2c). The GSN has been established by the GCOS to

[2] Information on these high quality data sets and access to them can be found at www.bom.gov.au/climate/change/datasets/datasets.shtml

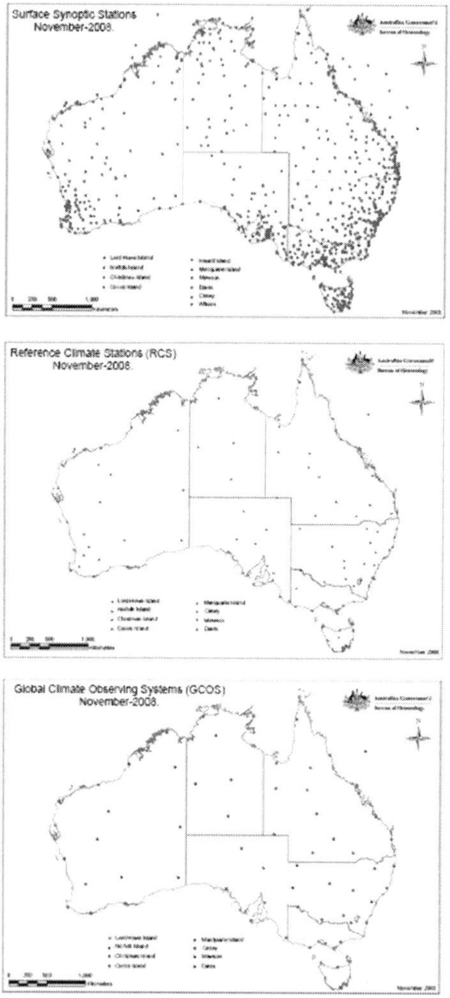

Figure 2 Australia's (a) surface meteorological network, (b) Reference Climate Station Network, (c) GCOS Surface Network.

primarily monitor and detect changes in temperature and atmospheric circulation at the Earth's surface.

The bureau also maintains a comprehensive database of historical station information or *metadata* to support the analysis and interpretation of the climate data. The database, with some records going back to the turn of the century, contains information concerning site location, exposure, instrumentation, maintenance, performance checks and inspections. Metadata are integral to understanding the quality of the climate data. This database was used to provide the data and photos discussed earlier in relation to the station relocation and data inhomogeneity at Yunta.

The high quality datasets are the basis for many of the key results indicating climate change trends across the Australian landscape. For instance, analyses show that Australia's average temperature has increased by 0.9 degrees since 1950 (Figure 3), although there have been significant variations across the continent. In addition to the change in average temperature, the number of hot days and nights has increased, whilst the number of cold days and nights has decreased.

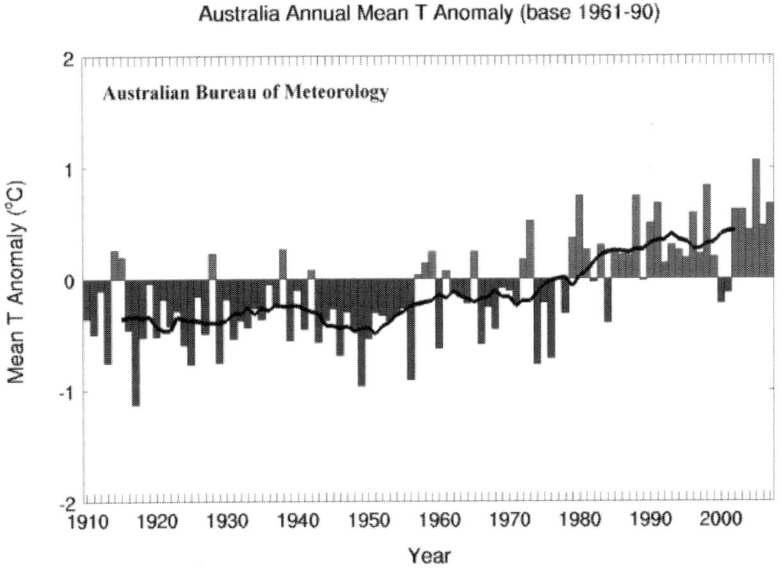

Figure 3 Australian average temperature changes over the past century.

Much of eastern and south-western Australia has seen declines in rainfall since 1950. Part of the explanation for this trend is the very wet period during the 1950s, particularly in New South Wales and Queensland. Recent years have been particularly dry, however, with the rainfalls in some areas comparable with the Federation Drought. Over the same period, north-west Australia has become considerably wetter with the changes most evident in the summer, rainy season. These signals are clear in the 1950–2007 trends in annual total rainfall (Figure 4).

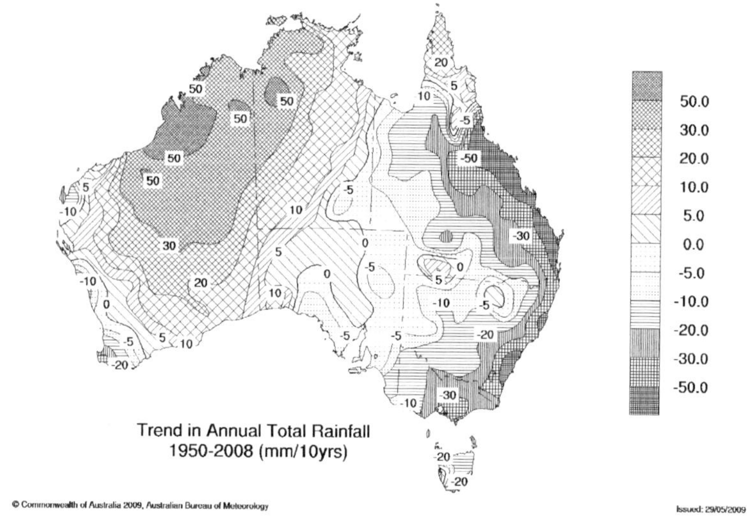

Figure 4 Trend in Annual Total Rainfall 1950–2007 (mm/decade).

The magnitude of the decline in rainfall over the last decade or so is evident in Figure 5, which shows the decile range of rainfall for the 10-year period November 1999–October 2008. The past five to 10 years mark one of the most severe drought periods in Australia's history with the worst declines centred over Australia's most productive zones in the Murray–Darling Basin and in the southern capital cities of Perth, Adelaide, Hobart, Melbourne and Sydney, with Brisbane also being adversely affected. Significantly, around

Perth, Melbourne and Brisbane there are areas where this period is noted as the driest on record. Almost all of Victoria is in decile 1, or *very much below average.*

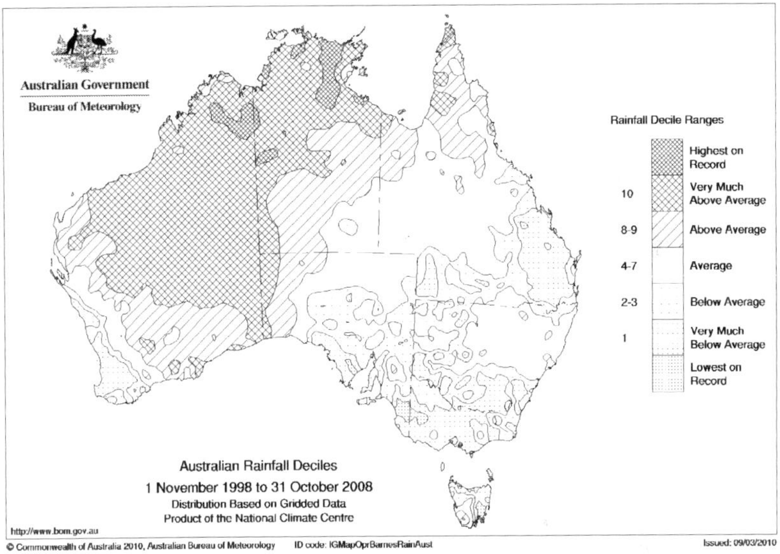

Figure 5 Australian rainfall deciles for the decade November 1998–October 2008.

One reason proposed for the changes in rainfall is the significant shift southward in the mean position of the subtropical ridge across Australia and the corresponding southward shift in mostly wintertime extra-tropical cyclones and rainfall bearing frontal systems. These findings are illustrated in Figure 6, which depicts the trend in anti-cyclone (high pressure) frequency between 1950 and 2006. The figure clearly shows a strong increase in the frequency of high pressure systems right across the south of Australia. High pressure systems are generally associated with stable, fine weather and reduced rainfall.

Also of significance are changes in the frequency and intensity of extremes of 'weather', which have can have severe consequences for life and property. These changes are evident in the rate at which records are broken over time, for example in maximum temperatures or overnight minimums, wind gusts, tropical cyclones, or storms with large hail. The Bureau of Meteorology regularly monitors and carefully verifies the breaking of records as part of its routine climate analysis operations, and occasionally issues Special Climate Statements on extreme events of particular significance.[3]

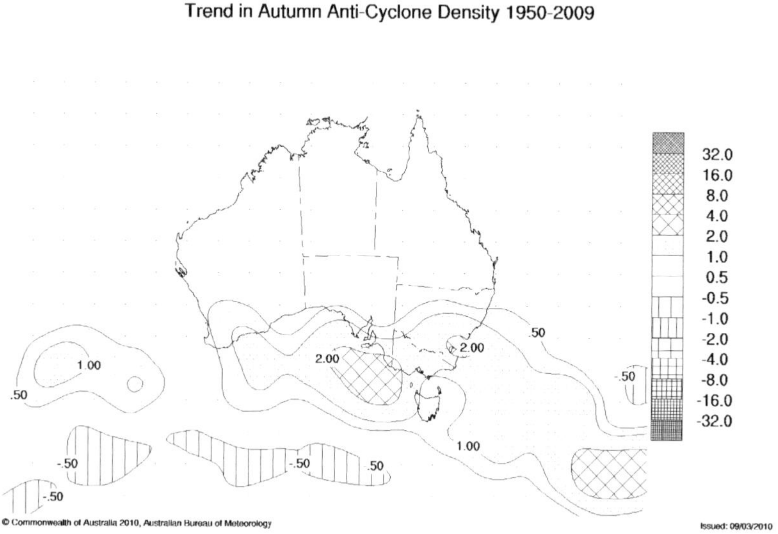

Figure 6 Trend in autumn anti-cyclone frequency 1950–2006 (systems/degree of latitude square/decade).

Upper air data

Australia operates 50 upper-air observation stations within its national meteorological network, including remote-island and Antarctic locations.

---

[3] Long-term trends in the occurrence of extreme events are also monitored and the results made available through the bureau's climate change website: www.bom.gov.au/silo/products/cli_chg/

Currently, 48 stations conduct meteorological balloon soundings, and the remaining two are Boundary Layer wind profilers.

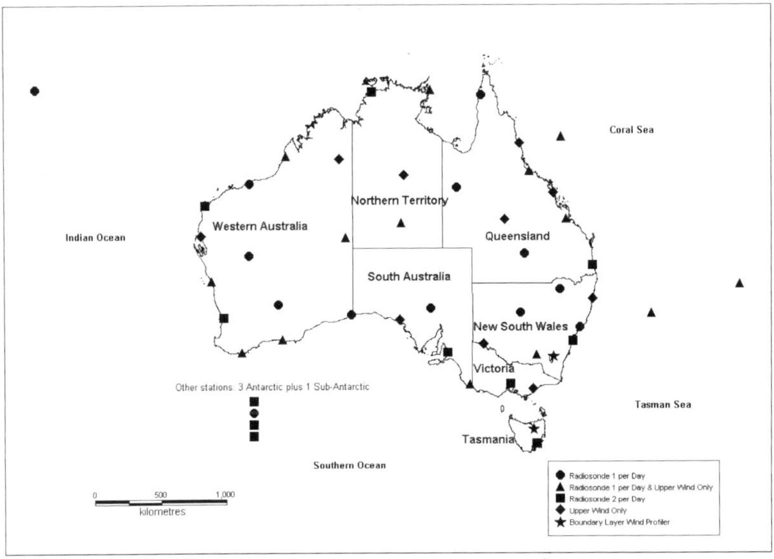

Figure 7 Australia's upper-air observation network of 50 stations (2008).

Thirty-eight of the stations conduct routine and non-routine soundings using balloon borne radiosondes[4] to report atmospheric pressure, temperature and relative humidity, as well as wind speed and direction up to balloon burst height, which typically occurs in the lower stratosphere (28–36 km). Full temperature and humidity soundings are made once or twice per day with wind speed and direction flights at up to six-hourly intervals.

The remaining 10 balloon operating stations conduct routine and non-routine soundings to report solely wind speed and direction, up to balloon

[4] A radiosonde is an instrument package that measure temperature and humidity and transmits the data back either directly or through a satellite relay. Wind speed and direction are measured using a tracking radar or increasingly through GPS technology.

burst height in the upper Troposphere (typically 12–18 km). Australia is evaluating wind profiler technology with a view to installing several (perhaps eight) additional profilers within the existing upper air network, during the period 2009–11. Some of these stations will be Stratosphere-Troposphere wind profilers, the remainder Boundary Layer wind profilers.

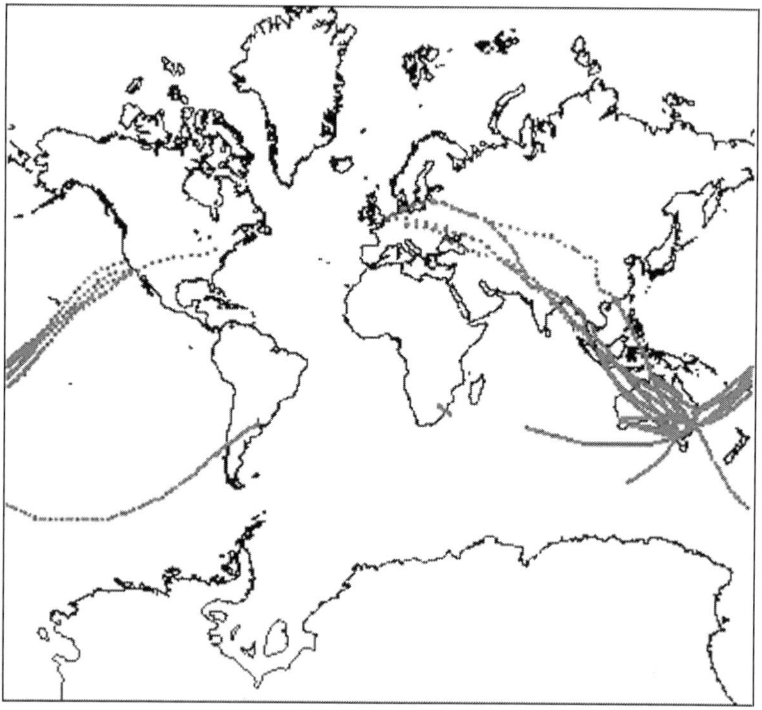

Figure 8 AMDAR upper-air data coverage for one 24-hour period during November 2008 from participating Qantas aircraft. Data from other aircraft operators (not shown) are also available to the Bureau of Meteorology.

Thirteen stations are equipped with automated radiosonde launching systems, which provide a fully automated upper air observing system by preparing and releasing the balloon train, tracking the radiosonde signal, and using the Global Positioning System (GPS) to derive wind information. These systems process and transmit the sounding data to a central location, where it is then verified manually.

Of the 48 balloon operating stations, 17 have been designated as part of the GCOS Upper-Air Network (GUAN), a network with relatively even distribution suitable for detecting the broad-scale patterns of climate change throughout the atmosphere. Soundings at these stations are conducted using stricter criteria, including better quality balloons to obtain data higher in the atmosphere (typically to 36 km). Nine of Australia's GUAN stations are located on the Australian mainland, with a further five located on remote islands and three in Antarctica.

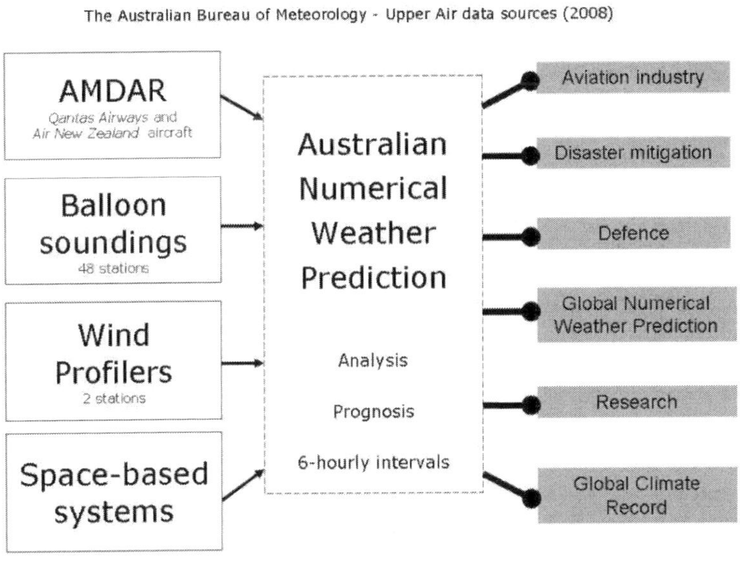

Figure 9 Observations of upper-air data and the uses that are made of them for different sectors.

The Australian upper-air network is supplemented by observations of temperature, wind and turbulence from new-generation commercial aircraft fitted with Aircraft Meteorological Data Relay (AMDAR) systems.

These observations are automatically transmitted from numerous aircraft in-flight, using very high frequency (VHF) air-to-ground communications. For Qantas aircraft these are transmitted to the Australian Bureau of Meteorology's communications centre where they are distributed internationally via the Global Telecommunications System (GTS). AMDAR-equipped aircraft are particularly valuable for providing coverage over the oceans and the remote continental interior. Like the balloon data, AMDAR data only cover specific locations and flight paths but do play a major part in building a better understanding of the atmosphere above the surface. However, due to the coarse resolution of the network and homogeneity issues associated with instrument changes, it is difficult to use Australia's upper-air network for the direct, routine monitoring of climate change.

Satellite-based observations of the upper atmosphere are an increasingly important component of Australia's composite observing system, together with ground-based and aircraft-based observations. Tropospheric and stratospheric temperatures and moisture content are derived routinely from radiance measurements obtained from instruments on the polar orbiting satellites of the US National Oceanic Atmospheric Administration (NOAA). Upper level winds are calculated by tracking cloud tracers on the visible, infrared and water vapour images of the Japanese Multifunctional Transport Satellite (MTSAT-1R) and experimentally from the Chinese satellite Fengyun-2C. Advanced infrared sounders, such as the Atmospheric Infrared Sounder on NASA's Aqua spacecraft and the Infrared Atmospheric Sounding Interferometer (IASI) on the operational European MetOp spacecraft, are providing temperature and moisture soundings with higher vertical resolution and better accuracy, than has been possible in the past.

Atmospheric constituent observations

Routine measuring of the chemical composition of the atmosphere is an important component of Australia's climate monitoring effort, which contributes to the wider international effort to understand the relationship between changing atmospheric composition and changes in global and regional climate. The WMO provides the international framework for countries to input the data for their region through the Global Atmosphere Watch (GAW) system. Australia's principal contribution to the GAW system is through the operation of the Cape Grim Baseline Air Pollution Station (CGBAPS) which was established in 1976 on the north-west tip of Tasmania. Cape Grim has gained an international reputation for the excellence of its scientific outputs and along with Mauna Loa in Hawaii is widely recognised as a premier station within the GAW, which has some 25 global background stations spread around the globe. Background stations sample air away from the influence of local human emission sources. A number of key atmospheric variables are monitored including chemical constituents such as carbon dioxide, methane, chlorofluorocarbons, ozone, particulate chemistry and physical characteristics such as radiation and particulate numbers and size distributions.

Air samples are also archived, so that in the future, the injection of newly relevant compounds can be investigated. CGBAPS has now produced a significantly long record. Quite consistent with other GAW stations around the globe, CGBAPS, for example, shows a steady increase in baseline carbon dioxide over the past 30 years. Within this steady increase there is a relatively consistent seasonal cycle which is understood to relate to the seasonally varying activity of biomass across the globe, a signal mostly emanating from the northern hemisphere.

In another contribution to GAW ozone monitoring activities, the Bureau of Meteorology operates Dobson spectrophotometers, measuring the total ozone in the atmospheric column and its vertical distribution, at four locations on the Australian mainland and at Macquarie Island. Weekly observations of the vertical distribution of ozone are also made using ozone sondes flown from Melbourne, Macquarie Island and at Davis Base, Antarctica. Figure 11 shows the average ozone partial pressure at Davis between 2003 and 2007 over the seasonal cycle. Notable is the clear depletion of ozone in the spring months. The bureau also routinely retrieves analyses and archives total ozone data derived from the TIROS Operational Vertical Sounder (TOVS) instruments on board NOAA polar orbiting satellites.

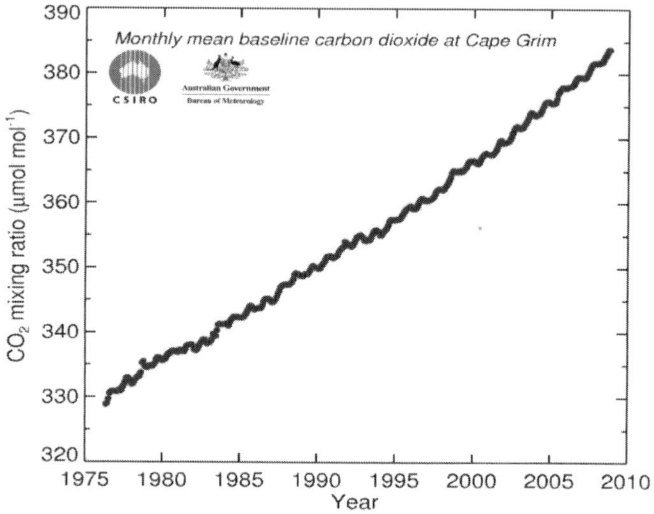

Figure 10 Monthly mean baseline carbon dioxide at the Cape Grim Baseline Air Pollution Station (CGBAPS) up to November 2008.

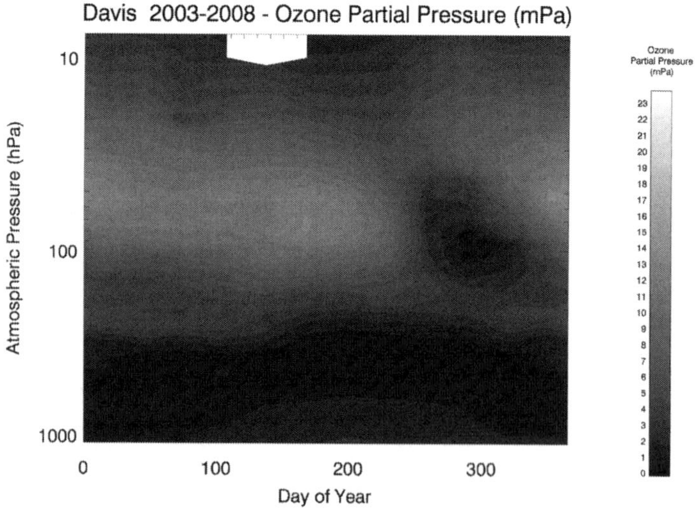

Figure 11 Ozone partial pressure at Davis between 2003 and 2007.

Australia has operated a global air sampling network of 5–20, mostly marine boundary layer, sites for varying periods since 1984 (Figure 12). In addition, flasks of air are collected on an opportunistic basis in support of regional and international campaigns. Shipboard sampling is done from Antarctic re-supply and CSIRO research vessels, and aircraft sampling is done over south-eastern Australia. Flask samples are analysed at the CSIRO's Global Atmospheric Sampling Laboratory (GASLAB).

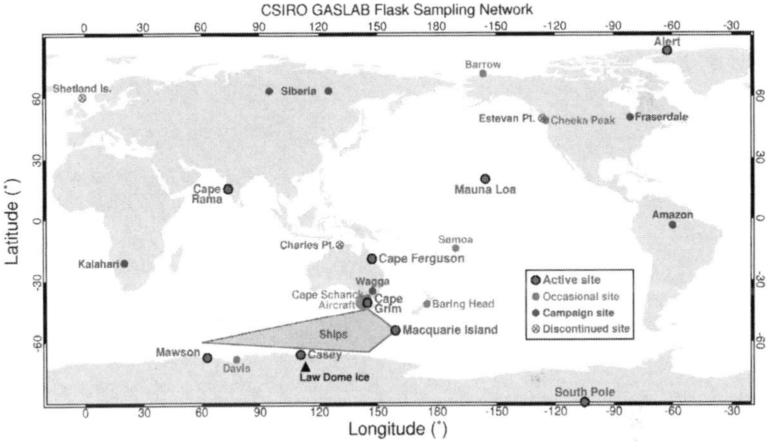

Figure 12 CSIRO GASLAB flask sampling network. Flasks are filled regularly at our current active sites and subsequently sent back to CSIRO in Aspendale for measurement in GASLAB (Global Atmospheric Sampling Laboratory).

The University of Wollongong monitors the long-term variability and change in stratospheric composition through ground-based remote sensing techniques involving solar infrared, visible and UV spectroscopy. Established in 1996, the Wollongong site contributes to the international Network for the Detection of Stratospheric Change (NDSC). Various agencies are involved in deriving concentrations and distributions of aerosols, such as dust, sulphate and volcanic ash, from a range of satellite instruments. The Bureau of Meteorology's Volcanic Ash Advisory Centre in Darwin uses volcanic ash detection techniques based on NOAA Advanced Very High Resolution Radiometer (AVHRR) data and GMS-5 imagery, combined with ground reports from vulcanological agencies, pilot reports, meteorological knowledge and numerical models, to track and forecast ash movements from volcanoes in Indonesia, Papua New Guinea and part of the Philippines.

Surface radiation

The Bureau of Meteorology operates a network of nine radiation stations monitoring direct, diffuse and global solar and terrestrial exposure, direct spectral irradiance, and, at two stations, UV measurements (Figure 13); three of which are WCRP Baseline Surface Radiation Network (BSRN) stations and the Alice Springs station is also a NDSC site.

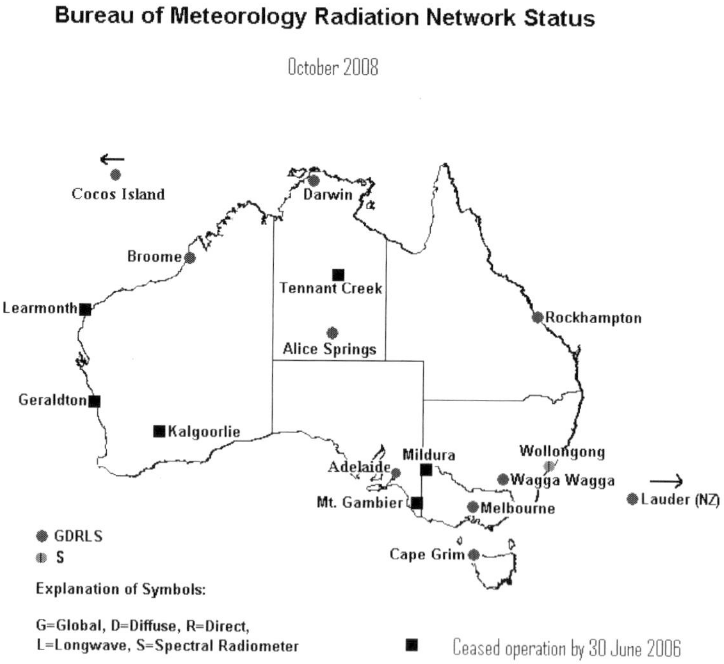

Figure 13 Australia's radiation observation sites.

These stations are intended to provide long-term, stable records for both climate studies and other applications such as agrometeorology or even new interests such as optimal sites for solar renewable energy locations. They also provide a calibration reference for the derivation of a nationwide solar exposure from satellite borne instruments, which provide a spatially extensive complement to the ground-based data, and are obtained using visible data from Japanese and USA-owned geostationary meteorological satellites. The software has recently been improved and older data are being back-processed to provide a long-term dataset covering around 15 years. In addition to the bureau's ground radiation stations, the bureau also hosts the WMO Region V Radiation Centre, which holds the Australian solar and terrestrial radiation standards.

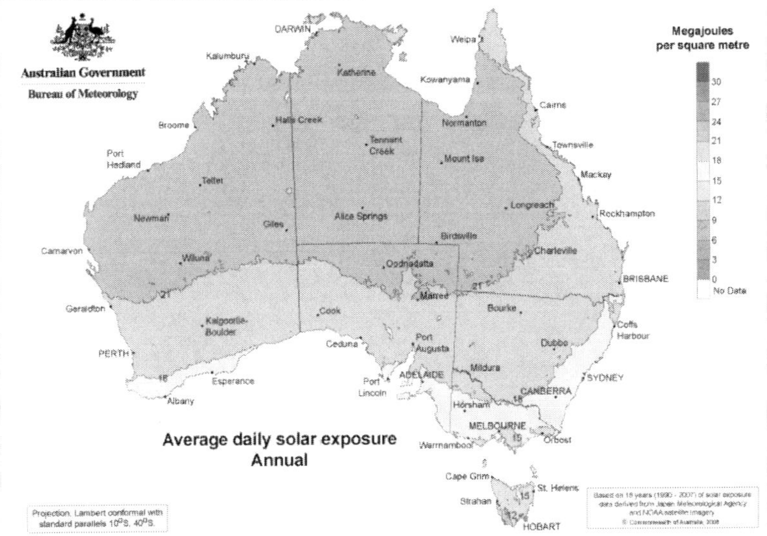

Figure 14 Average daily solar exposure (annual).

Hydrological observations

In addition to the rainfall networks operated by the Bureau of Meteorology, extensive networks of rainfall, streamflow, groundwater and climate stations are operated and maintained by the state and territory-based water agencies. These are a mix of public and private entities. The Bureau of Meteorology itself collects streamflow data to aid in its flood forecasting responsibilities. The vast array of different organisations collecting hydrological and related data, and the jurisdictional basis of their responsibilities, has led to a fragmented, inconsistent and incomplete understanding of Australia's water resources. The water resources assessment undertaken by the National Land and Water Resources Audit (NLWRA) and published as Australian Water Resources Assessment 2000 (AWRA) and the more recent Australian Water Resources 2005 undertaken by the National Water Commission used a subset of the networks to define a snapshot of the extent, quantity, use and quality of Australia's water resources. However, a more nationally consistent, hydrologically rigorous and regularly updated assessment is critical to improving the management of our limited water resources and thus a new approach was required. The federal government passed legislation in 2007 (*The Water Act 2007*) that charges the Bureau of Meteorology with responsibility to:

- issue standards for water information measurement and transmission
- gather water information and make it freely available via the web, with value-added analyses
- conduct national water resource assessments
- produce an annual national water account
- provide continuously updated water availability forecasts.

Whilst hydrological and related data are still observed by the array of organisations across Australia, legislation now requires that they provide the data to specified standards to the Bureau of Meteorology, which is responsible for generating national water accounts and making the data and information freely available via the internet. Whilst this new approach is still in its infancy, it provides an opportunity for a fundamental improvement in the understanding of Australia's water resources and the delivery of water information.

There remains the issue of development, which is occurring at an increasing rate in many river basins and is changing the hydrological characteristics of Australia's river basins. A limited number of stations, therefore, will be suitable for trend and climate analyses and only after considerable adjustment of the observed record to remove developmental impacts.

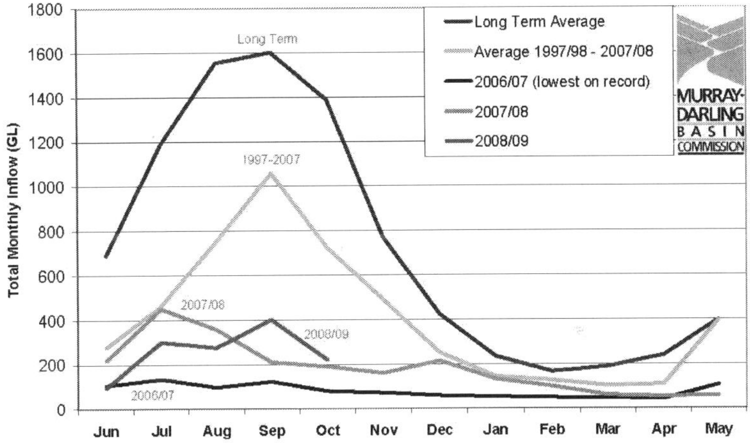

Figure 15 Total inflows to the River Murray System (including Menindee and excluding Snowy). The long-term average is compared with the past two years.

Over the past decade the Bureau of Meteorology has opened a number of new stations in more remote areas where there were insufficient data for water resources assessment and planning and drought monitoring. The river basins were selected using feedback from the state and territory water agencies to address their water management data needs and expected development in the basins. The technologies used have been a mix of manually read gauges where volunteer observers are available and automated data logging equipment in the more remote areas. Automated equipment has also been installed where volunteer observers are no longer available to ensure continuation of a valuable record. While the use of automated equipment improves the temporal resolution of data and increases the efficiency of data processing compared with manually read gauges, it has considerably higher initial capital cost and ongoing management and maintenance costs.

With Australia's growing population and changes in rainfall patterns, establishing adequate and reliable water resources has become even more critical than it was in the past. The combination of reductions in rainfall across much of the productive lands in the south of Australia, and a steady increase in temperature has seen a rapid decline in the inflows to catchments. In Figure 15, total inflows to the River Murray system over the long term are compared with recent inflows and they highlight the perilously low current values. Demand for water has never been higher yet the available water is severely limited. It is imperative that water use is sustainable amongst the mix of users which include individual households, businesses, agriculture as well as the natural environment; the challenge to do so is enormous. High-quality water information, including a good understanding of where the water is within the hydrological system and its quality, is paramount.

Summary

Monitoring of the atmosphere, and to a lesser extent the hydrosphere, has been comparatively well organised over a long period of time compared with other physical parameters; for example, the varying properties of the land surface. This outcome is by virtue of the fact that an operational agency, the Bureau of Meteorology, has been tasked with the collection and management of meteorological and related data virtually since Federation. There have also been international frameworks in place for long periods of time that have been established to organise the adequate collection and exchange of data. These frameworks have provided an effective basis for sharing global data to enable the forecasting of weather, and especially its manifestations of extremes, by national meteorological and hydrological services around the globe. Lessons learned from establishing networks to monitor the atmosphere have been valuable in establishing ocean monitoring networks in the most efficient manner and will be taken into account as efforts build to monitor the terrestrial domain more systematically.

Observing the world's climate is a very difficult and expensive process. The resources required to monitor climate in a sustained fashion far outweigh those for a one-off research project which will typically have a defined timeframe. However, it is absolutely vital to ensure that the monitoring effort is of the highest possible standard in order to expand understanding of the continuously changing climate. The climate record is essential for documenting the direct and indirect effects of climate change and therefore underpins efforts to determine areas of vulnerability and strategies for adapting to the changes.

Australia continues to work nationally and internationally to maintain and improve its monitoring effort, taking advantage of new technologies where possible and cooperating with international projects that target optimal deployment of effort. Amid the competition for resources, there must be efficiencies in providing the best cover possible with the available funds. Opportunities may evolve to see additional support for monitoring through the enormous challenge that climate change provides to Australia. It must be something that receives the highest priority.

Much of the information provided in this paper depicts activities within the Bureau of Meteorology. A great deal of information on bureau systems, products, services and data is available from the bureau's website, www.bom.gov.au.

Disclaimer

The views presented are those of the author and not the institution he works for.

Acknowledgements

The author would like to acknowledge the contributions of others in the editing and contribution of information and figures to this chapter. They include Michael Coughlan, Karl Monnik, Ross James, Graeme Ball, Gregory Smith, Bruce Forgan, Dean Collins, Paul Krummel, Matt Tully, Anthony Rea and Blair Trewin.

References

UNFCCC (2006). Report on a coordinated response from space agencies involved in global observations to the needs expressed in the Global Climate Observing System implementation plan. Available at: unfccc.int/resource/docs/2006/sbsta/eng/misc14.pdf.

WMO (1994). Guide to hydrological practices: data acquisition and processing analysis, forecasting and other applications .WMO-No 168. Geneva: World Meteorological Organization. Available at: www.bom.gov.au/hydro/wr/wmo/guide_to_hydrological_practices/WMOENG.pdf

WMO (1995). Manual on the Global Observing System. WMO-No. 544. Revised edition. Geneva: World Meteorological Organization. Available at: www.wmo.int/pages/prog/www/OSY/Manual/WMO544.RII.pdf

WMO (2007). Guide on the Global Observing System. WMO-No. 488, Third edition. Geneva: World Meteorological Organization. Available at: www.wmo.int/pages/prog/www/OSY/Manual/488_Guide_2007.pdf

6

Impact of climate change stressors on marine invertebrate life histories with a focus on the Mollusca and Echinodermata

Maria Byrne

*Abstract**

Global warming and increased atmospheric CO_2 are causing the oceans to warm and acidify. These stressors are likely to have deleterious impacts on marine biota, particularly on early life history stages. To address questions on future vulnerabilities, the thermo- and pH tolerance of fertilisation and development of sea urchins and bivalves is reviewed and placed within the setting of projected ocean change using data from single stressor studies of estuarine, coastal and shallow water species. Fertilisation in many urchin and bivalve species exhibit a broad tolerance being robust to warming or acidification beyond stressor values projected for 2100 (2100: ca. +2–4°C; pH −0.2–0.4 units, IPCC 2007). This is also the case for a number of other invertebrate taxa. It appears that fertilisation may not be a suitable indicator to discern the effects of climate change on marine life histories. Developmental stages, however, exhibit greater sensitivity to thermal and pH stress. Early embryos of some species are vulnerable within the range of near future warming. For embryos that reach the larval stage in a warm ocean, increased pCO_2 and acidification may have a negative effect on larval physiology due to hypercapnia and decreased bioavailability of the carbonate ions needed to produce their skeleton. Developmental bioassays with an initial focus

* Chapter originally submitted in March 2009.

on early endpoints (cleavage, gastrulation) provide a model system with which to assess and monitor the interactive effects of ocean warming and acidification. Animal development can fail at any stage and determination of the sensitivities of life stages to climate change stressors is essential. If embryogenesis fails due to warming then the question of compromised larval calcification may not be relevant. Marine life history stages are exposed to a plethora of stressors in addition to those from climate change and the interaction between stressors has potential for deleterious impacts, but these are difficult to predict. Most available data on effects of climate change stressors on marine life histories are for robust species used as model organisms for laboratory research and aquaculture. Studies to date indicate that the response of marine embryos and larvae to climate change stressors varies, even between closely related species, making it difficult to extrapolate broadly on potential climate change effects. Multifactorial experiments with marine embryos using methods adopted from ecotoxicology provide a framework to determine the potential effects of climate change on marine biota. Although it is unlikely that generic indicator species will be identified for monitoring impacts of ocean change, empirical data for a phylogenetically diverse suite of species from different habitats are required to assess this. A gloomy assessment is presented, but there is potential for in-built adaptive (genetic) and acclimation (phenotypic plasticity) capacity for some species. This, along with different tolerance thresholds, will influence species vulnerability and the potential winners and losers in the face of ocean regime change.

1. Introduction

Global warming and increased atmospheric CO_2 are causing the oceans to become warmer and acidify (Feely et al. 2004; IPCC 2007), the latter through the absorption of CO_2, which dissolves in seawater to form carbonic acid. This acid dissociates into protons and bicarbonate ions, thereby decreasing pH. Uptake of CO_2 is predicted to decrease global ocean pH by 0.2 to 0.4 units and increase sea surface temperature (SST) by a mean of 2°C by 2100

(Caldeira & Wickett 2003; Feely et al. 2004; IPCC 2007, Poloczanska et al. 2007; Fabry et al. 2008; but see Whooten et al. 2008). The extent of ocean warming and acidification will vary among regions, strongly influenced by differences in ocean circulation patterns, coastal processes, ocean chemistry and the inverse relationship between temperature and CO_2 uptake (Levitus et al. 2005; Blackford & Gilbert 2007; Poloczanska et al. 2007; McNeil & Matear 2008). The Earth has entered a new phase in its history, experiencing a pace of climate change far greater than the planet has experienced with respect to evolutionary time scales (Feely et al. 2004). Predicting the future state of our oceans and marine life in the face of climate change and other anthropogenic stressors is a major challenge (Pörtner et al. 2005; Przeslawski et al. 2008).

High pCO_2 in seawater and reduced pH have negative impacts on growth, reproduction and development of marine biota due to direct effects on metabolic processes caused by hypercapnia and decreased bioavailability of the carbonate ions needed for shell formation (Reviews: Burnett 1997; Fabry et al. 2008, Pörtner 2008; Widdicombe & Spicer 2008; Doney et al. 2009). Temperature affects everything organisms do through its pervasive physiological impact on all biological functions (Clark 2003). Gamete development, spawning and development are modulated by temperature (Giese & Pearse 1974; Pechenik 1989; Grubert & Ritar 2004a,b, 2005) and ocean warming is implicated in changes in the timing of reproduction and larval phenology (Hays et al. 2005; Edwards & Richardson 2004; Richardson 2008). Ocean warming increases susceptibility to disease and parasite infections in adult sea urchins, bivalves, abalone and other marine invertebrates (Alstatt et al. 1996; Friedman et al. 1997; Lester et al. 2007). This has negative impacts on fecundity and larval production. It is not known if ocean warming will increase incidence of disease in larvae and other developmental stages, although this seems likely.

Our understanding of the potential impacts of ocean warming, acidification and hypercapnia on marine biota is impeded by the scarcity of empirical data and prevalence of single stressor studies most of which use stressor levels well beyond values predicted by climate change models (Reviews: Przeslawski et al. 2008; Fabry et al. 2008; Dupont & Thorndyke

2009). Studies of echinoderms, crustaceans and molluscs show deleterious effects of CO_2-induced acidification in single stressor studies at levels predicted for 2300 (pH −0.7–1.2 units, ca. pH 6.0–7.4: Kurihara & Shirayama 2004; Bibby et al. 2007; Kurihara et al. 2004, 2007; Kurihara 2008; Clark et al. 2009).

The early life history stages (embryos and larvae) of most marine invertebrates occurs in the water column where climate change stressors are likely to have deleterious impacts on development. Understanding the vulnerabilities of the developmental stages integral to the persistence of marine populations is crucial as we endeavour to predict how marine populations and ecosystems will fare in the face of climate change. In a long history of research on the physiology of marine life histories, a plethora of studies document the role of temperature and pCO_2/pH on biological processes including gas exchange, sperm function, developmental dynamics, optimal culture conditions and teratogenic effects (Calabrese & Davis 1970; Kobayashi 1980; Burnett 1997; McEdward 1985; Brokaw 1990; O'Connor & Lawlor 2004; Carr et al. 2006). Due to their sensitivity to environmental perturbation, the gametes and embryos of marine invertebrates are used as key bioindicators for monitoring the environment. Ecotoxicological studies on the response of life history stages to thermal pollution (e.g. power plant effluent) and increased pCO_2/decreased pH (e.g. pore water, acid leachates) (Kennedy et al. 1974; Greenwood & Bennett 1981; Wright et al. 1983; Bay et al. 1993; Wilson & Hyne 1997; Carr et al. 2006; Byrne et al. 2008) provide data with which to assess vulnerabilities and tolerance limits to climate change stressors. Data on effects of increased pCO_2 on early life history stages are available in studies where CO_2 gas treatment of seawater is used as a standard method to investigate effects of acid pollution (e.g. Carr et al. 2006). Marine ecotoxicology has a new focus, to assess the impacts of climate change on marine life histories (Kurihara & Shirayama 2004; Kurihara et al. 2004, 2007; Byrne & Davis 2008; Havenhand et al. 2008; Byrne et al. 2009; Clark et al. 2009; Parker et al. 2009).

Most studies on the effects of temperature or acidification on marine invertebrates use sea urchins or bivalves as model organisms for aquaculture or environmental monitoring (Calabrese & Davis 1970; Rupp 1973;

Andronikov 1975; Greenwood & Bennett 1981; Mita et al. 1984; Roller & Stickle 1993; Carr et al. 2006; Gazeau et al. 2007; Havenhand et al. 2008; Kurihara et al. 2007; Byrne et al. 2009; Havenhand & Schlegel 2009; Kurihara 2008; Parker et al. 2009). These studies provide baseline data on the potential impacts of climate change stressors. Sea urchins and bivalves are an important component of near shore marine ecosystems and have received research emphasis due to their ecological and commercial significance. In this review the thermo- and pH tolerance of fertilisation and development of sea urchins and bivalves is reviewed using available data, from estuarine, coastal and shallow water species. To have comparative data on climate change stressors, the review focuses on species for which temperature and pH data are available and where ambient conditions are provided. The latter was essential to identify tolerance thresholds with respect to current conditions and to place these within the context of near-future scenarios (2100: ca. SST +2–4 °C; pH –0.2–0.4 units, IPCC 2007). The best data are for gamete and embryo (pre-larval) thermo- and pH tolerance (Table 1, 2) and recent studies have investigated the impact of near-future scenarios on larval development (Dupont et al. 2008; Parker et al. 2009).

To address questions on future vulnerabilities on marine life histories, the thermo- and pH tolerance of fertilisation and development of sea urchins and bivalves is placed within the setting of projected ocean change. Where available, data for other marine invertebrates are included. Gamete and embryotoxicity provide a model system to assess the potential impacts of climate change stressors on marine biota and risks to marine ecosystems . For marine invertebrate life histories this approach can be used to address a series of questions: 1) Can fertilisation occur in a warm and high pCO_2/acidic ocean? 2) If fertilisation is successful can the embryos develop? 3) Can these embryos reach the larval stage, make their skeleton and develop to the juvenile stage? Animal development can fail at any stage and determination of the comparative sensitivities of life stages to climate change stressors will provide insights as to where vulnerabilities lie. For instance if embryonic development fails due to warming then the question of comprised larval calcification due to ocean acidification may not be relevant.

2. *Echinoid and bivalve life histories in a changing ocean*

The direct effect of temperature on marine gametes and embryos is well documented. These effects include acceleration of sperm swimming speeds, modification of fertilisation kinetics, increased rate of development, reduced dispersal and induction or suppression of stress responses, until a lethal maximum is reached (Mita et al. 1984; McEdward 1985; Pechenik 1989; Hoegh-Guldberg & Pearse 1995; Hobday & Tegner 2002; Clark 2003; Staver & Strathmann 2002; Lee et al. 2004; Negri et al. 2007; O'Connor et al. 2007; Mumby et al. 2009; O'Donnell et al. 2009; Parker et al. 2009). The direct effects of increased pCO_2 (hypercapnia) and decreased pH on early life history of marine invertebrates are less well documented, although hypercapnia is well known to impair metabolic processes (Pörtner 2008; Widdicombe & Spicer 2008). The cellular control of fertilisation involves acidic conditions (e.g. acid egg jelly, egg acid secretion, low internal pH of sperm) and gonad tolerance of respiratory acidosis (Paucellier & Doree 1981; Holland & Gould-Somero 1982, Holland et al. 1984; Bookbinder & Shick 1986; Burnett et al. 2002). Studies on the effects of decreased pH and hypercapnia on adults report dissolution of skeleton and physiological stress (Knutzen 1981; Bamber 1990; Green et al. 2004; Michaelidis et al. 2005; Berge et al. 2006; Gazeau et al. 2007; Miles et al. 2007). In experiments in keeping with near future ocean acidification scenarios Gazeau et al. (2007) showed a linear decline in calcification in adult oysters and mussels with respect to increasing pCO_2.

Data on the effects of temperature and pH relevant to climate change scenarios are available for the gametes and embryos (pre-larval stages) of five genera and seven species of echinoids and six genera and seven species of bivalves (Table 1, 2). Although most of these studies were designed to address environmental (ecotoxicology) and aquaculture (hatchery) issues, they provide useful data on the thermo- and pH tolerance of marine life histories in response to individual stressors. With regard to ocean acidification, the most relevant data are from studies where experimental pH was adjusted through use of CO_2 gas (Carr et al. 2006). Some studies also provide data on pCO_2 conditions (Table 1, 2). Two studies provide data on the interactive effects of temperature and pH/pCO_2, one for echinoids and one for bivalves (Byrne et al. 2009; Parker et al. 2009).

There is considerable variation in the methodologies used in the studies listed in Tables 1 and 2. Some studies use single sire-dam crosses (e.g. Havenhand et al. 2008) while others use gametes pooled from multiple males and females (e.g. Byrne et al. 2009), a difference that influences fertilisation success (see Evans & Marshall 2005). For fertilisation some studies pretreat sperm in experimental seawater prior to fertilisation (Carr et al. 2006; Havenhand & Schlegel 2009), some pretreat eggs (e.g. Hägstrom & Hägstrom 1959; Fujisawa 1989), some pretreat both gametes (Byrne et al. 2009) while other studies placed both gametes directly in experimental water for fertilisation (e.g. O'Connor & Mulley 1977; Parker et al. 2009). For development some studies pretreated eggs in experimental seawater prior to addition of sperm followed by rearing in experimental conditions (e.g. Kurihara & Shirayama 2004), some pretreat eggs and sperm prior to fertilisation (Havenhand et al. 2008; Byrne et al. 2009), others placed both gametes directly in experimental water for fertilisation (Parker et al. 2009) while some studies place embryos fertilised in ambient conditions in experimental water for rearing (Carr et al. 2006; Kurihara et al. 2007; Clark et al. 2009). Despite the different methodologies used, general trends in the response of gametes and embryos to temperature and pH perturbations are similar among species (Table 1, 2). Experiments where gametes are pretreated in realistic scenario-based temperature/pCO_2 gas treatments (e.g. IPCC 2007; Poloczanska et al. 2007) prior to fertilisation, and the resulting embryos reared in these conditions, are most relevant to investigate the impact of ocean change on marine life histories (Byrne et al. 2009).

The potential interactive effects of climate change stressors also need to be placed in a climate and regionally relevant setting with respect to where species live. This is an essential consideration to inform risk assessment to local ecosystems. Researchers need to be cognisant of model projections for regional ocean change and the ecoclimate (microclimate) of the species being

Table 1 Data from single stressor studies of temperature and pH tolerance of sea urchin fertilisation and early development for species or genera where data for both factors are available. Where there are no data on ambient spawning temperature, location is indicative of thermal regime. Experimental pH was adjusted by treatment of seawater with CO_2 gas. Dev, development; Fert, fertilisation; ND, no data; SV, seasonally variable; Temp, temperature (pH scales not provided).

Species	**Study region & ambient spawning temp**	**Temp range (°C) Fert ≥ 75%**	**Upper Temp (°C) Fert < 70–75%**	**pH range Fert ≥ 75%**	**Low pH range Fert < 70–75%**	**Temp range (°C) early Dev ≥ 75%**	**Upper Temp (°C) early Dev. <70–75%**	**pH range early Dev ≥ 75%**	**Low pH early Dev < 70–75%**	**Reference**
Strongylo-centrotus purpuratus	W USA 13	13–25	30	7.3–8.2	7.2	13–20	25–30	7.5–8.4	7.4	Farmanfarmaian & Giese 1963; Bay et al. 1993
Hemicentrotus pulcherrimus	N Japan 5–19	0–30	ND	7.4–8.0[1]	7.0	5–24	> 24	7.6–8.1*[5]	7.4	Mita et al. 1984; Fujisawa 1989, 1995; Fujisawa & Shigei 1990; Kurihara & Shirayama 2004
Arbacia lixula	W Europe 14–19	15–22	> 32	ND	ND	ND	ND	ND	ND	Hagström & Hagström 1959

Species	**Study region & ambient spawning temp**	**Temp range (°C) Fert ≥ 75%**	**Upper Temp (°C) Fert < 70–75%**	**pH range Fert ≥ 75%**	**Low pH range Fert < 70–75%**	**Temp range (°C) early Dev ≥ 75%**	**Upper Temp (°C) early Dev. <70–75%**	**pH range early Dev ≥ 75%**	**Low pH early Dev < 70–75%**	**Reference**
Arbacia punctulata	E USA ND	ND	ND	7.0–8.6	< 7.0	ND	ND	7.5–8.6	7.2–7.4	Carr et al. 2006
Echinometra mathaei	Tropical Pacific 28	28–36	ND	7.6–8.1[2]	7.4	28–34	36	ND	ND	Rupp 1973; Kurihara & Shirayama 2004
Heliocidaris erythro-gramma	E Australia 19–24	18–28	>28	7.6–8.2[3]	ND	18–26	> 28	7.6–8.2[6]	> 7.6	Nguyen 2007; Byrne et al. 2009
Heliocidaris tuberculata	E Australia 17–26	17–28	>28	7.6–8.2[4]	ND	17–24	SV	7.6–8.2[7]	> 7.6	O'Connor & Mulley 1977; Byrne, unpub. data

*cleavage stages scored; pCO_2 ppm: 1365–2360, 2365–1000, 3230–700, 4230–700, 5365–100, 6365–690, 7365–690, pCO_2 ppm: [1]375–600, [2]375–600, [3]1000, [4]1000, [5]375–1000, [6]375–600, [6]375–600, [7]>1000.

Table 2 Data from single stressor studies of temperature and pH tolerance of bivalve fertilisation and early development for species where data for both factors are available. Where there are no data on ambient spawning temperature, location is indicative of thermal regime. Experimental pH was adjusted by treatment of seawater with CO_2 gas (*), acid (**) or method not provided (***). Dev, development; Fert, fertilisation; ND, no data; Temp, temperature (pH scales not provided).

Species	**Study region & ambient spawning temp**	**Temp range (°C) Fert ≥75%**	**Upper Temp (°C) Fert <70–75%**	**pH range Fert ≥75%**	**Low pH range Fert <70–75%**	**Temp range (°C) early Dev ≥75%**	**Upper temp (°C) early Dev <70–75%**	**pH range early Dev ≥75%**	**Low pH early Dev <70–75%**	**Reference**
Crassostrea gigas	E Japan 23 E Australia 26 Sweden 21–22	18–30	>30	7.8–8.2*[1]	7.7[3]	18–30	>30	7.7–8.2[5]	<7.7*[7]	Havenhand & Schlegel 2009; Kurihara et al. 2007; Kurihara 2008; Parker pers. comm.
Crassostrea virginica	E USA 25	20–32	ND	ND	ND	20–32	32	6.75–8.75	6.5–6.75**	Calabrese & Davis 1966, 1970; Wright et al. 1983

Species	Study region & ambient spawning temp	Temp range (°C) Fert ≥75%	Upper Temp (°C) Fert <70–75%	pH range Fert ≥75%	Low pH range Fert <70–75%	Temp range (°C) early Dev ≥75%	Upper temp (°C) early Dev <70–75%	pH range early Dev ≥75%	Low pH early Dev <70–75%	Reference
Mercenaria mercenaria	E USA 25	20–26	ND	ND	ND	18–30	31	7.0–8.75**	6.75**	Calabrese & Davis 1966,1970; Wright et al. 1983
Saccostrea glomerata	E Australia 26	26–30	>30	7.9–8.0*[2]	7.7[4]	22–26	30	7.0–8.0** 7.9–8.0**[6]	6.75** <7.8*[8]	Wilson & Hyne 1997; Parker et al. 2009; Parker pers. comm.
Adula californiensis	W USA 10–15	ND	ND	ND	ND	7–20	20	7.7–7.9	ND	Lough & Gonor 1973
Spisula solidissima	E Canada	8–20	22	7.5–8.0**	7.0	ND	ND	ND	ND	Clotteau & Dubé 1993
Placopecten magellanicus	E Canada	10–15	20	8.0–8.5***	7.5	8–12	15	7.5–8.5	<7.5	Desrosiers et al. 1996

considered. The latter plays a key role in determining threshold tolerance, phenotypic plasticity and adaptive capacity. In addition, determination of experimental pCO_2 and pH conditions, using the NBS scale or seawater scale will influence the parameter values provided in different studies (see Lewis & Wallace 1998) and so the scale used should be identified.

2.1 Fertilisation: influence of warming and acidification – single stressor studies

Reproduction of most marine invertebrates involves the shedding of eggs and sperm directly into the water column and this is where fertilisation occurs. Fertilisation is followed by development through planktonic embryo and larval stages with eventual metamorphosis and settlement to the sea floor by the juvenile stage. If climate change stressors have a negative effect on fertilization – the foundation stage for species persistence – this would have dire consequences for marine biota.

2.1.1 Temperature tolerance – fertilisation

The timing of gamete release and development in many marine invertebrates is often fine tuned with respect to annual seasonal temperature cycles (Giese & Pearse 1974; Pechenik 1989; Grubert & Ritar 2004a,b). Thus fertilisation in the sea often coincides with the narrow thermal range associated with natural spawning temperatures (e.g. winter, summer). Despite this, high fertilisation rates in sea urchins and bivalves can occur at temperatures well above ambient spawning season maxima (ca. +4–12°C, Table 1, 2, Fig. 1). A broad thermotolerance of fertilisation is reported for sea urchins (ca. nine species) as listed in Table 1 and other studies (Rupp 1973; Andronikov 1975; Horstadius 1975; Fujisawa 1989, 1995; Sewell & Young 1999). This is also the case for bivalves (ca. three species, Table 2, Andronikov 1975). For echinoderms, thermal tolerance of fertilisation is conveyed by protective maternal factors (e.g. heat shock proteins) and the thermal independent period post gamete binding (Fujisawa 1989; Sconzo et al. 1997; Yamada & Mihashi 1998). The magnitude of thermotolerance varies among species (Table 1, 2) and is likely to be species specific. A broad thermotolerance of fertilisation is also reported for asteroids, gastropods and corals (Davis 2000; Lee et al. 2004, Negri et al. 2007).

It appears that fertilisation in many species is robust to increased temperature (Table 1, 2) and, in many cases, is enhanced (Rupp 1973; Mita et al. 1984; Fujisawa 1989; O'Connor & Mulley 1977; Sewell & Young 1999; Davis 2000; Lewis et al. 2002; Negri et al. 2007; Parker 2009a,b). Thermal enhancement of fertilisation is likely due to positive effects on sperm swimming speeds and heightened sperm-egg collisions (Hagström & Hagström 1959; Mita et al. 1984).

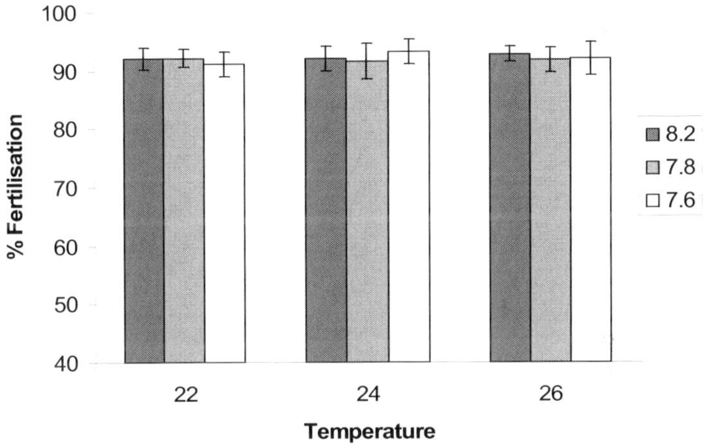

Figure 1 Effect of temperature and pH on the percentage of fertilisation of the sea urchin *Heliocidaris erythrogramma*. The 20°C and pH 8.2 treatment represented ambient conditions. At the upper warming (+6°C) and pCO_2 acidification (pH -0.4 units) scenarios predicted for 2070–2100, fertilisation of *H erythrogramma* was not affected. (Details of methods and experimental design in Byrne et al. 2009).

2.1.2 pH tolerance – fertilisation

A number of studies show high fertilisation rates in sea urchins across a broad pH range (Table 1, 2). The threshold for deleterious effects varies among species (Table 1). In *Strongylocentrotus purpuratus*, *Hemicentrotus pulcherrimus*, *Arbacia punctulata* and *Echinometra mathaei*, percent fertilisation dropped (<70%) at pH 7.2, 7.0, 7.0 and pH 7.4, respectively (Table 1). A similar result was obtained for *Heliocidaris erythrogramma* with high levels of fertilisation at the lowest experimental pH used (pH 7.6) (Figure 1) (but see Havenhand et al. 2008). Studies on the influence of pH using acid or samples of seawater from field sites known to have low pH show a similar tolerance of fertilisation in echinoids (Smith & Clowes 1924; Riveros et al. 1996; Kurihara & Shirayama 2004). The weight of evidence (Table 1) indicates that sea urchin fertilisation is impaired at pH below 7.4 (>1000 ppm CO_2), acidification levels well above projections for the ocean by 2100 (IPCC 2007).

For the bivalve *Saccostrea glomerata* (Table 2), a decrease from control pH 8.0 to pH 7.9 causes a significant reduction in percentage of fertilisation (Parker et al. 2009), indicating vulnerability of fertilisation in this species to near future ocean acidification. In contrast fertilisation in *Crossostrea gigas* was not negatively affected by pH 7.4–7.8 (Kurihara et al. 2007; Havenhand & Schlegel 2009). Fertilisation in *Spisula solidissima* is also unaffected by low pH 7.5 (Clotteau & Dubé 1993).

The gonads of marine invertebrates typically have a high internal pCO_2/low pH used to maintain sperm in a quiescent state prior to spawning. Hypercapnia has a well-known narcotic effect on sperm, reducing metabolism and swimming speed (Mohri & Yasumasu 1963; Ohtake 1976; Mita & Nakamura 2001). In nature, this is overcome by the stimulatory influence of compounds in the egg jelly coat on sperm metabolism (Ward et al. 1985; Brokaw 1990). A decrease in fertilisation in *H erythrogramma* (pH 8.1: 62%; pH 7.7: 51%) is attributed to decreased sperm motility (Havenhand et al. 2008). In contrast low pH (7.8)/high pCO_2 did not exert an inhibitory effect on sperm swimming in *Crossostrea gigas* (Havenhand & Schlegel 2009). As the ocean continues to uptake more CO_2, hypercapnia may be of concern for sperm function and fertilisation. Low pH has a negative effect on fertilisation in a polychaete species (Quesne & Bentley 2006).

2.2 Early (pre-larval) development: influence of warming and acidification – single stressor studies

Early embryogenesis (cleavage to gastrulation) is highly sensitive to environmental perturbation. Gastrulation is the first major morphogenetic event in development, generating the primary germ layers (ectoderm, endoderm, mesoderm) upon which all subsequent development depends. Vulnerability of early embryos to climate change stressors would have serious downstream effects on marine populations. Data on thermotolerance of early embryos are reported for a taxonomically diverse suite of sea urchins (ca. 20 species) in Table 1 and elsewhere (Rupp 1973; Andronikov 1975; Fujisawa 1989, 1995; Chen & Chen 1992; Sewell & Young 1999). For bivalves data are available for ca. eight species (Table 2, Lough & Gonor 1973; Kennedy et al. 1974; Andronikov 1975; Wright et al. 1983). Comparative data on temperature or pH tolerance on sea urchin embryos within the setting of climate change scenarios are available for six species where experimental pH was adjusted through use of CO_2 gas (Table 1). For bivalves, data are available for one species where seawater pH was adjusted with CO_2 gas, two species where acid was used and one where the method of pH adjustment was not stated (Table 2).

2.2.1 Temperature tolerance – early development

Temperature has a profound effect on early development dictating the pace of cell division and generation of the appropriate number of cells required for normal gastrulation. Depending on species, early development in sea urchins and bivalves is tolerant to +2–7°C warming. Thresholds for <70% normal development are ca. +4–10°C above ambient (Table 1, 2). Tolerance levels and thresholds for onset of deleterious effects vary among species. The ontogenetic pattern of thermotolerance also varies with some species being more vulnerable to increased temperature as early embryos (Kennedy et al. 1974; Andronikov 1975; Wright et al. 1983), while others are more vulnerable at a later stage of development (Lough & Gonor 1973). As shown for *Heliocidaris erythrogramma* (Figure 2) the thermotolerance range of early development (cleavage, gastrulation) is often narrower than that for fertilisation (Table 1, 2). Development in *H erythrogramma* was compromised at +4°C above ambient SST (ca. 50% normal) and failed at +6°C above ambient (Figure 2), indicating vulnerability to near future ocean warming (Byrne et al. 2009).

The embryos of some species such as the sea urchins *Echinometra mathaei* and *Hemicentrotus pulcherrimus* and the bivalves *Crassostrea gigas*, *Mercenaria mercenaria*, *Adula californiensis* and *Mulina lateralis* are highly thermotolerant (Table 1, 2; Lough & Gonor 1973; Kennedy et al. 1974). A broad thermotolerance of early development is also reported for gastropod, asteroid, coral and sponge species (Davis 2000; Lee et al. 2004; Negri et al. 2008; Whalen et al. 2008). Included in thermotolerant species are a number of hardy intertidal and shallow water species with broad latitudinal distributions used as for laboratory research (e.g. *Echinometra mathaei* and *Hemicentrotus pulcherrimus*) or aquaculture (*Crassostrea gigas*, *Mercenaria mercenaria*). Threshold thermotolerance for development (= biological zero points) have been determined for a number of bivalves and gastropods in hatchery culture and, depending on species, are generally >+4 above ambient indicating tolerance to near future warming (Table 1, 2; Grubert & Ritar 2004b).

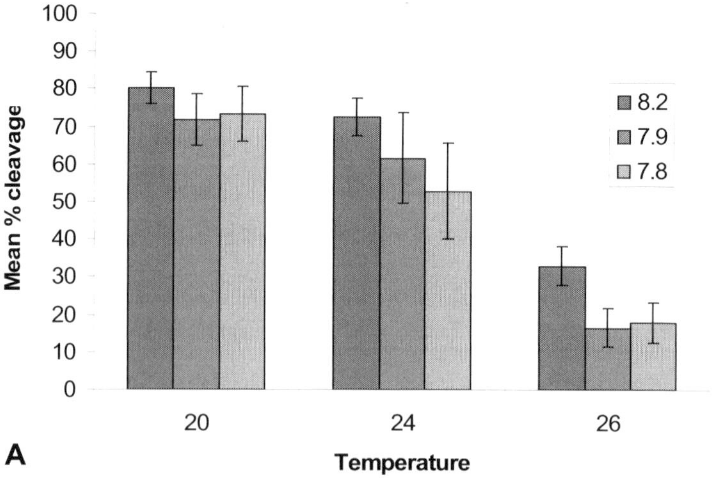

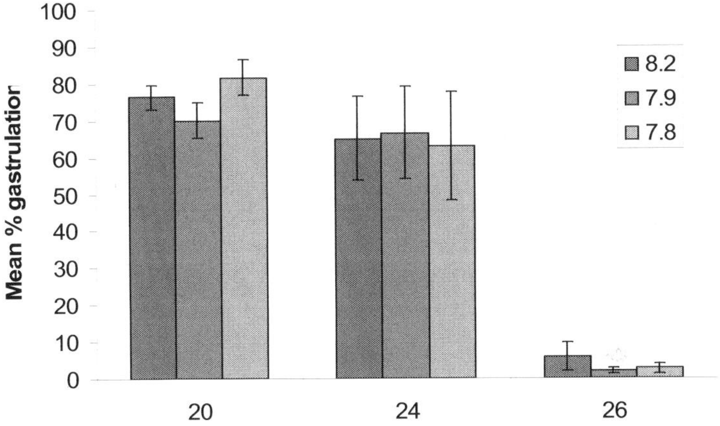

B

Figure 2 Effect of temperature and pH on early development of the sea urchin *Heliocidaris erythrogramma*. (A) Percentage of normal cleaving embryos – two hours post-fertilisation (B) percentage of normal gastrulae – 24–26 hours post-fertilisation. The 20°C and pH 8.2 treatment represented ambient conditions. Temperature had a significant effect on development, but pH did not (ANOVA: Cleavage: Temp; $F_{2,36}$=21.73, P<0.0001, pH: $F_{2,36}$=0.74, P=0.48; Gastrulation: Temp; $F_{2,36}$ =63.47, P<0.0001, pH: $F_{2,36}$ =0.28, P=0.76) and with no interaction between these factors (ANOVA: Cleavage: $F_{4,36}$ =0.27, P=0.89; Gastrulation: $F_{4,36}$ =0.47, P=0.76). The percentage of normal cleaving embryos decreased as temperature increased (SNK: 20°C>24°C>26°C) as was the percentage of normal gastrulae 24–26 hours post-fertilisation (SNK: 20°C=24°C>26°C). Temperature had a significant effect with develop-mental failure at the upper warming projected for 2100 for south-eastern Australian waters (+4–6°C) level, regardless of pH. (Details of methods and experimental design in Byrne et al. 2009.)

We have a paucity of data for a diversity of species, which may have a narrower developmental thermotolerance. Data on potential effects of ocean warming are particularly needed for subtidal species, which, due to their comparatively stable environment, would be expected to have a narrow thermal range and may be more susceptible to changed ocean conditions.

2.2.2 pH tolerance – early development

In single stressor studies of decreased pH/increased pCO_2 sea urchin embryos are negatively affected by pH below 7.4 (>1000 ppm CO_2) (Table 1). This level of acidification is more extreme than those projected for near future ocean acidification. For instance the percentage of normal *Strongylocentrotus purpuratus* embryos dropped to 70% at pH 7.2 (Bay et al. 2003). Studies on the influence of pH using acid or environmental samples to adjust pH show a similar tolerance of early development stages in other echinoderms to acidification (Smith & Clowes 1924; Riveros et al. 1996; Kurihara & Shirayama 2004). In bivalve development the threshold for low pH tolerance/high pCO_2 tolerance at ambient temperatures is pH 7.7 or lower (Table 2). In experiments where pH was adjusted by mineral acid treatment the threshold was pH 6.75–7.0 (Table 2). Early development of the scallop *Placopecten magellanicus* is not affected by reduced pH (pH 7.0) (Desrosiers et al. 1996).

While early embryos of some species are useful indicators to assess the deleterious effects of ocean warming, this stage of development seems relatively insensitive to ocean acidification.

2.3 Larval development: influence of warming and acidification – single stressor studies

For embryos that are able to reach the larval stage in a warm, acidic/high pCO_2 ocean, climate change stressors may have a negative impact on this crucial dispersive stage of marine animals. In most marine invertebrates the gastrula stage develops into a larval stage that remain in the plankton for a variable amount of time, depending on the species. Larval types are specific to the different marine invertebrate phyla.

2.3.1 Temperature and larval performance

As for early development, temperature has a profound effect on larval performance. Increased sea temperature causes perturbations in the timing of larval abundance, larval settlement and can generate a mismatch between trophic levels due to change in the temporal pattern of phytoplankton and zooplankton (Hart & Scheibling 1988; Garrido & Barber 2001; Philippart et al. 2003; Edwards & Richardson 2004; Hays et al. 2005; Richardson 2008). Temperature also has a significant impact on larval swimming and planktonic larval duration (PLD) (Hobday & Tegner 2002; O'Connor et al. 2007; Munday et al. 2009). Reduction in PLD as temperatures increase is documented for many species of marine invertebrates (Hobday & Tegner 2002; Staver & Strathmann 2002; Grubert & Ritar 2004; O'Connor et al. 2007; Whalen et al. 2008). While this may have a favourable effect through reduction of the vulnerable planktonic period when predation on larvae is very high (>90% of cohorts) (Rumrill 1990; Lopez et al. 1998; Lemare & Barker 1999), reduced dispersal reduces population genetic connectivity altering the source/sink dynamics of marine populations (Hobday & Tegner 2002; Mumby et al. 2009).

2.3.2 pH and larval performance

Most molluscs and many echinoderm species have calcifying larvae that form a skeleton or shell (Figure 4). The molluscan larval shell provides a protective shelter into which larvae withdraw when disturbed. In echinoderms, sea urchin and brittle star larvae produce delicate skeletal rods that support feeding structures and which also function in defence. Near-future ocean acidification may have a negative effect on development of these and other calcifying larvae due to the decreased bioavailability of the carbonate ions needed to produce a larval skeleton (Kurihara & Shirayama 2004; Dupont et al. 2008; Kurihara 2008; Clark et al. 2009; Parker et al. 2009b) and, at metamorphosis, the juvenile skeleton (Figure 3). Predicting the responses of calcification systems to ocean change is complicated because increased warming may ameliorate the negative effects of acidification on skeletogenesis through enhancement and stimulation of the cellular mechanisms underlying calcification (Reynaud et al. 2003; McNeil et al. 2004; Sosdian 2006; but see Anthony et al. 2008). Recent studies however, show that increased temperature does not compensate for the negative effect of acidification on calcification in bivalve larvae (Parker et al. 2009).

A study of tropical (*Tripneustes gratilla*), temperate (*Pseudechinus huttoni*, *Evechinus chloroticus*) and polar (*Sterechinus neumayeri*) sea urchins found that survival of echinopluteus larvae was only affected at pH levels < pH 7.0, but noted that one of the larval arm rods (post oral rod) was significantly smaller at pH 7.8 (Clark et al. 2009). Echinopluteus larvae of *Hemicentrotus pulcherrimus* reared at pH 7.8 and ambient temperature also have shorter arm skeletons (Kurihara & Shirayama 2004). Reduction in length of the larval arms would negatively impact feeding in echinoplutei (Soars et al. 2009). Development of the ophiopluteus larvae of two species has been investigated with respect to ocean acidification. *Ophiothrix fragilis* is negatively impacted by acidification (pH 7.7–7.9) with mortality, reduction in size and abnormal skeletogenesis (Dupont et al. 2008). In contrast skeleton formation in the ophioplutei of *Amphiura filiformis* is less sensitive with some larvae reaching the juvenile stage (Dupont & Thorndyke 2009). Interestingly, skeleton regeneration of adult *Amphiura filiformis* is not affected by low pH within the context of near future ocean acidification (Wood et al. 2008). For the sea cucumber *Holothuria spinifera* pH 7.8 is reported to be the most suitable pH for larval rearing and for *Cucumaria frondosa*, pH 7.5 was the threshold for deleterious effects (Hamel & Mercier 1996; Asha & Muthiah 2005).

Acidification has a negative effect on biomineralisation in mollusc larvae. A significant reduction in the shell length of oyster veliger larvae reared at ambient temperature occurs at pH 7.9 (Parker et al. 2009). At acidification levels projected for 2300 (pH 7.4) larvae of *Crassostrea gigas* failed to make a shell, although mussel larvae were able to make a shell, albeit an abnormal one (Kurihara et al. 2007; Kurihara 2008). Mollusc veligers appear to be highly sensitive to environmental perturbation with unshelled larvae observed in control temperature/pH conditions and in the presence of very low concentrations of toxicants at levels that did not impair fertilisation or early development (Ringwood 1992; Kurihara et al. 2007). In *Crassostrea gigas* 20% of veligers in control treatments lacked a shell (Kurihara et al. 2007). Juvenile bivalves are reported to have reduced recruitment in sediments prone to low pH 7.0–7.2 (Green et al. 2004).

The response of marine larvae to ocean acidification/increased pCO_2 (reduced carbonate availability, hypercapnia) varies greatly among species even between closely related species (Kurihara & Shirayama 2004; Dupont & Thorndyke 2009) making it difficult to extrapolate broadly on potential climate change effects. More empirical data are needed to place the prospects of calcification in the marine life stages with the most fragile skeletons in the context of climate change. For non-calcifying larvae there are less data on effects of pH. Regardless of possession of a shell however, it is important to recognise that the negative effects of hypercapnia on growth and metabolism may compromise larval success.

3. Invertebrate life histories in a warm, acidified ocean – interactive effects of climate change stressors on fertilisation and development: multiple stressor studies

Recent multifactorial experiments with the sea urchin *Heliocidaris erythrogramma* and the bivalve *Saccostrea glomerata* provide the only data on the interactive effects of ocean warming and acidification on invertebrate life histories within a climate-relevant setting (Byrne et al. 2009; Parker et al. 2009). The distributions of these species overlap with the southeastern Australia climate change hot spot where, due to changes in ocean circulation, disproportionate warming has occurred over the past 60 years (+2.3°C) with projection for a further +2–3°C increase in SST by 2070 (Poloczanska et al. 2007). Although ocean uptake of pCO_2 will cause a simultaneous reduction in pH, marine biota in this region are presently, and for the foreseeable future, particularly vulnerable to ocean warming.

For *Heliocidaris erythrogramma* ocean warming (ca. +4°C) exerts a negative effect on development and there is no detectable effect of (CO_2 gas induced) acidification (pH 7.6) (Figure 2). There is also no interactive effect of temperature and pH on fertilisation and early development (Figure 1, 2). For this sea urchin near future ocean warming is the most serious contemporary climate change teratogen (Figure 3). Embryos may not reach the skeleton forming stage in a warm ocean regardless of pH (Byrne et al. 2009).

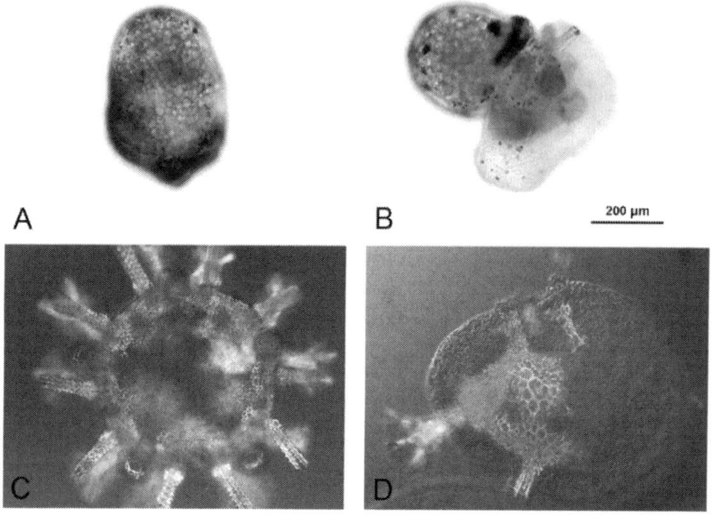

Figure 3 Effect of temperature and pH on larval development of the sea urchin *Heliocidaris erythrogramma.* (A) Normal larva three days post-fertilisation. (B) Thermal-effects phenotype characteristic of this species (see Kobayashi 1983), larvae reared at increased (+4°C) temperature (C) Normal juvenile – five days post-fertilisation (5 dpf). (D) Juvenile (5 dpf) reared at 24°C and pH 7.6 exhibiting abnormal calcification.

In contrast, simultaneous stressor experiments with the oyster *Saccostrea glomerata* showed vulnerabilities of fertilisation and development to both pH (mineral acid induced) and temperature (Parker et al. 2009). There was also a negative interaction between these factors on fertilisation success of this species. In the range of optimal rearing temperatures (22–26°C), pH $\leq$7.9 decreased fertilisation rates, impaired larval development and decreased larval shell length (Parker et al. 2009).

4. Perspectives and future monitoring

4.1 Perspectives

The major climate change stressors set to alter the global ocean are temperature and pH/pCO_2 and these will change simultaneously. While it is important to understand organism tolerance thresholds for these stressors in isolation, it is also imperative to understand their interactive effects. It is well known that a single factor may be sublethal, but when combined with other stressors can have additive, synergistic or antagonistic effects (Hoffmann et al. 2003; Przeslawski 2004; Przeslawski et al. 2005, 2008; Crain et al. 2008). For instance, the potential that ocean warming and acidification will alter the speciation of trace elements (Byrne et al. 1988), making the toxic elements (e.g. Cu, Zn) that interfere with skeleton development and other developmental processes more bioavailable (Byrne et al. 1988; Hunt & Anderson 1989; Ringwood 1992; Hollows et al. 2007), is a real concern.

Data are needed on biotic responses to ocean change (warming, acidification, hypercapnia) within the setting of current local conditions, species ecoclimates, multiple stressors, contemporary climate change and future environmental scenarios for local biota (Poloczanska et al. 2007; Fabry et al. 2008; Przeslawski et al. 2008).

Although the data are sparse it appears that ocean change may cause recruitment failure of a suite of invertebrates with flow-on ecosystem effects (Hobday & Tegner 2002; Przeslawski et al. 2008). Relatively small perturbations in recruitment can translate to large alterations of adult populations, a particularly serious consideration for keystone species (Uthicke et al. 2009). The climate change driven range extension of the sea urchin *Centrostephanus rodgersii*, a keystone species, caused a dramatic ecosystem shift from kelp to barrens habitat in Tasmanian waters resulting in a marked change in biodiversity (Ling et al. 2008). The southern migration of *C rodgersii* was facilitated by change in the southerly flow of the East Australian Current and probably also by warm conditions favourable for its larvae. In this case ocean warming facilitated range extension of *C rodgersii*. The distributional change of this species in Tasmania was readily detected because of its large size and its role as an ecosystem engineer (Andrew & Byrne 2007; Ling et al. 2008). It is likely that changes in abundance and range (contraction/extension) of a plethora of other species accompanied the southern extension of *C rodgersii*, but these changes may be difficult to detect. Climate change driven alteration in biodiversity and range shifts have occurred in many marine regions (e.g. Hobday & Tegner 2002; Mueter & Litzow 2009). Many marine species appear to be living near their thermal threshold, particularly at range limits and will have no alternative but to move into cooler regions (e.g. depth, latitude) through larval or adult migration to survive (Hobday & Tegner 2002; Metzger et al. 2007; Przeslawski et al. 2008). Some species may have a capacity for phenotypic plasticity facilitating tolerance and adaptation to altered local conditions. This may be expressed in changes in the timing of gametogenesis, spawning and larval behaviour (Garrido & Barber 2001; Grubert & Ritar 2004a; McDonald 2004), with flow on effects.

With regard to near future ocean warming, early development of some species may be thermotolerant to a +2°C change but not a +4°C change. On the other hand species such as *Crassostrea gigas* are highly thermotolerant across entire life history. This oyster has a broad thermotolerance range (ca. 20°C range) and pH tolerance range and thus may be a comparative winner in the climate change stakes (Kurihara et al. 2007; Parker et al. 2009).

Our knowledge of environmental tolerance thresholds of marine invertebrates is largely based on studies of widely distributed and hardy species used as model organisms for laboratory research and aquaculture, most of which are from shallow water habitats (Tables 1, 2). Through their evolutionary adaptation to environmentally variable estuarine or intertidal conditions these species may show greater tolerance to ocean warming, acidification and hypercapnia than their deeper water counterparts. The immersion-emersion thermal fluctuation in shallow water habitats is well characterised, as is their extreme pH fluctuations (tide pools pH 6.4–10.1; estuaries pH 7.1–8.3), due to the diurnal pattern of respiration and photosynthesis (Truchot & Duhamel-Jouve 1980; Morris & Taylor 1983; Ringwood & Keppler 2002; Marshall et al. 2008; Whooten et al. 2008). However, physiological studies indicate that some species may already be living close to their thermal limits e.g. intertidal molluscs and crabs (Helmuth et al. 2006; Metzger et al. 2007). We have a paucity of data on life history tolerance thresholds for the vast diversity of marine species many of which are likely to be vulnerable to changing ocean conditions and which may already be on the edge of existence due to anthropogenic stressors, particularly pollution and UV (Przeslawski 2004; Harley et al. 2006; Przeslawski et al. 2005, 2008; Crain et al. 2008). As marine species differ in their pH, thermo- and pCO_2 tolerance, data are needed on a phylogenetically diverse species from a range of habitats to inform predictions on how species may fare in the face of climate change stressors and potential for survivorship. It is likely that negative effects of ocean change on marine life histories will be identified as more species are investigated.

4.2 Future monitoring

Knowledge of the relative vulnerabilities and thresholds of different life history stages is essential for monitoring the effects of ocean change and for assessment of risk from climate change. Determination of the thermal and pH/pCO_2 tolerances of gametes, embryos and larvae is crucial as we endeavour to predict how marine species may respond to climate change. The life history stages of the species in this review exhibited different tolerances to increased temperature and acidification, even among closely related sympatric species with larvae in the plankton at the same time (Table 1, 2; Calabrese & Davis 1970; Dupont & Thorndyke 2009). Such differences will influence comparative vulnerabilities to climate change. Although research focus on the impact of climate change stressors on marine life histories is recent, ecotoxicology has a long history of use in environmental monitoring and provides a framework for investigation of the effects of ocean change on marine biota.

Marine invertebrate developmental stages provide a model system to identify vulnerabilities, to assess how species may respond to climate change at different parts of their latitudinal range and to monitor the interactive effects of climate change stressors on marine biota. They can be used to determine potential synergistic negative, positive or neutral effects of climate change and other stressors using the multifactorial approach (Przelawski et al. 2005, 2008; Byrne & Davis 2008; Byrne et al. 2009; O'Donnell et al. 2009).

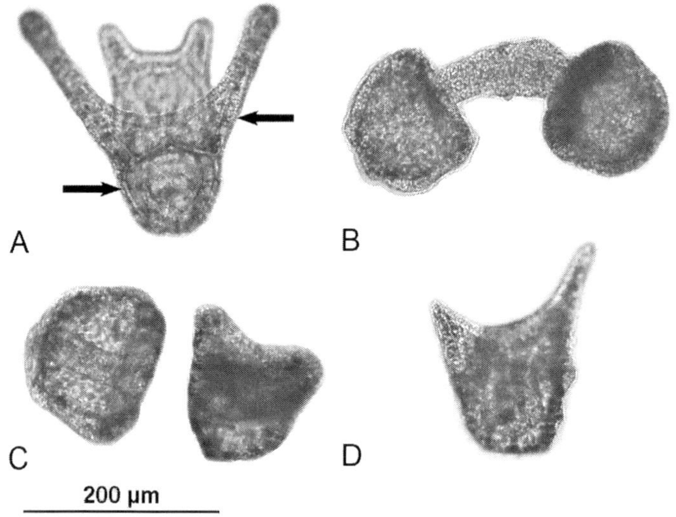

Figure 4 Developmental bioassays with the sea urchin *Heliocidaris tuberculata*. (A) Normal early echinopluteus larva (70 h) with distinct larval arms supported by thin skeletal rods. Right arrow, postoral rod; Left arrow, body rod. (B–D) Abnormal phenotypes reared in test (toxic leachate) solutions. (B) Abnormal dumbbell shaped larvae with no traces of skeleton (C) Development arrested at gastula (left) and early larval (right) stages (D) Larva with an abnormal skeleton (d) (from Byrne et al. 2008).

4.2.1 Choice of bioassays

Routinely used endpoints in marine toxicity tests include fertilisation, early cleavage, gastrulation and larval development, with scoring based on presence of fertilised eggs, arrested or delayed development, reduction in size and abnormal phenotypes (Figure 3, 4). Bioassays based on calcifying stages provide a means to quantify the effects of ocean change on skeletogenesis. Calcification bioassays in molluscs and echinoderms are typically scored at 24–72 hrs at the echinopluteus and veliger larval stage, respectively (Figure

4). Abnormal skeleton in plutei (Figure 4; Kurihara & Shirayama 2004; Dupont et al. 2008) or shell formation in veliger larvae is readily scored (Cherr et al. 1990; Conroy et al. 1996). To detect subtle impacts of climate changes on life history, use of a broad range of stressor levels to determine lowest observable effect levels (LOEC approach ca. 10%) may provide an indication of tipping points. It has also been suggested that when no effect of stressors are detected power analysis is needed to have confidence in non-significant results (Havenhand & Schlegel 2009).

With regard to fertilisation bioassays, variable experimental conditions (e.g. gamete source, gamete age, concentration, vessel size, volume) strongly influence fertilisation rates (e.g. Clotteau & Dubé 1993; Benzie & Dixon 1994; Encena et al. 1998; Hollows et al. 2007; Song et al. 2009). This presents a significant challenge to generation of comparative data. To reduce inter-test variation and to facilitate detection of effects, it is important to use optimal fertilisation conditions (e.g. sperm to egg ratio) predetermined for each test species for conditions used by individual laboratories (Dinnel et al. 1987; Clotteau & Dubé 1993; Song et al. 2009; Byrne et al. in press). Ecologists are often interested in maternal, paternal or genetic effects and recent fertilisation studies have used gametes from single males and females in experimental dam-sire crosses to address these aspects of fertilisation in the sea (e.g. Evans & Marshall 2005). However spawning in the field generally involves multiple males and females (Pennington 1985; Himmelman et al. 2008) and this is known to enhance fertilisation success (Evans & Marshall 2005). If fertilisation bioassays are to inform the debate on the impact of climate change on fertilisation in the sea, use of gametes from multiple males and females in experiments designed to investigate the interactive effects of ocean warming and acidification (and other stressors e.g. UV) are most realistic. That said, the weight of evidence (Table 1, 2) indicates that fertilisation is not particularly sensitive to climate change stressors and may not be a suitable test endpoint for assessment of risk to marine biota due to climate change. The potentially confounding influence of different findings in fertilisation bioassays with the same species (cf. Havenhand et al. 2008; Byrne et al. 2009) indicates that application of fertilisation tests is problematic.

In contrast, developmental stages are more sensitive to climate change stressors with respect to near future scenarios (Table 1, 2; Figure 3) and may be more useful to determine the interactive effects of climate change stressors on marine life. Due to the stereotypic nature of invertebrate cleavage, gastrulation and larval form, developmental bioassays are relatively easy to score and are also less prone to inter-test variation. Efficient generation of a large database for a diverse suite of species through use of short-term screening tests will be facilitated by focusing in three major developmental endpoints, cleavage (2–4 hr), gastrulation (20–24 hr) and the first larval stage (24–72 hr). This approach would quickly answer key questions: 1) can embryos faced with ocean warming, hypercapnia and acidification develop to the gastrula stage and, if so 2) can these gastrulae produce a normal larva? Considering the dependence of later life stages on the success of embryogenesis, application of the ecotoxicology approach would provide a rapid assessment of life stage vulnerability and identify species for which long term rearing experiments should be considered. Longer-term tests can be used to follow cohorts of embryos through the cleavage/gastrulation mortality bottleneck to determine the presence of robust sub-populations of larvae. In addition, tests using embryos from the latitudinal range limits of species present an opportunity to determine the presence of cold and warm adapted developmental genotypes. This would provide useful data to model future range contractions/extensions.

While it is unlikely that generic bioassay indicator species will be identified for monitoring impacts of climate change stressors on marine life histories, more empirical data from diverse species are needed to determine if this is the case. For logistical reasons in the long term it may become evident that it will be most practical to focus on the response of developmental stages of commercially (e.g. aquaculture, fisheries) or ecologically (e.g. keystone species, ecological engineers) important species.

5. Climate adaptation

The strong influence of adult thermal history, particularly maternal acclimatisation and egg thermal imprinting on the thermotolerance of fertilisation and development indicates the potential for adaptive phenotypic plasticity with respect to prevailing temperatures (O'Connor & Mulley 1977; Fujisawa 1989, 1995; Johnson & Babcock 1994; Bingham et al. 1997). The potential that adaptive phenotypic plasticity may help buffer the negative effects of climate change is also suggested for corals (Wooldridge et al. 2005; Edmunds & Gates 2008). In addition, widely distributed species may also be able to keep up with a warming world by microevolution through immigration of propagules with genotypes adapted to warmer environments (Visser 2007). This latter response may be particularly important along coastlines such as that of eastern Australia where a major long-coast current facilitates dispersal of propagules from lower to higher latitudes (e.g. *Centrostephanus rodgersii*). The prediction that propagules from warm adapted parents may provide a source of pre-adapted phenotypes/genotypes for conspecific populations at higher latitudes can be addressed through multifactorial experiments on the impact of climate change stressors on life history stages of populations located in different latitudes.

The inbuilt pre-adaptive tolerance and flexibility (polytopy) of marine invertebrate life histories to environmental stressors is suggested to provide a mechanism to increase the probability of persistence of some species in geological time (Reviews Palmer 1994; Hadfield & Strathmann 1996; Davis 2000). Those species lacking polytopy are likely to be vulnerable to extinction as the oceans change. Although the future for many marine invertebrates appears gloomy in the face of climate change, the potential for inbuilt adaptive (genetic) and acclimation (phenotypic plasticity) capacity of some species warrants attention. Determination of these capacities will assist in identification of species that are potential winners and losers in the face of ocean change. These data are needed to inform debate on how climate change may alter marine ecosystems and the services they provide (e.g. fisheries, water quality) for the wellbeing of human populations.

Acknowledgements

Supported by the Australian Research Council. Thanks to Dr A Davis, University of Wollongong for reading and commenting on the manuscript. Dr A Hobday (CSIRO) and an anonymous reviewer also provided helpful comments on the manuscript. M Ho, H Nguyen N Soars, E Woolsey and E Wong provided assistance. This represents contribution no. 0027 from the Sydney Institute of Marine Science.

References

Andrew NL & Byrne M (2007). *Centrostephanus*. In J Lawrence (Ed), *The biology and ecology of edible urchins,* (pp191–204). Amsterdam: Elsevier Science.

Alstatt JM, Ambrose RF, Engle JM, Haaker PL, Lafferty KD & Raimondi PT (1996). Recent declines of black abalone *Haliotis cracherodii* on the mainland coast of central California. *Mar Ecol Prog Ser,* 142, 185–92.

Anthony KRN, Kline DI, Diaz-Pulido G, Dove S & Hoegh-Guldberg O (2008). Ocean acidification causes bleaching and productivity loss in coral reef builders. *Proc Natl Acad Sci USA*, 105, 17442–46.

Andronikov VB (1975). Heat resistance of gametes of marine invertebrates in relation to temperature conditions under which the species exist. *Mar Biol,* 30, 1–11.

Asha PS & Muthiah P (2005). Effects of temperature, salinity and pH on larval growth, survival and development of the sea cucumber *Holothuria spinifera* Theel. *Aquaculture*, 250, 823–29.

Bamber RN (1990). The effects of acidic seawater on three species of lamellibranch mollusk. *J Exp Mar Biol Ecol*, 143, 181–91.

Bay S, Burgess R & Nacci D (1993). Status and applications of echinoid (Phylum echinodermata) toxicity test methods. In WG Landis, JS Hughes & MA Lewis (Eds), *Environmental toxicology and risk assessment*, (pp281–302). Philadelphia: American Society for Testing and Materials.

Benzie JAH & Dixon P (1994). The effects of sperm concentration, sperm: egg ratio and gamete age on fertilisation success in crown-of-thorns starfish (*Acanthaster planci*) in the laboratory. *Biol Bull*, 186, 139–52.

Berge JA, Bjerkeng B, Pettersen O, Schaanning MT & Oxnevad S (2006). Effects of increased sea water concentrations of CO_2 on growth of the bivalve *Mytilus edulis* L. *Chemosphere*, 62, 681–87.

Bingham BL, Bacigalupi M & Johnson LG (1997). Temperature adaptations of embryos from intertidal and subtidal sand dollars (*Dendraster excentricus*, Eschscholtz). *Northwest Science*, 71, 108–14.

Blackford JC & Gilbert FJ (2006). pH variability and CO_2 induced acidification in the North Sea. *J Mar Sys*, 64, 229–41.

Bookbinder LH & Shick JM (1986). Anaerobic and aerobic energy metabolism in ovaries of the sea urchin *Strongylocentrotus droebachiensis*. *Mar Biol*, 93, 103–10.

Brokaw CJ (1990). The sea urchin spermatozoon. *BioEssays*, 12, 449–52.

Burnett LE (1997). The challenges of living in hypoxic and hypercapnic aquatic environments. *Am Zool*, 37, 633–40.

Burnett LE, Terwilliger N, Carroll A, Jorgensen D & Scholnick D (2002). Respiratory and acid-ase physiology of the purple sea urchin, *Strongylocentrotus purpuratus*, during air exposure: presence and function of a facultative lung. *Biol Bull*, 203, 42–50.

Byrne M & Davis A (2008). The acid test: Responses of benthic invertebrates to climate change. *Aust Antarctic Mag*, 15, 8–9.

Byrne M, Oakes DJ, Pollak JK & Laginestra E (2008). Toxicity of landfill leachate to sea urchin development with a focus on ammonia. *Cell Biol Toxicol*, 24, 503–12.

Byrne M, Ho M, Selvakumaraswamy P, Nguyen HD, Dworjanyn SA & Davis AR (2009). Temperature, but not pH, compromises sea urchin fertilisation and early development under near-future climate change scenarios. *Proc Roy Soc B*, 276, 1883–88.

Byrne M, Soars NA, Selvakumaraswamy P, Dworjanyn SA & Davis AR (2009). Sea urchin fertilization in a warm, acidified and high pCO_2 ocean across a range of sperm densities. *Mar Env Res* (in press).

Byrne RH, Kump LR & Cantrell KJ (1988). The influence of temperature and pH on trace metal speciation in seawater. *Mar Chem* 25, 163–81.

Calabrese A & Davis HC (1966). The pH tolerance of embryos and larvae of *Mercenaria mercenaria* and *Crassostrea virginica. Biol Bull,* 131, 427–36.

Calabrese A & Davis HC (1970). Tolerances and requirements of embryos and larvae of bivalve mollusks. *Helgol Wissens Meeres*, 20, 553–64.

Caldeira K & Wickett ME (2003). Anthropogenic carbon and ocean pH. *Nature*, 425, 365.

Carr RS, Biedenbach JM & Nipper M (2006). Influence of potentially confounding factors on sea urchin porewater toxicity tests. *Arch Environ Contam Toxicol*, 51, 573–79.

Chen CP & Chen BY (1992). Effects of high-temperature on larval development and metamorphosis of *Arachnoids placenta* (Echinodermata, Echinoidea). *Mar Biol*, 112, 445–49.

Cherr GN, Shoffner-McGee J & Shenker JM (1990). Methods for assessing fertilisation and embryonic/larval development in toxicity tests using the California mussel (*Mytilus californianus*). *Environ Toxicol Chem,* 9, 1137–45.

Clark D, Lamare M & Barker M (2009). Response of sea urchin pluteus larvae (Echinodermata: Echinoidea) to reduced seawater pH: a comparison among a tropical, temperate, and a polar species. *Mar Biol,* 156, 1125–37.

Clarke A (2003). Costs and consequences of evolutionary temperature adaptation. *Trends Ecol Evol*, 18, 573–81.

Clotteau G & Dubé F (1993). Optimization of fertilisation parameters for rearing surf clams (*Spisula solidissima*), *Aquaculture,* 114, 339–53.

Conroy PT, Hunt JW & Anderson BS (1996). Validation of a short-term toxicity test endpoint by comparison with longer-term effects on larval red abalone *Haliotis rufescens. Environ Toxicol Chem,* 15, 1245–50.

Crain CM, Kroeker K & Halpern BS (2008). Interactive and cumulative effects of multiple human stressors in marine systems. *Ecology Letters*, 11, 1304–15.

Davis M (2000). The combined effects of temperature and salinity on growth, development, and survival for tropical gastropod veligers of *Strombus gigas*. *J Shellfish Res*, 19, 883–89.

Desrosiers RR, Désilets J & Dubé F (1996). Early developmental events following fertilisation in the giant scallop *Placopecten magellanicus*. *Can J Fish Aquat Sci*, 53, 1382–92.

Doney SC, Fabry VJ, Feely RA & Kleypas JA (2009). Ocean acidification: the other CO_2 problem. *Ann Rev Mar Sci*, 1, 169–92.

Dinnel PA, Link JM & Stober QJ (1987). Improved methodology for a sea-urchin sperm cell bioassay for marine waters. *Arch Environ Contam Toxicol*, 16, 23–32.

Dupont S, Havenhand J, Thorndyke W, Peck L & Thorndyke M (2008). Near-future level of CO_2-driven ocean acidification radically affects larval survival and development in the brittlestar *Ophiothrix fragilis*. *Mar Ecol Prog Ser*, 373, 285–94.

Dupont S & Thorndyke MC (2009). Impact of CO_2-driven ocean acidification on invertebrates early life-history – what do we know, what we need to know and what we can do. *Biogeosciences Discuss*, 6, 3109–31.

Edmunds PJ & Gates RD (2008). Acclimatization in tropical coral reefs. *Mar Ecol Prog Ser*, 361, 307–10.

Edwards M & Richardson AJ (2004). Impact of climate change on marine pelagic phenology and trophic mismatch. *Nature*, 430, 881–84.

Encena VC, Capinpin EC & Bayona NC (1998). Optimal sperm concentration and time for fertilisation of the tropical abalone, *Haliotis asinina* Linné 1758. *Aquaculture*, 165, 347–52.

Evans JP & Marshall DJ (2005). Male-by-female interactions influence fertilisation success and mediate the benefits of polyandry in the sea urchin *Heliocidaris erythrogramma*. *Evolution*, 59, 106–12.

Fabry VJ, Seibel BA, Feely RA & Orr JC (2008). Impacts of ocean acidification on marine fauna and ecosystem processes. *ICES J Mar Sci*, 65, 414–32.

Farmanfarmaian A & Giese AC (1963). Thermal tolerance and acclimation in the western purple sea urchin, *Strongylocentrotus purpuratus. Physiol Zool*, 36, 237–343.

Feely R, Sabine CL, Lee K, Berelson W, Kleypas J, Fabry VJ & Millero FJ (2004). Impact of anthropogenic CO_2 on the $CaCO_3$ systems in the oceans. *Science*, 305, 362–66.

Friedman CS, Thomson M, Chun C, Haaker PL & Hedrick RP (1997). Withering syndrome of the black abalone, *Haliotis cracherodii* (Leach): water temperature, food availability, and parasites as possible causes. *J. Shellfish Res,* 16, 403–11.

Fujisawa H (1989). Differences in temperature dependence of early development of sea urchins with different growing seasons. *Biol Bull*, 176, 96–102.

Fujisawa H (1995). Variation in embryonic temperature sensitivity among groups of the sea urchin, *Hemicentrotus pulcherrimus*, which differ in their habitats. *Zool Sci*, 12, 583–89.

Fujisawa H & Shigei M (1990). Correlation of embryonic temperature sensitivity of sea urchins with spawning season. *J Exp Mar Biol Ecol*, 136, 123–39.

Garrido CL & Barber BJ (2001). Effects of temperature and food ration on gonad growth and oogenesis of the green sea urchin, *Strongylocentrotus droebachiensis. Mar Biol*, 138, 447–56.

Gazeau F, Quiblier C, Jansen JM, Gattuso JP, Middelburg JJ & Heip CHR (2007). Impact of elevated CO_2 on shellfish calcification. *Geophys Res Lett*, 34, L07603.

Giese AC & Pearse JS (1974). Introduction: general principles. In AC Giese & JS Pearse (Eds), *Reproductive biology of invertebrates.* Vol. 1, (pp1–49). New York: Academic Press.

Green MA, Jones ME, Boudreau CL, Moore RL & Westman BA (2004). Dissolution mortality of juvenile bivalves in coastal marine deposits. *Limnology and Oceanography*, 49, 727–34.

Greenwood PJ & Bennett T (1981). Some effects of temperature-salinity combinations on the early development of the sea urchin *Parechinus angulosus* (Leske) fertilisation. *Mar Biology Ecol*, 51, 119–31.

Grubert MA & Ritar A (2004a). Temperature effects on the dynamics of gonad and oocyte development in captive wild-caught of blacklip (*Haliotis rubra*) and greenlip (*H. laevigata*) abalone. *Inv Rep Dev*, 45, 185–96.

Grubert MA & Ritar A (2004b). The effect of temperature on the embryonic and larval development of blacklip (*Haliotis rubra*) and greenlip (*H. laevigata*) abalone. *Inv Rep Dev* 45, 197–203.

Grubert MA & Ritar A (2005). The effect of temperature and conditioning interval on the spawning success of wild-caught blacklip (*Haliotis rubra*, Leach 1814) and greenlip (*H. laevigata*, Donovan 1808) abalone. *Aquaculture Res*, 36, 654–65.

Hadfield MG & Strathmann MF (1996). Variability, flexibility and plasticity in life histories of marine invertebrates. *Oceanologica Acta*, 19, 323–24.

Hagström BE & Hagström B (1959). The effect of decreased and increased temperatures on fertilisation. *Exp Cell Res*, 16, 174–83.

Hamel JF & Mercier A (1996). Early development, settlement, growth and spatial distribution of the sea cucumber *Cucumaria frondosa* (Echinodermata: Holothuroidea). *Can J Fish Aquat Sci*, 53, 253–71.

Harley CDG, Hughes AR, Hultgren KM, Miner BG, Sorte CJB, Thornber CS et al. (2006). The impacts of climate change in coastal marine systems. *Ecology Letters*, 9, 228–41.

Hart MW & Scheibling RE (1988). Heat waves, baby booms and the destruction of kelp beds by sea urchins. *Mar Biol*, 99, 167–76.

Havenhand JN, Butler FR, Thorndyke MC & Williamson JE (2008). Near-future levels of ocean acidification reduce fertilisation success in a sea urchin. *Current Biology*, 18, 651–52.

Hays GC, Richardson AJ & Robinson C (2005). Climate change and marine plankton. *Trends Ecol Evol*, 20, 338–44.

Helmuth B, Mieszkowska N, Moore P & Hawkins SJ (2006). Living on the edge of two changing worlds: forecasting the responses of rocky intertidal ecosystems to climate change. *Ann Rev Ecol Evol Syst*, 37, 373–404.

Himmelman JH, Dumont CP, Gaymer CF, Vallières C & Drolet D (2008). Spawning synchrony and aggregative behaviour of cold-water echinoderms during multi-species mass spawnings. *Mar Ecol Prog Ser*, 361, 161–68.

Hobday AJ & Tegner MJ (2002). The warm and the cold: influence of temperature and fishing on local population dynamics of red abalone. *California Cooperative Oceanic Fisheries Investigations Report*, 43, 74–96.

Hollows CF, Johnston EL & Marshall DJ (2007). Copper reduces fertilisation success and exacerbates allee effects in the field. *Mar Ecol Prog Ser*, 333, 51–60.

Hoegh-Guldberg O & Pearse JS (1995). Temperature, food availability, and the development of marine invertebrate larvae. *Amer Zool*, 35, 415–25.

Holland LZ & Gould-Somero M (1982). Fertilisation acid of sea urchin eggs: evidence that it is H^+ not CO_2. *Dev Biol*, 92, 549–52.

Holland LZ, Gould-Somero M & Paul M (1984). Fertilisation acid release in *Urechis* eggs. I. The nature of the acid and the dependence of acid release and egg activation on external pH. *Dev Biol*, 103, 337–42.

Hörstadius S (1975). A note on the effect of temperature on sea urchin eggs. *J Exp Mar Biol Ecol*, 18, 239–42.

IPCC (2007). Intergovernmental Panel on Climate Change 2007. *The fourth assessment report of the IPCC*. Cambridge, UK: Cambridge University Press.

Johnson LG & Babcock RC (1994). Temperature and the larval ecology of the crown-of-thorns starfish, *Acanthaster planci*. *Biol Bull*, 187, 304–08.

Kennedy VS, Roosenburg WH, Zion HH & Castagna M (1974). Temperature – time relationships for survival of embryos and larvae of *Mulinia lateralis* (Mollusca: Bivalvia). *Mar Biol*, 24, 137–45.

Knutzen J (1981). Effects of decreased pH on marine organisms. *Mar Poll Bull*, 12, 25–29.

Kobayashi N (1980). Comparative sensitivity of various developmental stages of sea urchins to some chemicals. *Mar Biol*, 58, 163–71.

Kurihara H (2008). Effects of increased atmospheric CO2 on sea urchin early development. *Mar Ecol Prog Ser, 274*, 161–69.

Kurihara H & Shirayama Y (2004). Effects of increased atmospheric CO_2 on sea urchin early development. *Mar Ecol Prog Ser*, 274, 161–69.

Kurihara H, Shimode S & Shirayama Y (2004). Sub-lethal effects of elevated concentration of CO_2 on planktonic copepods and sea urchins. *J Oceanogr*, 60, 743–50.

Kurihara H, Kato S & Ishimatsu A (2007). Effects of increased seawater pCO_2 on early development of the oyster *Crassotrea gigas*. *Aquat Biol*, 1, 91–98.

Lamare MD & Barker MF (1999). In situ estimates of larval development and mortality in the New Zealand sea urchin *Evechinus chloroticus* (Echinodermata: Echinoidea). *Mar Ecol Prog Ser,* 180, 197–211.

Lee CH, Ryu TK & Choi JW (2004). Effects of water temperature on embryonic development in the northern Pacific asteroid, *Asterias amurensis*, from the southern coast of Korea. *Invert Reprod Dev*, 45, 109–16.

Lester SE, Tobin ED & Behrens MD (2007). Disease dynamics and the potential role of thermal stress in the sea urchin, *Strongylocentrotus purpuratus*. *Can J Fish Aquat Sci*, 64, 314–23.

Levitus S, Antonov J & Boyer T (2005). Warming of the world ocean. *Geophys Res Lett*, 32, L02604.

Lewis C, Olive PJW, Bentley MG & Watson G (2002). Does seasonal reproduction occur at the optimal time for fertilisation in the polychaetes *Arenicola marina* L. and *Nereis virens* Sars? *Invert Reprod Dev*, 41, 61–71.

Lewis E & Wallace DWR (1998). *Program developed for CO_2 system calculations; ORNL/CDIAC-105*. Oak Ridge, Tennessee: Oak Ridge national Laboratory US Department of Energy.

Ling SD, Johnson CR, Frusher SD & King CK (2008). Reproductive potential of a marine ecosystem engineer at the edge of a newly expanded range. *Global Change Biol*, 14, 1–9.

López S, Turon X, Montero E, Palacín C, Duarte CM & Tarjuelo I (1998). Larval abundance, recruitment and early mortality in *Paracentrotus lividus* (Echinoidea). Interannual variability and plankton-benthos coupling. *Mar Ecol Prog Ser,* 172, 239–51.

Lough RG & Gonor JJ (1973). A response-surface approach to the combined effects of temperature and salinity on the larval development of *Adula californiensis* (Pelecypoda: Mytilidae). II Long-term larval survival and growth in relation to respiration. *Mar Biol*, 22, 295–305.

Marshall DJ, Santos JH, Leung KMY & Chak WH (2008). Correlations between gastropod shell dissolution and water chemical properties in a tropical estuary. *Mar Env Res*, 66, 422–29.

McDonald K (2004). Patterns of embryonic motility: effects of size and environmental temperature on vertical velocities of sinking and swimming in echinoid blastulae. *Biol Bull* 207, 93–102.

McEdward LR (1985). Effects of temperature on the body form, growth, electron transport system activity and development rate of an echinopluteus. *J Exp Mar Biol Ecol*, 93, 169–81.

McNeil BI, Matear RJ & Barnes DJ (2004). Coral reef calcification and climate change: The effect of ocean warming. *Geophys Res Lett*, 31, L22309.

McNeil BI & Matear R (2008). Southern Ocean acidification: A tipping point at 450-ppm atmospheric CO_2. *Proc Natl Acad Sci USA*, 105, 18860–64.

Metzger R, Sartoris FJ, Langenbuch M & Pörtner HO (2007). Influence of elevated CO_2 concentrations on thermal tolerance of the edible crab *Cancer pagurus*. *J Thermal Biol*, 32, 144–51.

Michaelidis B, Ouzounis C, Paleras A & Pörtner HO (2005). Effects of long-term moderate hypercapnia on acid-base balance and growth rate in marine mussels *Mytilus galloprovincialis*. *Mar Ecol Prog Ser*, 293, 109–18.

Miles H, Widdicombe S, Spicer JI & Hall-Spencer J (2007). Effects of anthropogenic seawater acidification on acid-base balance in the sea urchin *Psammechinus miliaris*. *Mar Poll Bull*, 54, 89–96.

Mita M, Hino A & Yasumasu I (1984). Effect of temperature on interaction between eggs and spermatozoa of sea urchin. *Biol Bull*, 166, 68–77.

Mita M & Nakamura M (2001). Energy metabolism of sea urchin spermatozoa: the endogenous substrate and ultrastructural correlates. In M Jangoux & JM Lawrence (Eds), *Echinoderm studies*, Vol. 6, (pp85–110). Lisse: AA Balkema.

Mohri H & Yasumasu I (1963). Studies on the respiration of sea-urchin spermatozoa. *J Exp Biol*, 40, 573–86.

Morris S & Taylor AC (1983). Diurnal and seasonal variation in physico-chemical conditions within intertidal rock pools. *Est Coastal Shelf Sci*, 17, 339–55.

Mueter FJ & Litzow MA (2008) Sea ice retreat alters the biogeography of the Bearing Sea continental shelf. *Ecological Applications*, 18, 309–20.

Mumby PL, Leis JM, Lough JM, Paris CB, Kingsford MJ, Berumen ML & Lambrechts J (2009). Climate change and coral reef connectivity. *Coral Reefs* (in press).

Negri AP, Marshall PA & Heyward AJ (2007). Differing effects of thermal stress on coral fertilisation and early embryogenesis in four Indo Pacific species. *Coral Reefs*, 26, 759–63.

Nguyen HD (2007). Effect of elevated temperatures on HSP70 expression and development in congeneric rocky shore sea urchins. BSc Thesis, University of Sydney.

O'Connor C & Mulley JC (1977). Temperature effects on periodicity and embryology, with observations on the population genetics, of the

aquacultural echinoid *Heliocidaris tuberculata*. *Aquaculture*, 12, 99–114.
O'Connor MI, Bruno JF, Gaines SD, Halpern BS, Lester SE, Kinlan BP & Weiss JM (2007). Temperature control of larval dispersal and the implications for marine ecology, evolution, and conservation. *Proc Natl Acad Sci USA*, 104, 1266–71.
O'Connor WA & Lawler NF (2004). Salinity and temperature tolerances of embryos and juveniles of the pearl oyster, *Pinctada imbricata* Röding. *Aquaculture*, 229, 493–506.
O'Donnell MJ, Hammond LM & Hoffman GE (2009). Predicted impact of ocean acidification on a marine invertebrate: elevated CO_2 alters response to thermal stress in sea urchin larvae. *Mar Biol*, 156, 439–46.
Ohtake H (1976). Respiratory behaviour of sea-urchin spermatozoa II: sperm-activating substance obtained from jelly coat of sea-urchin eggs. *J Exp Zool*, 198, 313–22.
Palmer AR (1994). Temperature sensitivity, rate of development, and time to maturity: geographic variation in laboratory reared *Nucella* and a cross-phyletic overview. In WH Wilson Jr, SA Stricker & GL Shinn (Eds), *Reproduction and development of marine invertebrates*, (pp177–94). Baltimore: Johns Hopkins University Press.
Parker LM, Ross PM & O'Connor WA (2009a). The effect of ocean acidification and temperature on the fertilisation and embryonic development of the Sydney rock oyster *Saccostrea glomerata* (Gould 1850). *Glob Change Biol*, 15, 2123–36.
Peaucellier G & Dorée M (1981). Acid release at activation and fertilisation of starfish oocytes. *Dev Growth Differ*, 23, 287–96.
Pechenik JA (1989). Environmental influences on larval survival and development. In AC Giese, JS Pearse & VB Pearse (Eds), *Reproductive biology of invertebrates*. Vol. 5, (pp551–608). Palo Alto, California: Blackwell.
Pennington JT (1985). The ecology of fertilisation of echinoid eggs: the consequences of sperm dilution, adult aggregation, and synchronous spawning. *Biol Bull*, 169, 417–30.

Philippart CJM, van Aken HM, Beukema JJ, Bos OG, Cadee, GC & Dekker R (2003). Climate-related changes in recruitment of the bivalve *Macoma balthica. Limnology and Oceanography*, 48, 2171–85.

Poloczanska ES, Babcock RC, Butler A, Hobday AJ, Hoegh-Guldberg O, Kunz TJ et al. (2007). Climate change and Australian marine life. *Oceanog Mar Biol Ann Rev*, 45, 407–78.

Pörtner HO (2008). Ecosystem effects of ocean acidification in times of ocean warming: a physiologist's view. *Mar Ecol Prog Ser*, 373, 203–17.

Pörtner HO, Langenbuch M & Reipschläger A (2005). Biological impact of elevated ocean CO_2 concentrations: lessons from animal physiology and earth history. *J Oceanog*, 60, 705–18.

Przeslawski R (2004). A review of the effects of environmental stress on embryonic development within intertidal gastropod egg masses. *Molluscan Res*, 24, 43–63.

Przeslawski R, Davis AR & Benkendorff K (2005). Synergies, climate change and the development of rocky shore invertebrates. *Glob Change Biol*, 11, 515–22.

Przeslawski R, Ahyong S, Byrne M, Wörheide G & Hutchings P (2008). Beyond corals and fish: the effects of climate change on non-coral benthic invertebrates of tropical reefs. *Glob Change Biol*, 14, 2773–95.

Quesne WJF & Bentley MG (2006). Effects of acidification on fertilisation success in free spawning marine invertebrates. *Geophys Res Abstr*, 8, 02658.

Reynaud S, Leclercq N, Romaine-Lioud S, Ferrier-Pagés C, Jaubert J & Gattuso JP (2003). Interacting effects of CO_2 partial pressure and temperature on photosynthesis and calcification in a scleractinian coral. *Glob Change Biol*, 9, 1660–68.

Richardson AJ (2008). In hot water: zooplankton and climate change. *ICES Journal of Marine Science*, 65, 279–95.

Ringwood AH (1992). Comparative sensitivity of gametes and early developmental stages of a sea urchin species (*Echinometra mathaei*) and a bivalve species (*Isognomon californicum*) during metal exposures. *Arch Environ Contam Toxicol*, 22, 288–95.

Ringwood AH & Keppler CJ (2002). Water quality variation and clam growth: is pH really a non-issue in estuaries? *Estuaries* 25, 910–07.

Riveros A, Zuñiga M, Larrain A & Becerra J (1996). Relationships between fertilisation of the Southeastern Pacific sea urchin *Arbacia spatuligera* and environmental variables in polluted coastal waters. *Mar Ecol Prog Ser*, 134, 159–69.

Roller RA & Stickle WB (1993). Effects of temperature and salinity acclimations of adults on larval survival, physiology, and early development of *Lytechinus variegatus* (Echinodermata: Echinoidea). *Mar Biol*, 116, 583–91.

Rumrill SS (1990). Natural mortality of marine invertebrate larvae. *Ophelia*, 32, 163–98.

Rupp JH (1973). Effects of temperature on fertilisation and early cleavage of some tropical echinoderms, with emphasis on *Echinometra mathaei*. *Mar Biol*, 23, 183–89.

Sconzo G, Amore G, Capra G, Giudice G, Cascino D & Ghersi G (1997). Identification and characterization of a constitutive HSP75 in sea urchin embryos. *Biochem Biophys Res Commun*, 234, 24–29.

Sewell MA & Young CM (1999). Temperature limits to fertilisation and early development in the tropical sea urchin *Echinometra lucunter*. *J Exp Mar Biol Ecol*, 236, 291–305.

Smith HW and Clowes GHA (1924). The influence of hydrogen ion concentration on the fertilisation process in *Arbacia*, *Asterias* and *Chaetopterus* eggs. *Biol Bull*, 47, 333–34.

Soars NA, Prowse TAA & Byrne M (2009). Overview of phenotypic plasticity in echinoid larvae, '*Echinopluteus transverses*' type vs. typical echinoplutei. *Mar Ecol Prog Ser*, 383, 113–25.

Song YP, Suquet M, Quéau I & Lebrun L (2009). Setting of a procedure for experimental fertilisation of Pacific oyster (*Crassostrea gigas*) oocytes. *Aquaculture*, 287, 311–14.

Sosdian S (2006). Strontium to calcium ratios in the marine gastropod *Conus ermineus*: growth rate effects and temperature calibration. *Geochem Geophys Geosys*, 7, Q11023.

Staver JM & Strathmann RR (2002). Evolution of fast development of planktonic embryos to early swimming. *Biol Bull*, 203, 58–69.
Truchot JP & Duhamel-Jouve A (1980). Oxygen and carbon dioxide in the marine intertidal environment: diurnal and tidal changes in rockpools. *Respiration Physiology*, 39, 241–254.
Uthicke S, Schaffelke B & Byrne M (2009). A boom and bust phylum? Ecological and evolutionary consequences of large population density variations in echinoderms. *Ecol Monogr*, 79, 3–24.
Visser ME (2008). Keeping up with a warming world; assessing the rate of adaptation to climate change. *Proc R Soc B*, 275, 649–59.
Ward GE, Brokaw CJ, Garber Dl & Vacquier VD (1985). Chemotaxis of *Arbacia punctulata* spermatozoa to resact, a peptide from the egg jelly layer. *J Cell Biol*, 101, 2324–29.
Whalan S, Ettinger-Epstein P & de Nys R (2008). The effect of temperature on larval pre-settlement duration and metamorphosis for the sponge, *Rhopaloeides odorabile*. *Coral Reefs*, 27, 783–86.
Widdicombe S & Spicer JI (2008). Predicting the impact of ocean acidification in benthic biodiversity: What can animal physiology tell us? *J Exp Mar Biol Ecol*, 366, 187–97.
Wilson SP & Hyne RV (1997). Toxicity of acid-sulfate soil leachate and aluminium to embryos of the Sydney Rock Oyster. *Ecotoxicol and Env Safety*, 37, 30–36.
Whooten JT, Pfister CA & Forester JD (2008). Dynamic patterns and ecological impacts of declining ocean pH in a high-resolution multi-year dataset. *Proc Natl Acad Sci USA*, 105, 18848–853.
Wood HL, Spicer JI & Widdicombe S (2008). Ocean acidification may increase calcification rates, but at cost. *Proc R Soc B*, 275, 1767–73.
Wooldridge S, Done T, Berkelmans R, Jones R & Marshall P (2005). Precursors for resilience in coral communities in a warming climate: a belief network approach. *Mar Ecol Prog Ser*, 295, 157–69.
Wright DA, Kennedy VS, Roosenburg WH, Castagna M & Mihursky JA (1983). Temperature tolerance of embryos and larvae of five bivalve

species under simulated power plant entrainment conditions: A synthesis. *Mar Biol*, 77, 271–78.
Yamada K & Mihashi K (1998). Temperature-independent period immediately after fertilisation in sea urchin eggs. *Biol Bull*, 195, 107–11.

7
Climate impact on hypersalinity in an Australian coastal bay

Joachim Ribbe

*Abstract**

Recent work on episodes of hypersalinity in a subtropical east Australian coastal bay is reviewed. Hypersalinity is a response of coastal bays and estuaries to the local freshwater balance which fluctuates due to variable rainfall, river run-off and evaporation. Longer lasting episodes of hypersaline conditions are a likely consequence of a drying and warming climate. The response of the coastal ocean to changes in the hydrological cycle has been observed in several Australian coastal systems, including the response to persistent drought conditions resulting in increased salinity and more frequent hypersaline conditions. In addition to observing hypersaline conditions in Hervey Bay off the central Queensland coast, first modelling experiments indicate that the frequency of hypersaline conditions has increased during the last 20 years. Coastal regions are important nurseries for Australian fisheries and the likely impact of persistent below average rainfall upon marine environmental conditions is yet to be assessed. This work indicates that routine monitoring of coastal salinity is essential in order to evaluate future impacts of climate change.

1. Coastal Vulnerability to Climate Change

Climate change and variability are impacting upon natural and human coastal systems due to changes in frequency and intensity of rainfall. In particular during the last 50 years, much of coastal east, south, and south-

* Chapter originally submitted in March 2009.

west Australia has experienced significant declines in rainfall with decreases exceeding 200–250 mm (e.g. Shi et al. 2008a). These trends are likely to persist into the future and are associated with large scale changes in the ocean and atmosphere circulation. Rainfall variability has been attributed to persistent changes in phenomena such as the Southern Annular Mode, the El Niño – Southern Oscillation and the Indian Ocean Dipole which drive much of Australia's interannual rainfall variability. It is likely that some of the changes are linked to global climate change (e.g. Cai et al. 2005; Murphy & Ribbe 2004; Shi et al. 2008a; Shi et al. 2008b; Ummenhofer et al. 2009). The Australian coastal marine systems are of immense economic value with fisheries, for example, being the fifth largest contributor to the Australian economy (e.g. ABARE 2005). Yet, our understanding of the coastal ocean's vulnerability and response to climate change is rated as poor in a study commissioned by the Australian Government (Voice et al. 2006).

The freshwater balance composed of evaporation, rainfall and river run-off is one of the key drivers of marine environmental conditions in Australian coastal systems such as estuaries and bays and directly controls the salinity within these systems. It is likely that a change in the freshwater balance due to reduced rainfall impacts upon these environments. An indicator of possible changes in coastal systems due to reduced freshwater supplies is the occurrence of coastal hypersaline zones. These are already observed in many Australian coastal systems due to a climatological freshwater balance that is negative, meaning the evaporation is larger than the supply of freshwater. This leads to elevated coastal salinity with lower salinity in the open ocean which exchanges water with the coastal systems.

Hypersalinity has been observed in Australian coastal systems of Western Australia (Burling et al. 1999), the Northern Territory and northern Australia (Wolanski 1986), South Australia (de Silva Samarasinghe & Lennon 1987; Petrusevics et al. 2009) and Queensland (Ribbe 2006; Benfer et al. 2007). A possible response of Australian coastal systems to persistent climatic changes (i.e. shift to low rainfall regime) comes from two recent studies by Lee et al. (2007) and Gräwe et al. (2009), both studies finding from observations and modelling that drought conditions are likely to lead to more frequent and persistent hypersaline conditions.

In this chapter, recent work is reviewed that has been carried out to better understand the impact of a variable climate and changes in freshwater supplies upon the hydrography of Hervey Bay which is an east Australian bay off the central coast of Queensland. An analysis of hydrographic observations shown in this chapter and obtained since Ribbe (2006) in combination with the historical freshwater balance indicates that continual drought conditions along eastern Australia favour the persistence and maintenance of inverse estuarine-like conditions in Hervey Bay. These are only reversed during severe flood events as observed in June 2008. During this storm event the Mary River, located in the south of Hervey Bay, discharged about three times the monthly climatology mean (~300,000 megalitres). Due to a predominantly cyclonic circulation within the Bay (Ribbe et al. 2008; Gräwe et al. 2009), freshwater is rapidly advected along the western coast and leaves the Bay. Consequently, the bay is expected to return to inverse estuarine-like conditions following events of high river run-off. Additional information from modelling studies is discussed that indicates inverse-like conditions are likely to have been enhanced during the last two decades. Salinity is not regularly and continuously monitored in Australian coastal waters, however, from this work it is projected that many Australian estuaries and confined larger bays are likely to experience increased levels of salinity due to reduced rainfall; as such operational salinity monitoring along Australia's coastal line should be established.

2. Hervey Bay – a coastal system

Hervey Bay is located on the continental shelf of subtropical southeast Queensland, Australia. It is situated to the south of the Great Barrier Reef Marine Park and covers an area of about 4000 km^2 with an average depth of about 15 m (Figure 1). The main exchange of water with the open shelf and ocean occurs in the north where the shelf is about 80–90 km wide. In the east, the bay is separated from the open ocean by Fraser Island which is the world's largest sand island. In the south, it is connected with the open ocean via a narrow, about 90 km long and shallow system (average depths of less than 2 m) of channels, the Great Sandy Strait and an eastward facing gap of about 2 km width south of Fraser Island. The Mary, Burrum, Elliott and Burnett rivers drain into the bay. The Mary and Burnett rivers are the main sources for river water discharges with the Burrum and Elliott having only a small catchment area. The Mary River mouth is located at the northern end of the Great Sandy Strait and freshwater is drained into the southern parts of Hervey Bay and southward into the Strait itself. The mouth of the Burnett River is at the north of the study area. The Bay came into existence about 7000–8000 years ago with the flooding of the coastal regions due to global sea level rise that followed the termination of the last ice.

The climate of the Hervey Bay region is classed as subtropical, with predominantly high monsoonal summer and low winter rainfalls, and the region is characterised by significant interannual rainfall variability (e.g. Murphy & Ribbe 2004). The key climate drivers controlling the rainfall delivering weather systems into the Hervey Bay region are primarily the Australasian Monsoon circulation and the Easterly Trade Winds while interannual rainfall variability is primarily a result of the El Niño – Southern Oscillation (Murphy & Ribbe 2004). The climatological freshwater water balance (E-P-R) due to evaporation (E), precipitation (P) and total river run-off (R) is shown in Figure 2. The data is obtained as averages from three rainfall weather observation locations, i.e. Bundaberg in the north west, Sandy Cape in the north of Fraser Island, and Urangan in the south of Hervey Bay and for about the last 100 years. Monthly river flow discharges are provided by the Queensland State Government's Department of Natural Resources. For this comparison, river run-off is converted in a virtual rainfall equivalent using a surface area for Hervey Bay of about 4000 km^2, i.e. river run-off is assumed to be equally distributed across the surface of the bay.

The annual freshwater balance for the bay is dominated by a net freshwater loss of about 113 mm resulting from a total annual supply of freshwater of 1425 mm due precipitation and river run-off and a total evaporative loss of about 1538 mm. In terms of seasons, the southern hemisphere summer and autumn seasons are dominated by a negative balance, i.e. freshwater supply is larger than loss, while the winter and spring seasons are dominated by a positive balance, i.e. evaporation is larger than freshwater input (Figure 2b). Recent observations indicate that the climate of

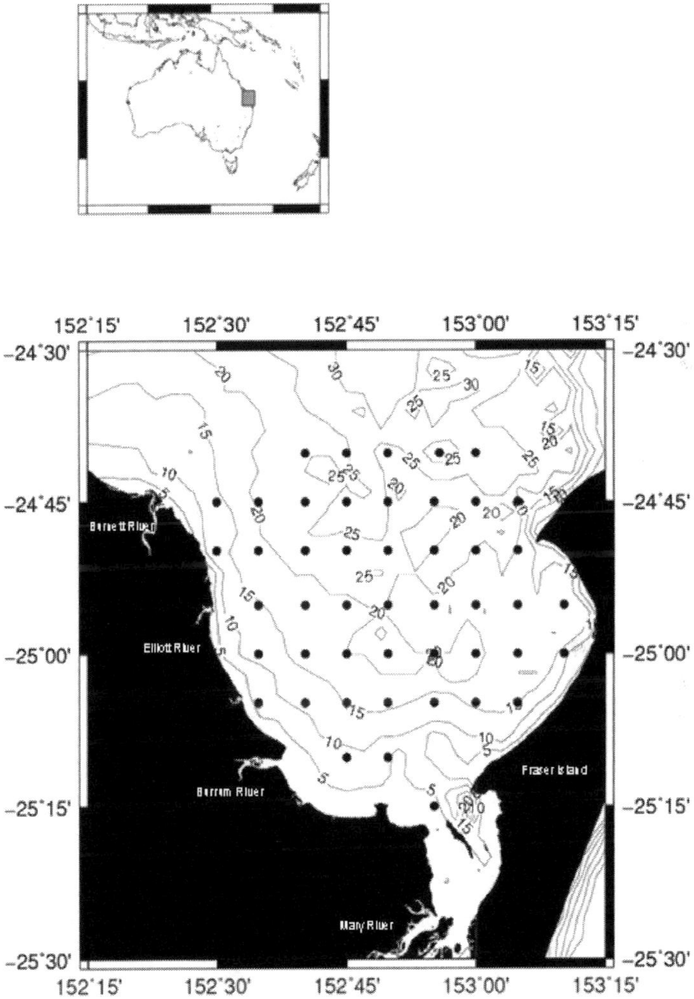

Figure 1 Map of Australia, location of Hervey Bay, and bathymetry of Hervey Bay [m].

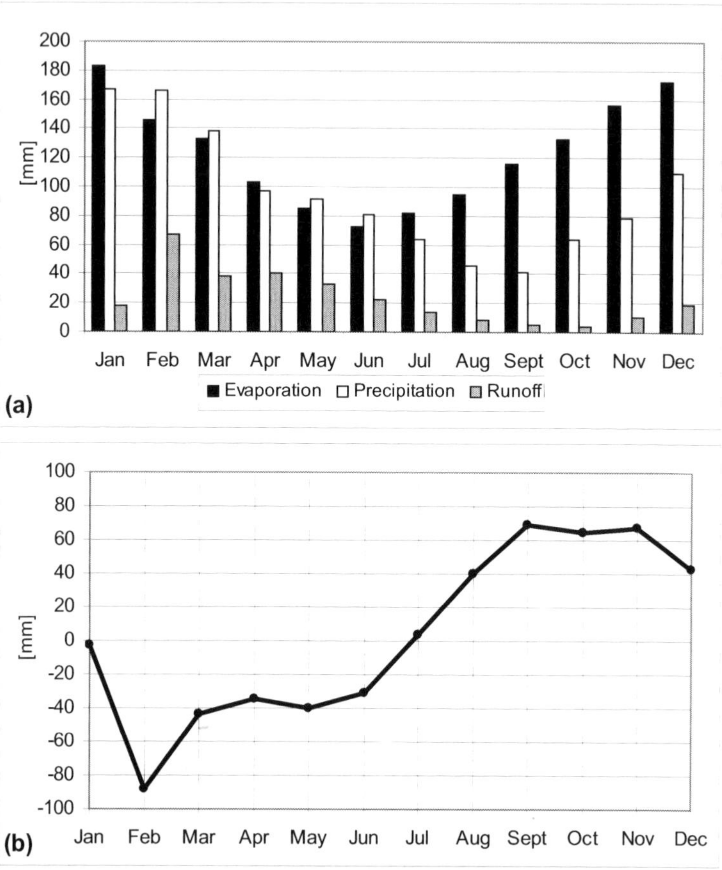

Figure 2 a) Climatological rainfall, run-off, and evaporation for Hervey Bay in [mm/month]. The total annual gain/loss due to rainfall, river run-off and evaporation is 1146 mm, 279 mm, and 1538 mm, i.e. the annual freshwater balance is dominate by a net loss of about 113 mm due to evaporation. b) Climatological monthly freshwater balance. Data source: Clewett et al. (2003).

the region is rapidly changing and is characterised by significant downward trends in rainfall (Shi et al. 2008), an upward trend in surface air temperature (Beer et al. 2006) and shifts in ocean climate zones (Lough 2008).

Systematic research into the circulation, hydrographic characteristics of Hervey Bay, and the impact of variable climate and freshwater flows is lacking. Many previous studies of the region support the status of Hervey Bay as an important marine ecosystem and whale sanctuary, focus upon its sedimentary history, and its importance for fisheries (Preen 1995; Preen et al. 1995; Preen & Marsh 1995; Moss & Kocovski 1998; Chaloupka et al. 1999; Campbell & McKenzie 2004; Boyd et al. 2004; Ward et al. 2003). The studies report little if anything at all about physical conditions of the bay that are clearly the key determinants of marine ecological systems and interact with biological, chemical and sedimentary processes.

Some new insight into the physical characteristics of Hervey Bay comes from recent work by Ribbe (2006), Ribbe et al. (2008) and Gräwe et al. (2009). Ribbe (2006) reported the first hydrographic survey completed for the bay in September 2004. The survey identified a hypersaline zone along the western region of the bay. Ribbe (2006) linked the observed salinity distribution to freshwater supplies arguing that hypersalinity in Hervey Bay is likely to be a climatological feature during September which has not been reported or observed previously. This is clearly supported by climatological data shown in Figure 2 with evaporation exceeding freshwater supplies by at least a factor of two during September, but also by actual data during the September 2004 survey, with an almost diminished freshwater supply during that month due to persistent drought conditions.

Hervey Bay can possibly be also classed as an inverse estuary with inflowing surface water of lower salinity above a layer of outflowing sub-surface water of high salinity away from the hypersaline zone. This high salinity water is referred to as Hervey Bay Water (Ribbe 2006). It is generated within the bay, exported to the northeast of Fraser Island and possibly advected with East Australian Current (EAC) southward, a feature that has been observed by Middleton et al. (1994) within the EAC to the southeast of Fraser Island or perhaps entrained in a quasi-permanent EAC recirculation north-west along the shelf break (Griffin et al. 1987; Burrage et al. 1996).

The formation and maintenance of a hypersaline zone within Hervey Bay is aided by the water renewal process within the bay. From simple freshwater /salinity budget modelling Ribbe (2006) obtained a basin-average water renewal times scale of about 90 days. The subsequent application of a three-dimensional ocean circulation model provided insight into the spatial variation of this time scale (Ribbe et al. 2008) and was found to vary spatially from a few tens of days to more than 100 days within the interior of the bay (Ribbe et al. 2008). The region is dominated primarily by north-easterly and south-easterly trade winds which force a cyclonic or clockwise circulation within Hervey Bay. A more detailed account of the processes that drive the generation of hypersalinity and the density-driven export of saline Hervey Bay Water is provided by Gräwe et al. (2009).

Since Ribbe (2006) reporting on only one hydrographic survey, a total of four additional surveys were completed during August and December 2007 and May and June 2008. The observations from these surveys are presented and analysed in this paper and discussed in the context of the regional freshwater balance for the period from 2004 to 2008. It provides additional insight into the response of a subtropical large bay to variable freshwater supplies and provides further evidence of hypersalinity within Hervey Bay.

The hydrographic surveys reported in this chapter were carried out during a period characterised by persistent drought conditions along the central Queensland coast as well as other regions of Australia (e.g. Shi et al. 2008a). The resulting freshwater balance for the period January 2004 to June 2008 is shown in Figure 3. It is dominated by evaporative loss of water for most of the period. Climatic variability and possible change in the region lead to significant departures from the climatology mean (Figure 2), clearly affecting the supply of freshwater from river run-off and rainfall into the bay.

3. *Observing hypersalinity*

A total of four surveys were conducted in 2007 and 2008 complementing the only other survey for the region reported previously by Ribbe (2006). All cruises commenced and terminated in Bundaberg Port and utilised the fisheries research vessel FRV *Tom Marshall* which is managed by the

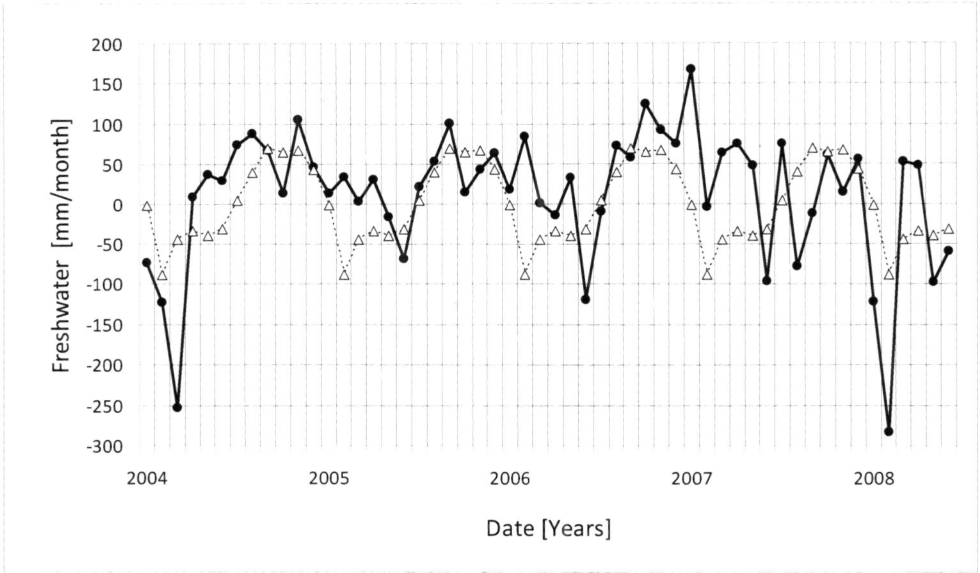

Figure 3 Monthly freshwater balance (mm/month) for the period January 2004 to June 2008 (solid line) and climatological water balance (dashed lines). The period is dominated by evaporation with only few instances where the supply of freshwater is larger than the loss due to evaporation. Data source: Clewett et al. (2003), Australian Bureau of Meteorology and Queensland Government.

Queensland Department of Primary Industries and Fisheries (DPI&F). The FRV *Tom Marshall* is a research catamaran cruising with an average speed of about 17–18 knots between sample locations which allows for a rapid quasi-synoptic survey of the bay. The distance between sample locations in all surveys reported here is about 5 nm in both longitudinal and latitudinal direction. The surveys took place during late winter (28–30 August) and early summer (4–7 December) in 2007 and during later autumn (26–27 May) and early winter (16–18 June) in 2008. This is a total of 12 observing days during which a total of 269 CTD profiles were recorded. The data was recorded using a Sea-Bird Electronics SBE 19 plus SEACAT Profiler and Falmouth Scientific Instrument CTD in August 2007. A full account of these surveys is provided by Ribbe (2008).

Depth-averaged temperature, salinity and density observations are shown in Figures 4–6 for all hydrographic surveys conducted into Hervey Bay. Included for comparison is the information from the previously referred to September 2004 survey (see Ribbe 2006 for a detailed discussion). Temperature ranges from a minimum of about 20.6 °C to a maximum of about 21.2°C in September 2004 (Figure 4). For all the following surveys values are: 19.4°C to 20.8°C in August 2007, 24.4°C to 25.6°C in December 2007, 20.4°C to 23.0°C in May 2008 and 19.8°C to 21.4°C in June 2008 (Figure 4). In all cases, a southwest to northeast temperature gradient is observed with the warm water in the northeast being a signature of open ocean water entering the bay. Only during the southern hemisphere summer month of December 2007, the temperature gradient is reversed with the warmer water found within the bay. This is most likely a result of the much more rapid warming of the shallow bay region compared to that of the open ocean. In all other seasons rapid cooling overnight dominates the heat gain during the day in shallow water. In contrast, the deeper ocean's heat capacity delays rapid cooling and establishes a temperature gradient with decreasing temperature toward the southwest. A remnant of the warming of shallow regions can also be observed during September 2004 and August 2007 where the shallow western region of the bay is found to be warmer.

In all but one of the surveys conducted, a hypersaline zone within the bay extending toward the western shore characterises Hervey Bay (Figure 5).

Maximum depths-averaged salinity within the hypersaline zone is about 36.8 psu in September 2004, 36.3 psu in August 2007, 36.2 psu in December 2007, and 35.9 psu in May 2008. In the latter case, this is just to the north of the bay. For this particular case, the survey commenced further in the north in order to chart the northern extent of the likely existing hypersaline zone outside Hervey Bay. Unfortunately, due to deteriorating weather conditions, the survey had to be terminated before Hervey Bay proper could be covered. Yet, the data shown in Figure 5 indicate elevated salinity and hypersalinity within the bay. In June 2008, no hypersaline zone within the bay is found. Depth-averaged salinity is lower throughout the bay with values smaller than 34.4 psu along the western boundary. Although these are depth-averaged salinity values, actual observed values for salinity within the hypersaline zone are of similar magnitude since the hypersaline zone, in fact the bay as a whole is vertically well mixed. The observed horizontal salinity gradients between the northeast oceanic region and the southwest within the bay are about 1 psu in September 2004, 1 psu in August 2007, 0.9 psu in December 2007, and likely to be larger than 0.7 psu in May 2008.

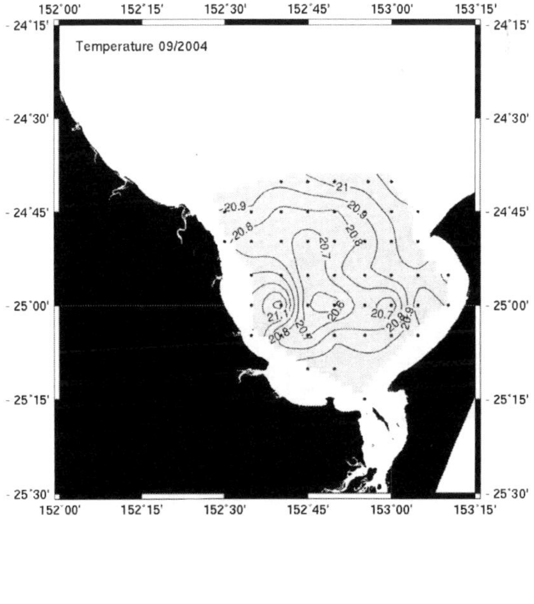

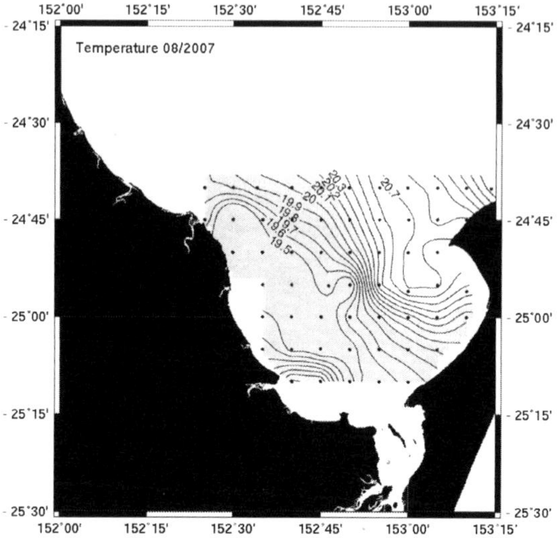
Temperature 08/2007
152°00'
152°15'
152°30'
152°45'
153°00'
153°15'
24°15'
24°30'
24°45'
25°00'
25°15'
25°30'
20.7
19.9
19.8
19.7
19.6
19.5

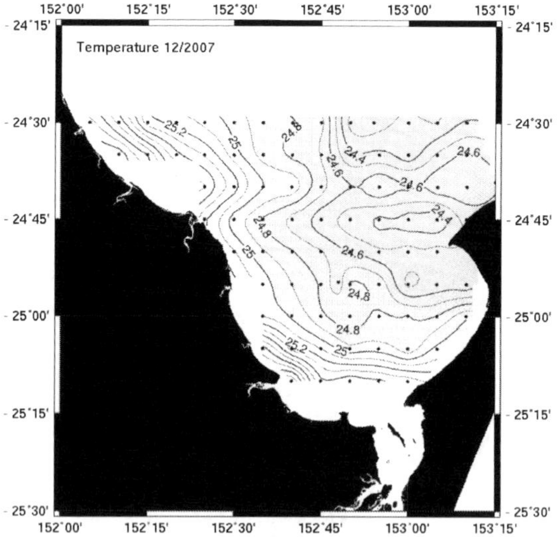
Temperature 12/2007
152°00'
152°15'
152°30'
152°45'
153°00'
153°15'
24°15'
24°30'
24°45'
25°00'
25°15'
25°30'
25.2
25
24.8
24.4
24.6
24.6
24.6
24.4
24.8
25
24.6
24.8
24.8
25.2
25

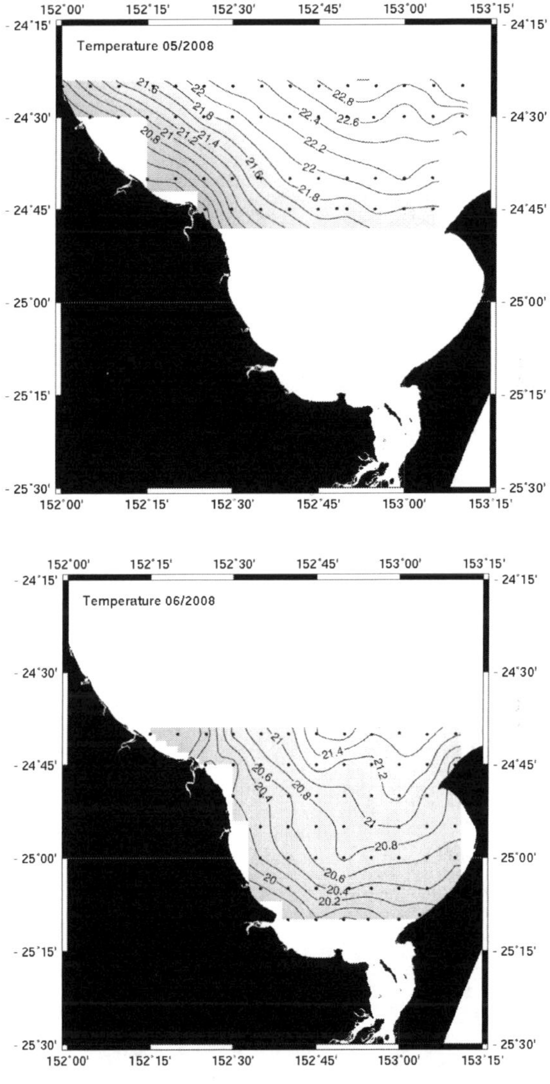

Figure 4 Depth-averaged temperature [°C] distribution from observations in a) September 2004, b) August 2007, c) December 2007, d) May 2008 and e) June 2008. Contour interval is 0.1°C[3].

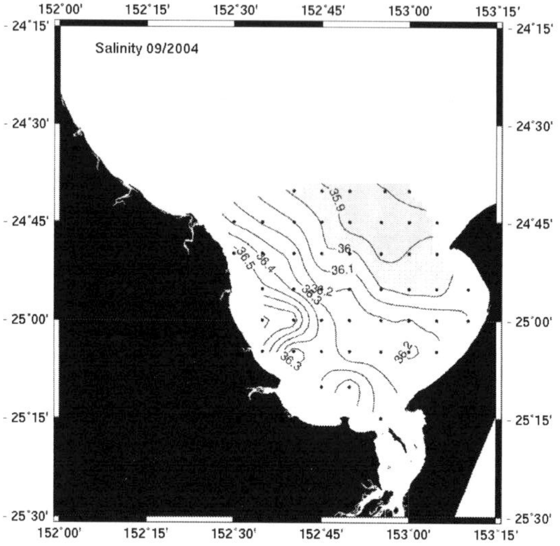
Salinity 09/2004
152°00'
152°15'
152°30'
152°45'
153°00'
153°15'
24°15'
24°30'
24°45'
25°00'
25°15'
25°30'
35.9
36
36.1
36.2
36.3
36.4
36.5

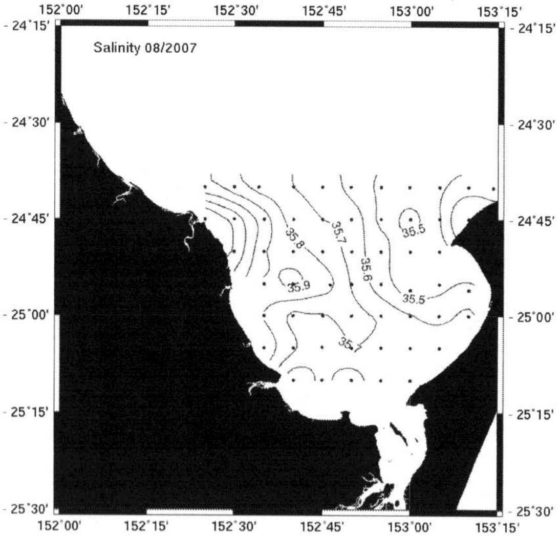
Salinity 08/2007
152°00'
152°15'
152°30'
152°45'
153°00'
153°15'
24°15'
24°30'
24°45'
25°00'
25°15'
25°30'
35.5
35.6
35.7
35.8
35.9

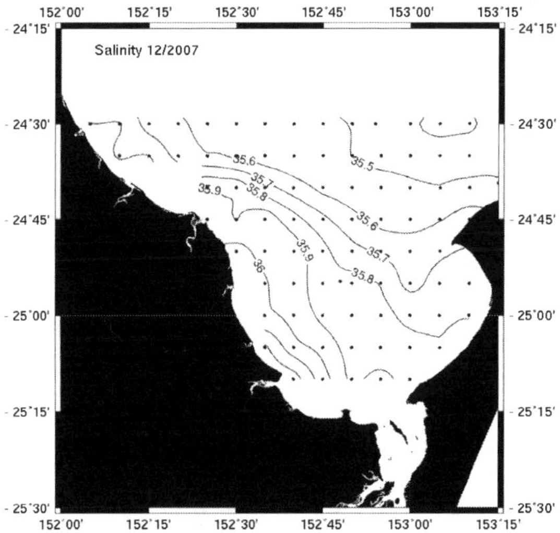

Salinity 12/2007
152°00'
152°15'
152°30'
152°45'
153°00'
153°15'
24°15'
24°30'
24°45'
25°00'
25°15'
25°30'
35.5
35.6
35.7
35.8
35.9
36

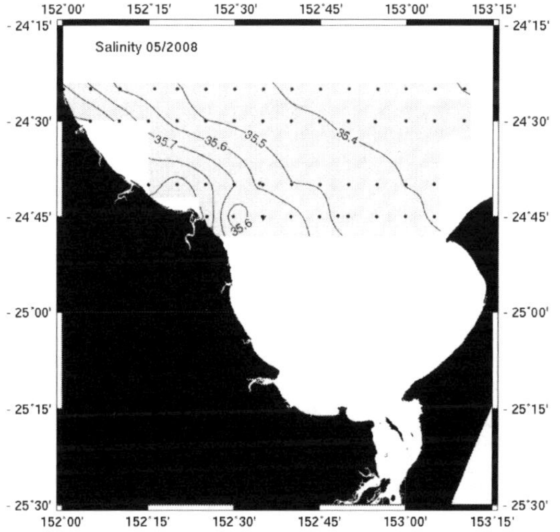

Salinity 05/2008
152°00'
152°15'
152°30'
152°45'
153°00'
153°15'
24°15'
24°30'
24°45'
25°00'
25°15'
25°30'
35.7
35.6
35.5
35.4

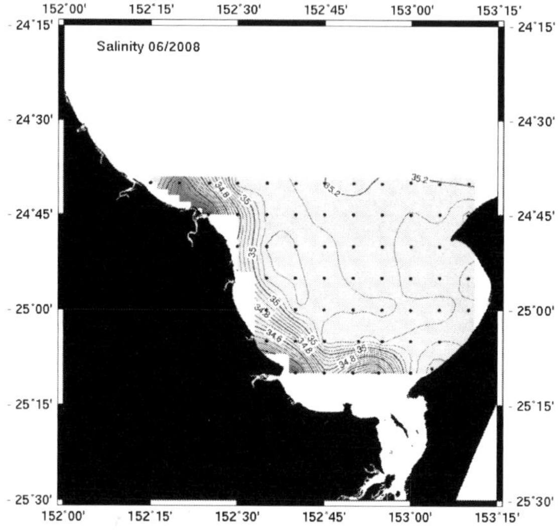

Figure 5 Depth-averaged salinity [psu] distribution from observations in a) September 2004, b) August 2007, c) December 2007, d) May 2008 and e) June 2008. Contour interval is 0.1°C[3].

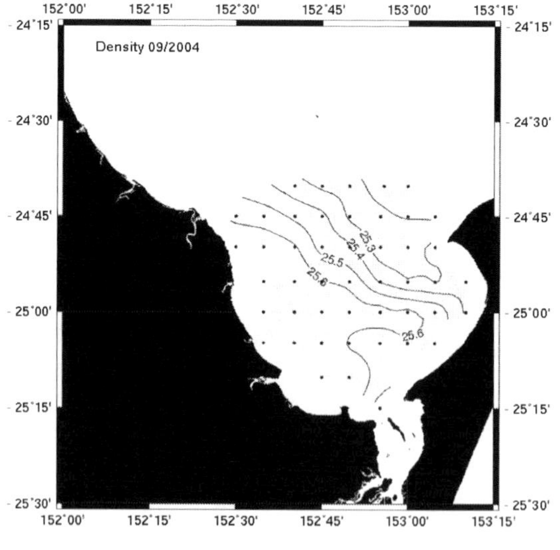

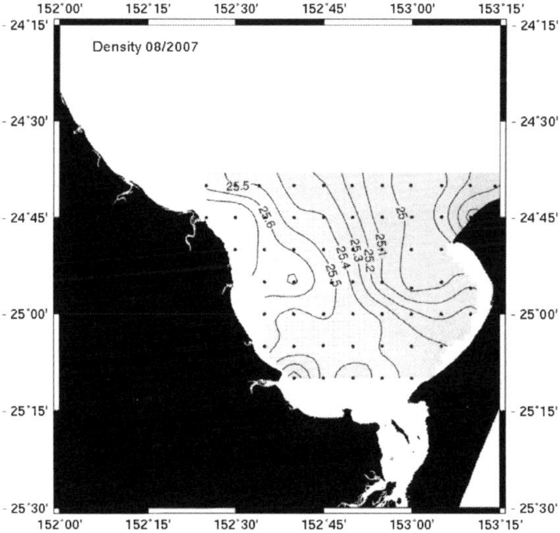
Density 08/2007
152°00'
152°15'
152°30'
152°45'
153°00'
153°15'
24°15'
24°30'
24°45'
25°00'
25°15'
25°30'
25.5
25.6
25.4
25.3
25.2
25.1
25
25.5

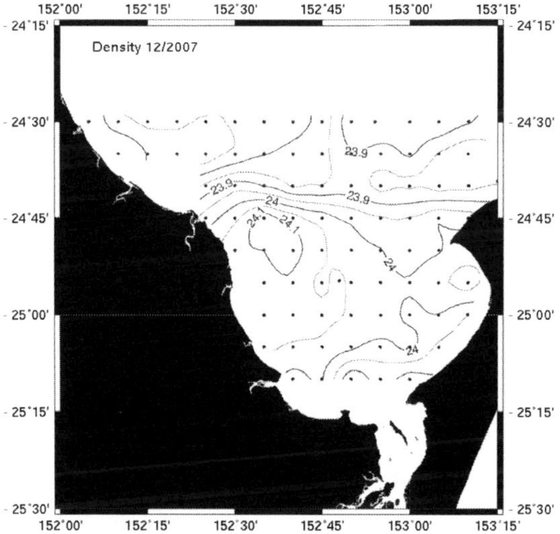
Density 12/2007
152°00'
152°15'
152°30'
152°45'
153°00'
153°15'
24°15'
24°30'
24°45'
25°00'
25°15'
25°30'
23.9
23.9
23.9
24
24.1
24.1
24
24

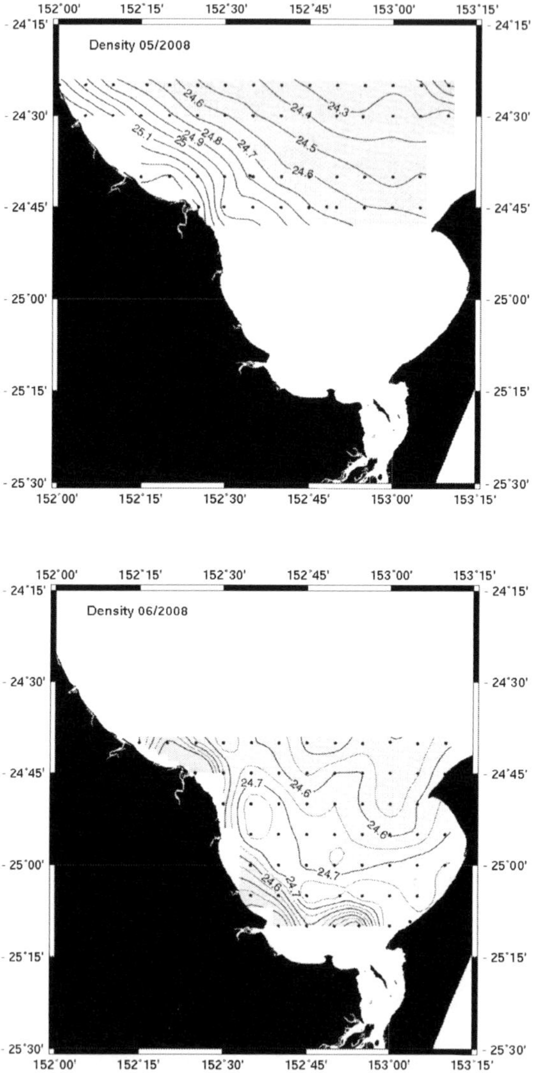

Figure 6 Depth-averaged density [kg·m^{-3}] distribution from observations in a) September 2004, b) August 2007, c) December 2007, d) May 2008 and e) June 2008. Contour interval is 0.1°C^{3}.

The establishment of and maintenance of a hypersaline zone within the bay is aided by bathymetry, the residual circulation within the bay, and the local freshwater balance shown in Figure 3. Simple modelling carried out by Ribbe (2008) and with a more complex model configuration by Gräwe et al. (2009) indicate that predominant north-easterly to south-easterly trade winds establish a cyclonic circulation that is associated with basin-average water renewal or flushing time scales in the order of about 89 days (Ribbe et al. 2006). The physical characteristics (bathymetry and circulation) of the bay combined with climatic conditions (freshwater balance dominated by evaporation) indicate that a mean salinity gradient of about 0.9 psu is the likely signature of a hypersaline zone within Hervey Bay during periods of persistent low freshwater supplies.

The distribution of density observed in all five surveys is characterised by a high density region that in four out of five instances is linked with zones of hypersalinity (Figure 6). In all these cases, the depth-averaged density maximum values within the hypersaline zone are about 25.8 $kg{\cdot}m^{-3}$ during September 2004, 25.8 $kg{\cdot}m^{-3}$ during August 2007, 24.1 $kg{\cdot}m^{-3}$ during December 2007, and 25.4 $kg{\cdot}m^{-3}$ during May 2008. In June 2008, depth-averaged density is largest in what could be referred to as a transition zone between lower density regions toward the west and the open ocean. Maximum depth averaged density within this transition zone is larger than about 24.8 $kg{\cdot}m^{-3}$. This transition region is likely to be a remnant of a hypersaline zone found in May 2008 and seemingly to extend into the bay along the western region of the bay. This region is now dominated by low density water resulting from freshwater river run-off that is entering the bay as a result of the end of May/early June 2008 storm and following flooding event.

Although as indicated above, wind and tidal mixing are usually sufficient to homogenise the water column throughout (Gräwe et al. 2009) some indication of a vertical salinity (and density) stratification within the hypersaline zone is observed to occur (Figure 7). Examples of the vertical salinity distribution along several west to east transects across the bay are shown for the September 2004, August 2007 and December 2007 surveys. The eastern region of the bay is homogenised with only little or no stratification at all. However, within the western zone of hypersalinity, changes in salinity between the surface and bottom layer are about 0.2–0.5 psu.

The establishment and maintenance of coastal hypersaline zones is driven by climatological/physical factors such as evaporation, precipitation, ocean circulation and is aided by the bathymetry or physical shape of the coastal system. A physical but unobservable quantity that characterises a confined coastal region such as a bay or large estuary and its communication with the open ocean is the flushing time scales. This is the time required to renew water within the bay with water derived from the open ocean. This renewal process imposes also an upper limit to salinity within hypersaline coastal systems. The flushing time scale τ (days) can be assessed from a simple relationship (Wolanski 1986; de Silva Samarasinghe & Lennon 1987; Ribbe 2006):

$$\tau = \frac{H \cdot (S_2 - S_1)}{S_1 \cdot e}$$

with H (m) as the mean depth of the bay (about 15 m), $S_2 - S_1$ the salinity gradient across the bay with S_2 being the salinity within the hypersaline zone and S_1 the salinity of incoming oceanic water, e is the annual climatological evaporation rate (m/d). The total annual climatological loss of water due to evaporation is about 1540 mm (from data in Figure 2) which is a daily evaporation rate of about 0.42 cm/day. The observations show that an average salinity gradient of about 0.9 psu (S_1 ~35.4 psu, S_2 ~36.3 psu) characterises hypersaline conditions within Hervey Bay. This leads to a mean or climatological flushing time scale of about 91 days, or about three months. This is very similar to the 89 days estimate provided by Ribbe (2006); the difference here is that the flushing time scale is assessed on the basis of four hydrographic surveys and a mean salinity gradient of 0.9 psu characterising hypersaline conditions within Hervey Bay.

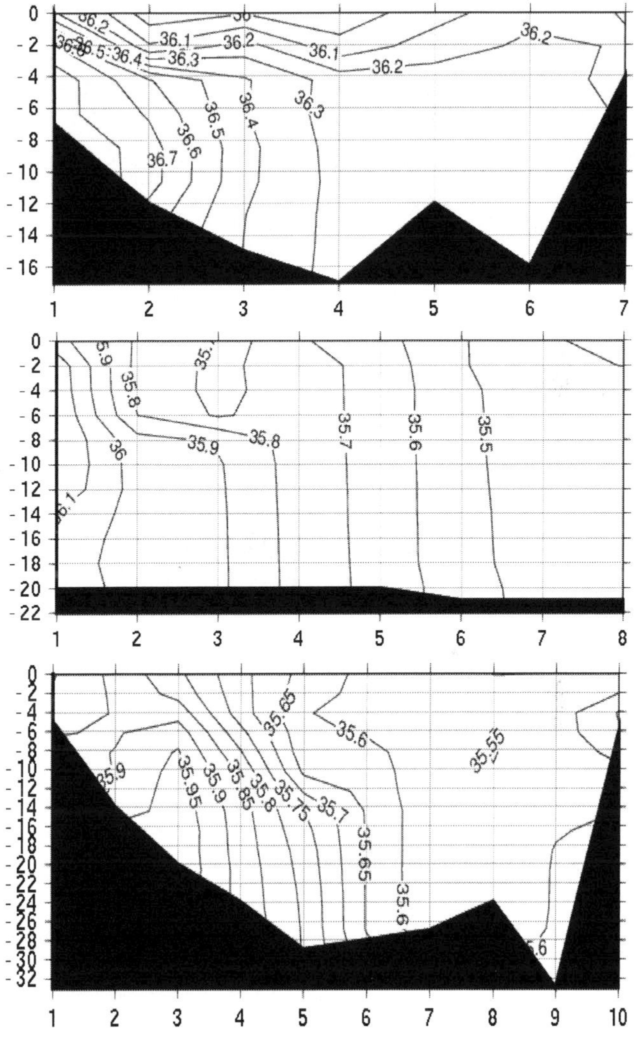

Figure 7 Vertical west to east salinity distribution in Hervey Bay during (top) September 2004 at 24° 55' S, (middle), August 2007 at 24° 50' S and (bottom) December 2007 at 24° 45' S.

4. Future hypersalinity in coastal systems

There is some indication that continued climatic changes associated with a reduction in rainfall and river run-off could lead to more frequent and persistent hypersaline conditions in coastal systems. Two recent studies found some evidence that this is likely to be the case. Lee et al. (2007) examined salinity data recorded continuously from 1947 to 2006 in Port Phillip Bay, Victoria. They identified continued and increased hypersaline conditions since 1997 as a most likely response to the persistent drought conditions in eastern Australia and associated reduction in precipitation and river run-off. Lee et al. (2007) further concluded from projections based on climate change scenarios from the Intergovernmental Panel on Climate Change that hypersaline conditions within the bay will likely be more prevalent. In another study, Gräwe et al. (2009) applied an ocean general circulation model to the region of Hervey Bay and generated a hindcast of the salinity distribution within the bay for the period January 1990 to December 2007. For this period, the authors find that the salinity flux out of the bay increases by about 25%. Furthermore, hypersaline conditions are more frequent with an upward trend of about three days per year in the number of days characterised by hypersalinity. This is likely to be direct consequence of persistent drought conditions in coastal Queensland.

Salinity is a key environmental variability that impacts upon the aquatic health of marine coastal systems. There is evidence emerging from observational data and modelling studies that climatic changes leading to reduced rainfall and river run-off is impacting coastal systems which are of immense economic value to society. Routine operational salinity monitoring should be encouraged and realised in order to assess the impact of projected future climatic changes on coastal systems.

Acknowledgements

The author would like to thank the Burnett Mary Regional Group in Bundaberg, Queensland, the Hanse Institute for Advanced Study, Germany, and the University of Southern Queensland for supporting this work.

References

ABARE (2005). *Australian fisheries statistics 2004.* Canberra: Australian Bureau of Agricultural and Resource Economics.

Beer T, Borgas M, Bouma W, Fraser P, Holper P & Torok S. (2006). Atmosphere. Theme commentary prepared for the 2006 Australia State of the Environment Committee, Department of Environment and Heritage, Canberra.

Benfer NP, King A & Lemckert CJ (2007). Salinity observations in a subtropical estuarine system on the Gold Coast, Australia, *Journal of Coastal Research,* 50, 646–51.

Burling MC, Ivey GN & Pattiaratchi CB (1999). Convectively driven exchange in a shallow coastal embayment. *Continental Shelf Research,* 19, 1599–616.

Burrage DM, Steinberg CR, Skirving JW & Kleypas JA (1996). Mesoscale circulation features of the Great Barrier Reef region inferred from NOAA satellite imagery. *Remote Sensing of Environment,* 56, 21–41.

de Silva Samarasinghe JR & Lennon GW (1987). Hypersalinity, flushing, and transient salt-wedges in a tidal gulf – an inverse estuary. *Estuarine Coastal Shelf Science,* 24, 483–98.

Cai W, Shi G, Cowan T, Bi D & Ribbe J (2005). The response of the southern annual mode, the East Australian current, and the southern mid-latitude ocean circulation to global warming. *Geophysical Research Letters,* 32, L23706, doi: 10.129/2005GL024701.

Clewett JF, Clarkson NM, George DA, Ooi SH, Owens DT Partridge IJ & Simpson GB (2003). Rainman StreamFlow version 4.3: a comprehensive climate and streamflow analysis package on CD to assess seasonal forecasts and manage climate risk. QIO3040, Department of Primary Industries, Queensland.

Gräwe U, Wolff J-O & Ribbe J (2009). Mixing, hypersalinity and gradients in Hervey Bay. *Ocean Dynamics,* doi: 10.1007/s10236-009-0195-4.

Griffin DA, Middleton JH & Bode L (1987). The tidal and longer-period circulation of Capricornia, southern Great Barrier Reef, *Australian Journal of Marine and Freshwater Research,* 38, 461–74.

Lee R, Martinez G, Harrison S, Bosserelle C & Black K (2007). Examining patterns of climatic variance and trend in Port Phillip Bay. Greenhouse 2007 Conference, Sydney, Australia.

Lough J (2008). Shifting climate zones for Australia's tropical marine ecosystems. *Geophysical Research Letters*, 35, L14708.

Middleton JH, Coutis P, Griffin DA, Macks A, McTaggart A, Merrifield MA & Nippard GJ (1994). Circulation and water mass characteristics of the southern Great Barrier Reef. *Australian Journal of Marine and Freshwater Research*, 45, 1–18.

Murphy BF & Ribbe J (2004). Variability of southeast Queensland rainfall and its predictors. *International Journal of Climatology*, 24, 703–21.

Petrusevics P, Bye JAT, Fahlbusch V, Hammat J, Tippins DR, & van Wijk E (2009). High salinity winter outflow from a mega-inverse estuary: the Great Australian Bight. *Continental Shelf Research*, 29, 371–80.

Ribbe J (2006). A study into the export of saline water from Hervey Bay, Australia. *Estuarine, Coastal and Shelf Science*, 66, 550–58.

Ribbe J, Wolff J-O, Staneva J & Gräwe U (2008). Assessing water renewal time scales for marine environments from three dimensional modelling: a case study for Hervey Bay, Australia. *Environmental Modelling and Software*, 23, 1217–28.

Shi G, Cai W, Cowan T, Ribbe J, Rotstayn L, & Dix M (2008a). Variability and trend of the northwest Western Australia Rainfall: observations and coupled climate modelling. *Journal of Climate*, 21, 2938–59.

Shi G, Ribbe J, Cai W & Cowan T (2008b). Interpretation of Australian summer and winter rainfall projections. *Geophysical Research Letters*, 35, L02702, doi: 10.1029/2007GL032436.

Ummenhofer CC, England MH, McIntosh PC, Meyers GA, Pook MJ, Risbey JS et al. (2009). What causes Southeast Australia's worst droughts? *Geophysical Research Letters*, 36, L04706.

Voice M, Harvey N & Walsh K (2006). Vulnerability to climate change of Australia's coastal zone: analysis of gaps in methods, data and system thresholds. Report to the Australian Greenhouse Office, Department of the Environment and Heritage, Canberra, Australia.

Wolanski E (1986). An evaporation-driven salinity maximum zone in Australian tropical estuaries. *Estuarine Coastal Shelf Science*, 22, 415–24.

8

Indonesian Throughflow: PACific Source Water INvestigation (PACSWIN) An international ocean climate program

Y You, T Rossby, W Zenk, AG Ilahude, M Fukasawa, R Davis, D Hu, D Susanto, PL Richardson, C Villanoy, C-T Liu, JH Lee, R Molcard, WW Pandoe, M Koga, T Qu, RA Fine, A Gabric, R Robertson, Y Masumoto, S Riser, H Hasumi, P Sigray and T Lee

*Abstract**

Our understanding of the role of the Indonesian Throughflow (ITF) in the global climate system has improved due to progress made on observations and modelling in the last decade or so. The International Nusantara Stratification and Transport (INSTANT) program from 2004–06 measured the ITF simultaneously at several of the primary straits and showed a 20–30% increase of the ITF transport under relatively weak El Niño – Southern Oscillation (ENSO) conditions. This suggests that prior ITF transports could be underestimated. Further simultaneous observations encompassing a complete ENSO cycle are required for a more accurate estimate of the ITF transport. Climate monitoring requires cost-effective long-term measurement of the ITF on decadal or even centennial time scales. Currently, no such sustainable monitoring program exists for the Indonesian seas and adjacent region.

---

* Chapter originally submitted in April 2009.

To address this need, a coordinated international program is necessary. Modelling the role of the ITF in global climate has primarily been along the line of switching the ITF on or off. The ITF has been established as affecting tropical wind stress, thermocline depth and precipitation in ocean general circulation and coupled atmosphere and ocean models. Teleconnection of the climate impact of the ITF with higher latitudes results from the westward shift of the western Pacific warm pool inducing changes in atmospheric deep convection. However, the role of the ITF in the global climate system has not been well addressed by the models, due to two distinct obstacles. First, there has been insufficient resolution of the complicated bottom topography and numerous narrow straits in the Indonesian seas. Second, the tidal mixing which blends all Pacific source waters into one Indian Ocean water-mass has been poorly resolved. As a consequence of mixing, the major ITF in the thermocline layer actually cools and freshens the eastern Indian Ocean. These observational, modelling and climate inadequacies provide the rationale for this newly proposed international ocean climate program, the Indonesian Throughflow: PACific Source Water INvestigation (PACSWIN), which will focus on the variability of the ITF passing through the Indonesian seas and straits. This new program aims to bridge the data gap in the Indonesian seas left by the current Argo program and to resolve various source waters (mainly of Pacific origin) and their teleconnections and pathways, thus supplementing the INSTANT moorings completed by the end of 2006.

1. Introduction

The United Nation (UN) Climate Change Conference, held in Bali, Indonesia, in December 2007, created a roadmap to continue the goals of the Kyoto Protocol after it ends in 2012. The conference sent strong messages about the critical need to effectively monitor climate change and provide accurate information of global warming to society. Nowhere is this more evident than in the Indonesian Throughflow (ITF) region between the Pacific and Indian Oceans. The ITF, draining water from the tropical Pacific into the Indian Ocean, plays an important role in Australian and global climate. It is reasonable to expect that the potential dramatic climate change associated with the weakening of the ocean circulation in the northern and tropical North Atlantic could also affect the ITF and cause subsequent regional climate change in southeast Asia and Australia because the ITF is a choke point of the global ocean circulation affecting earth's climate.

The Indonesian Throughflow: PACific Source Water INvestigation (hereafter PACSWIN) program is an international coordinated ocean climate monitoring program proposed to address the role of the ITF in regional and global climate change. Geographically, PACSWIN intends to cover the Indonesian seas and adjacent regions (Figure 1) including the entrance of the Pacific source waters east of Mindanao, north of Papua New Guinea (PNG) and the far western equatorial Pacific as well as the exit of the ITF in the eastern Indian Ocean (see dashed square in Figure 1; note that acronyms for currents and water masses used in the paper are defined in the caption of Figure 1). The aims are twofold: first, to improve identification of the various source waters from the Pacific and Indian oceans through the use of extended tracers including temperature, salinity, oxygen, nutrients and biological and chemical tracers and, second, to investigate the Indo-Pacific water mass exchange via the ITF linkage with current system east of Mindanao, north of PNG, the equatorial current and that in the eastern Indian Ocean. The second goal will be to coordinate activities with two other major international programs in the western tropical-subtropical Pacific, SPICE (Southwest Pacific Ocean Circulation and Climate Experiment) (Ganachaud et al. 2007) and NPOCE (Northwestern Pacific Ocean Circulation Experiment) (Hu, personal communication, 2007) as well as the global Argo (Array for Real-time Geostrophic Oceanography) program (Roemmich et al. 2009). We want to illuminate how various Pacific source waters are transformed into a unique mixture of the Indonesian sea water and how the new Indian Ocean sources, such as the transformation of Antarctic Intermediate Water (AAIW) to Indonesian Intermediate Water (IIW), are formed by tidal and vertical mixing processes within the Indonesian seas. A water mass moving from its formation region to another place carries an index of climate change signals.

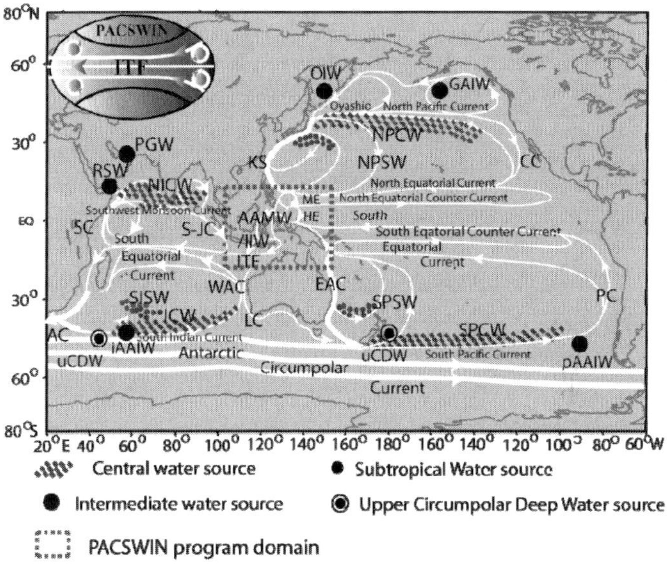

Figure 1 Schematic Indo-pacific ocean circulation and source water formation that are associated with the Indonesian Throughflow. Abbreviations are used for names of currents and eddies (in italics): ITF (Indonesian Throughflow), EAC (East Australia Current), WAC (West Australia Current), LC (Leeuwin Current), S-JC (Sumatra-Java Current), KS (Kuroshio), SC (Somali Current), AC (Agulhas Current), PC (Peru Current), CC (California Current), ME (Mindanao Eddy) and HE (Halmahera Eddy), and of water masses: uCDW (upper Circumpolar Deep Water), iAAIW (Indian Ocean Antarctic Intermediate Water), pAAIW (Pacific Ocean Intermediate Water), ICW (Indian Central Water), SISW (South Indian Subtropical Water), NICW (North Indian Central Water), RSW (Red Sea Water), PGW (Persian Gulf Water), AAMW/IIW (Australasian Mediterranean Water and Indonesian Intermediate Water), SPCW (South Pacific Central Water), SPSW (South Pacific Subtropical Water), NPCW (North Pacific Central Water), NPSW (North Pacific Subtropical Water), OIW (Okhotsk Intermediate Water) and GAIW (Gulf Alaska Intermediate Water). The proposed PACSWIN program area is marked as a dashed square.

PACSWIN is designed to better quantify the role of the ITF in global climate change.

Secondly, a climate-sensed teleconnection of the water-mass inventory of the Indonesian seas has to be established to link local water masses with their remote source formation regions. The Indonesian seas lie in a tropical region where no water mass can be formed locally by air/sea interaction. All water masses have to come from subtropical and high latitudes of the Pacific Ocean. The remote source waters at tropical and subtropical regions are marked in Figure 1. They are mainly high salinity and oxygen waters initially formed by Ekman subduction since they are ventilated at the tropical and subtropical convergence zones. Those intermediate waters are originally formed in the subpolar regions (see big black dots in Figure 1). However, significant transformation of these prototype source waters is necessary before they can cross the subtropical convergence zone at about 40°S and enter the subtropical latitudes. In the past, the contribution of marginal seas to the ITF has not been well accounted. PACSWIN has established a monitoring component of marginal seas input such as at the Kalimantan Strait where the South China Sea transports a significant amount of water from and into the Indonesian seas during reversal monsoons.

The PACSWIN initiative is motivated by four factors: 1) the need for continuous monitoring of the ITF across the major Indonesian straits subsequent to International Nusantara Stratification and Transport (INSTANT) program, 2) the paucity of Argo profiles in the region prompting consideration of a different monitoring strategy, 3) the inability of the INSTANT program to resolve the various source waters and their pathways and 4) the focus of the present wider global climate issue on the North Atlantic concerning the weakening or even halting of the meridional overturning circulation (MOC) but ignoring the role of the ITF in global circulation. Recent observational programs monitoring the MOC in the North Atlantic indicate high current variability that must be incorporated into any predictive models (Church 2007; Kanzow et al. 2007; Cunningham et al. 2007).

2. Aims and objectives

PACSWIN aims to bridge the data gap in the Indonesian seas left by the current Argo program, and to resolve various source waters (mainly of Pacific origin) and their teleconnection with the high latitude formation region and pathways and hence the climate. This will supplement the INSTANT moorings completed by the end of 2006. PACSWIN will provide a major international ocean climate monitoring network in the Indo and western Pacific for at least the coming decade and beyond. The objectives of PACSWIN are

- to strategically establish a long-term and cost-effective climate monitoring program including the priority components of moorings, submarine cables and modified floats, permitting data from the Indonesian seas to complement the Argo array in compliance with other global ocean monitoring programs such as that in the North Atlantic for the MOC
- to obtain accurate estimates of the ITF transport, heat and freshwater fluxes through the Indonesian seas
- to elucidate the various source waters entering the Indonesian seas, especially of Pacific origin, their pathways, transformations and teleconnections
- to identify forcing mechanisms such as wind, ENSO, monsoon and tides, etc, which affect the ITF and heat and fresh water fluxes
- to apply high resolution ocean and climate models to better simulate and predict the ITF and its climate impact.

PACSWIN also includes some components of topical studies aimed at resolving some important dynamical issues

- the association of ITF with the Pacific and Indian Ocean equatorial current system, low latitude boundary currents, eddies of Mindanao, Halmahera and eastern Indian Ocean
- the forcing of wind and equatorial waves including Rossby and Kelvin waves

- the interior mixing, topographic interaction and water-mass transformation within the Indonesian seas
- the various time-scales of intraseasonal, seasonal, interannual, annual and decadal variability and their interplay
- the dynamic correlation of the ITF with the Indian Ocean dipole, Pacific decadal Oscillation and North Pacific Gyre Oscillation.

3. Scientific background

Argo is designed to continuously monitor ocean climate, especially temperature and salinity in the upper 2000 m of the global ocean covered by ~3000 free-drifting floats (The Argo Science Team 1998). Argo started in 1999 as part of the Global Climate Observing System (GCOS), Global Ocean Observing System (GOOS), Global Ocean Data Assimilation Experiment (GODAE) and Climate Variability and Predictability Experiment (CLIVAR). Figure 2 shows the updated state of the Argo floats as of 15 March 2009, with a total number of 3325 floats now deployed. The important role of Argo in monitoring climate change and validating climate models has been increasingly recognised. However, there is a gap in the Indonesian seas due to the limit of the designed diving depth of 2000 m for open ocean, (see marked square area in Figure 2). Therefore, the Indonesian seas need a different strategy to monitor changes in ocean climate.

The Indonesian seas play a crucial role in global ocean circulation and climate. The Indonesian archipelago is the only low latitudinal connection of tropical oceans between the western equatorial Pacific and eastern Indian Ocean. Flow passing through the archipelago is the well-known ITF guided by geographically complicated basins and straits. From the Snellius expedition, van Riel (1932) was the first to suggest that the transport within the East Indies (now Indonesian) waters (ITF) was directed to the Indian Ocean (Pariwono et al. 2005). In the Indian Ocean, the inflow of the ITF can be identified as a jet-like flow westward along 12°S reaching the coast of eastern Africa, feeding the Agulhas Current from both sides of Madagascar and exiting in the southwest Indian Ocean. A small fraction of the escaped

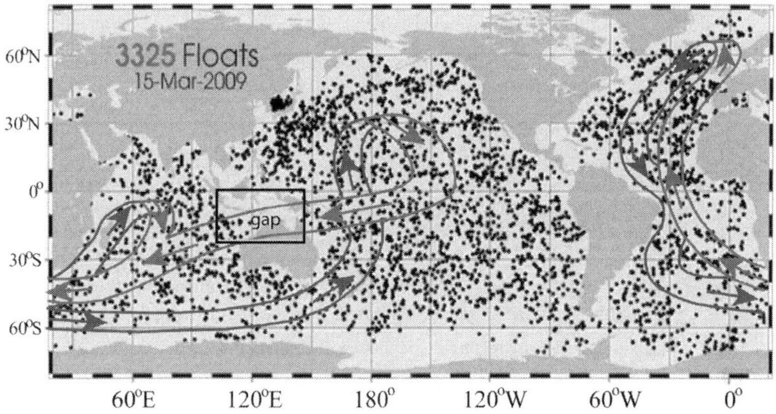

Figure 2 Argo stations on 15 March 2009. Note the gap in the Indonesian seas (outlined in square). Superimposed is the great ocean conveyor belt with the cold water route and warm water route.

ITF water from the Agulhas Retroflection continues its journey to the South Atlantic via Agulhas rings/eddies, joining the global ocean circulation in the Benguela Current. Within the Indian Ocean, the ITF is a significant component of the northern and southern Indian Ocean gyre circulation. The former switches between cyclonic and anticyclonic circulation under monsoonal forcing. The water mass mixture of the Indonesian seas from various thermocline source waters can be identified as a low temperature and salinity tongue in the Indian Ocean, and is called Australasian Mediterranean Water (AAMW) (You & Tomczak 1993), a nomenclature reserved for its unique water mass property while the ITF is named for the surface current. In terms of basin-scale description of water-mass structure in the Pacific and Indian Oceans, AAMW is a sink resulting from the mixing of North and South Pacific thermocline waters predominantly from North Pacific sources and transformed through the "Indonesian Mix-Master" (IMM) (Gordon 2005) to be a new source of Indian Ocean thermocline water. The same scenario of water mass transformation is retained for the intermediate water layer. The Indonesian Intermediate Water (IIW) is characterised by a silicate maximum in the Indian Ocean (You 1998) and originates from the Indonesian seas as a mixture of Pacific AAIW and local deep water in the Banda Sea. The mixed IIW is deepened as a result of vertical mixing and buoyancy decrease. The return water of the North Pacific sources is through lower Circumpolar Deep Water (lCDW) just above the Antarctic Bottom Water (AABW),) which is characterised by an oxygen minimum and maxima in the nutrients. The lCDW is transported to the North Pacific via the deep western boundary currents and upwells in the subpolar latitudes of the northern North Pacific, forming a limb of warm water in the great ocean conveyor belt circulation (see overlapped sketch for cold and warm route in Figure 2) (Gordon 1986). Another route by which water mass characteristics entering the ITF are changed before entry was identified by Godfrey et al. (1996), and McCreary et al. (2007). The latter showed (in a model) that some of the salty water entering the Pacific south of Australia joins the Tsuchiya Jets, is upwelled in the east Pacific and returns west in the South Equatorial Current.

The great conveyor belt is sensitive to the freshening of the northern North Atlantic by ice-sheet and glacial melting in Greenland and the Arctic as well as increased precipitation and run-off caused by global warming. The potential dramatic climate change associated with the alteration of ocean circulation in the northern North Atlantic may affect the ITF and may cause regional climate change in southeast Asia and Australia because the ITF is a choke point of the global thermohaline circulation (THC). In the southern hemisphere, two global scale water masses, AAIW and AABW, are formed near Antarctica and contribute to the THC as well. In the Indonesian seas, only AAIW contributes to the ITF because AABW lies too deep.

The Indonesian seas link two warm pools – the western Pacific and eastern Indian Ocean. The former regulates the global heat budget and controls atmospheric circulation. The latter guides the Indian monsoon circulation and African drought through the Indian dipole mode (Saji et al. 1999). The Indonesian archipelago contains many shallow and deep straits and interior basins (Figure 3) (Gordon et al. 2006; Gordon 2005). The Sulawesi Sea (or Celebes Sea) and Banda Sea are among the largest and deepest. The unique geographical conditions and strong tides permit blending of incoming Pacific source waters into a different water mass. The primary source of the ITF is from the North Pacific thermocline and intermediate water carried by the Mindanao Current into the Sulawesi Sea, through the Makassar Strait into the Flores Sea and Banda Sea and finally to the eastern Indian Ocean via either side of Timor with a small amount flowing out from the Lombok Strait (Wyrtki 1961; Gordon 1986; Murray & Arief 1988; Gordon & Fine 1996; Meyers 1996).

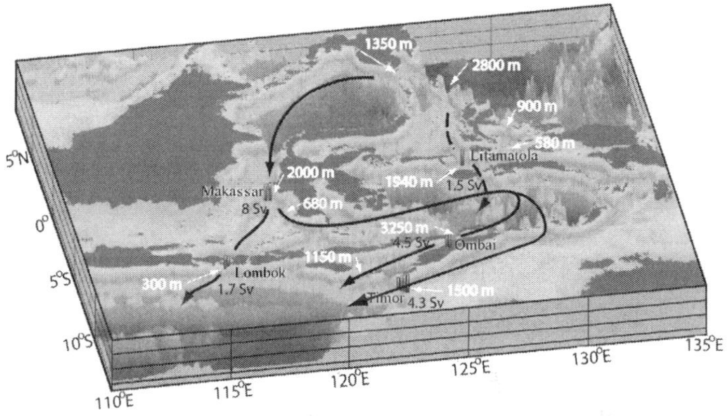

Figure 3 Three-dimentional view of the topography of the Indonesian seas. The ITF paths are indicated by a solid line in the west and a dashed line in the east. The sill depth, transport and the 11 INSTANT moorings at the major straits are marked (2 at Makassar, 2 at Lombok, 4 at Timor, 2 at Ombai and 1 at Lifamatola Straits). Results are based on figure 1 of Gordon et al. (2006) (with permission from Gordon in June 2009). According to Gordon, the Makassar Strait transport has an uncertainty of 1–2Sv.

Water masses from North and South Pacific sources are most distinguishable by their salinities with a low salinity and high temperature for the North Pacific source and a high salinity and low temperature for the South Pacific source (Figure 4) (also see Ilahude & Gordon 1996). Through vertical mixing in the Banda Sea and Timor Sea, the North and South Pacific thermocline waters are blended into a vertically rather homogeneous water mass (see Figure 4). For the intermediate water layer, NPIW at about 300 m (or 26.5σ_θ) is well mixed and is no longer identifiable in the Timor Sea. AAIW enters the eastern Indonesian seas at about 600–1000m (or 27.1–27.2 σ_θ) but cannot be identified in the Banda Sea. A replaced IIW at 1000m (or σ_θ=31.96) is observed in the eastern Indian Ocean South Equatorial Current (SEC) (Talley & Sprintall 2005). The Indonesian seas act as a sink for various Pacific water masses and a source for the Indian Ocean (Table 1).

Table 1 Inventory of water masses in the Indonesian seas, their teleconnection with the Pacific source formation and transformation to new source of the Indian Ocean.

Water mass	Property feature	Source formation region	Transformed Indian new source water mass and property feature
NPSW	high S (34.65), T (15°C) O_2 (3.0 ml/l), low nutrients	shallow subtropical North Pacific	AAMW low S (34.6), T (14°C)
SPSW	high S (34.75), T (14°C), O_2 (2.6 ml/l), low nutrients path: via Halmahera Sea	shallow subtropical South Pacific	AAMW low S (34.6), T (14°C) high silicate (35 mol kg^{-1})
NPIW	S minimum (34.4) path: via Makassar and Maluku Sea	Okhotsk Sea and Gulf of Alaska	AAMW low S (34.6), T (14°C) high silicate (35 mol kg^{-1})
AAIW	S minimum (34.56) low nutrients path: via Maluku Sea, Seram Sea, Banda Sea to Timor Sea	southeast South Pacific	IIW silicate maximum (80 μmol kg^{-1})
uCDW	high S (32.62) and nutrients low T (2.6°C) and O_2 (2.6 ml/l) path: via Maluku Sea, Seram Sea to Banda Sea	circumpolar region of the Southern Ocean	none

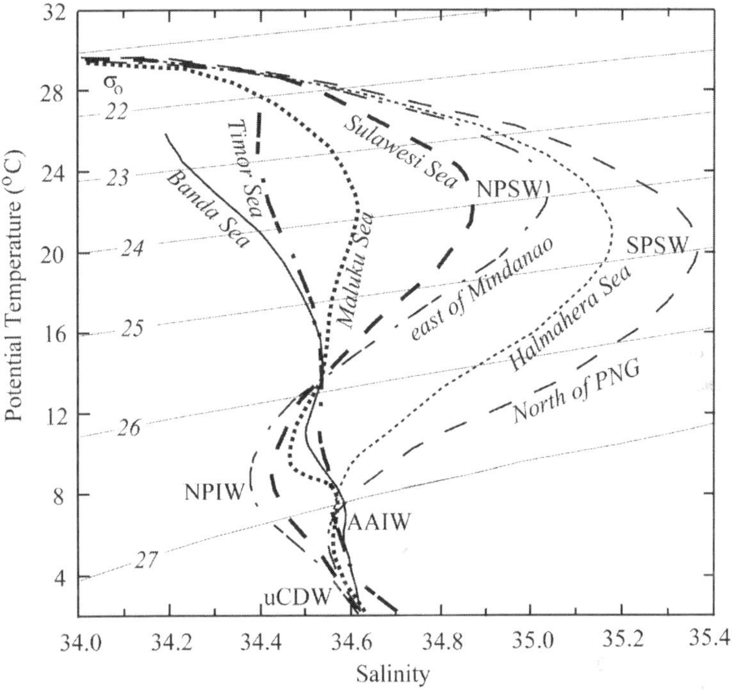

Figure 4 A θ-S diagram showing the Pacific source waters at the eastern entrance and their transformation in the interior of the Indonesian seas, the north of PNG (dashed line), Halmahera Sea (dotted line), east of Mindanao (dot dashed line), Sulawesi Sea (thick dashed line), Maluku Sea (thick dotted line), Banda Sea (solid line) and Timor Sea (thick dot-dashed line).

When the water mass cannot be identified through potential temperature θ and S becoming obscured, nutrients and chemical tracers help to identify the various Pacific source waters. Gordon and Fine (1996) used chlorofluorocarbons (CFC-11 and CFC-12) to examine water mass stratification within the Indonesian seas. However, CFC-11 and CFC-12 do not delineate vertical water-mass structure. Oxygen generally has a high value in the South Pacific source and a low value in the North Pacific source below the upper thermocline on isopycnal surfaces mapped by Hautala et al. (1996) to follow the major water mass cores. This situation is reversed in the upper thermocline. The North Pacific source shows relatively higher oxygen content. Within the Indonesian seas, Ffield and Gordon (1992) discovered that oxygen curves could not be used to distinguish between sources, suggesting that oxygen consumption may have to be considered. However, silicate is a major signal to identify the transformed Indonesian source waters for the Indian Ocean. Nutrients have so far been very poorly sampled in the Indonesian seas.

The thermohaline stratification and water-mass structure of the Indonesian seas (referring to Figure 4 and Table 1) can be summarised as upper thermocline water with a salinity maximum (>34.75) at 130 m (or $24.5\sigma_\theta$) (SPSW), main thermocline water with relative high salinity (~34.65) and oxygen (3.0 ml l^{-1}) but no extrema at 220 m (NPSW), a salinity minimum (<34.4) at 300 m (or $26.5\sigma_\theta$) (NPIW), a relative low salinity (~34.56) and oxygen (2.35 ml l^{-1}) but no extrema at 1000 m (or $27.25\sigma_\theta$) (AAIW) and deep water with a relative high salinity (~32.62) below 1500 m (or larger than $27.4\sigma_\theta$) (Ilahude & Gordon 1996; Ffield & Gordon 1992; Hautala et al. 1996). Note that the sill depths in the Maluku Sea and Lifamatola Strait are sufficient to allow passage of upper Circumpolar Deep Water (uCDW) from the Southern Ocean into the Banda Sea (see Figures 3 and 4). The pathways of the remote Pacific sources into the Indonesian seas are not well resolved, especially in the eastern route since the western route is dominated by North Pacific sources and is shallower with a sill depth of ~600 m (implying fewer water types involved). Both North and South Pacific sources flow into the eastern route. A teleconnection of the Indonesian water types with the remote Pacific source waters has yet to be established in extended property fields including nutrients and chemical tracers. For example, it is still unclear whether or not there is a shortcut by the North Pacific source from the Mindanao Current (MC) southward along a line east of the Sangihe Ridge to the Maluku Sea, and a circulatory path by South Pacific source to the North Pacific through the equatorial current system (Godfrey 1996), and how much South Pacific source water successfully crosses the Lifamatola Passage, although evidence is building up that South Pacific origin does get into the Halmahera and Maluku Seas (Cresswell & Luick 2001; Luick & Cresswell 2001; Talley & Sprintall 2005). The model results of Zenk et al. (2005) show that South Pacific waters not only feed the Halmahera Sea and Maluku Sea but probably also the southern Sulawesi Sea though the flow is strongly seasonally dependent.

The ITF transport (Figure 3) is based on the estimates by Gordon (2001, 2005). The new estimate from the INSTANT program has not yet been published but suggests a value 20–30% higher. The main throughflow path passes through the Makassar Strait within the thermocline layer with a transport of 8 Sv (1 Sv=10^6 m^3 s^{-1}) with an uncertainty of 1–2 Sv. The continuation of the southward transport is limited by a shallow Dewakang Sill of 680 m. About 1.7 Sv of throughflow water escapes to the Indian Ocean through the 300 m deep Lombok Strait. Most of this water flows eastward to the Banda Sea and enters the Indian Ocean over the deep gaps of the Lesser Sunda Arc on either side of Timor with almost the same amount of transport: 4.5 Sv through Ombai Strait into the Sawu Sea and 4.3 Sv into the Timor Sea. The easterly route contributes to a small portion of 1.5 Sv through Lifamatola Passage via the Halmahera and Maluku Seas, consisting of mainly South Pacific sources. This gives a total mean ITF transport of 10.5 Sv, close to the mean of previous range of 2–22 Sv (Table 2) (Godfrey 1996; Fieux et al. 1994). Improved estimates are possible through recent mooring programs such as Arus Lintas Indonesia (ARLINDO) (Gordon & Susanto 1999). Using the World Ocean Circulation Experiment (WOCE) and other sections, Talley and Sprintal (2005) derive a range of 3–7 Sv for the IIW transport through the eastern route which, for a single water mass, is much larger than the estimate of 1.5 Sv for the entire strait throughflow.

Table 2 Inventory of the Indonesian Throughflow transport estimates.

Method	Transport (Sv)	Author(s)
Geostrophy, top 300 m	1.7	Wyrtki (1961)
Geostrophy, 11 unclosed Indian Ocean sections	10.0	Godfrey and Golding (1981)
Inverse, 43°S and 28°S Pacific section	<<10.0	Wunsch et al. (1983)
Freshwater budget, Pacific and Indian Oceans	14.0	Piola and Gordon (1984)
Tritium budget, top 300 m	5.0	Fine (1985)
Inverse, Timor Strait	7.0	Fu (1986)
Salinity budget, West Pacific	<5.0	Toole et al. (1988)
Current meter, Lombok Strait	1.5	Murray and Arief (1988)
Geostrophy from Levitus data, Australia – Sumatra	12.0	Godfrey (1989)
Inverse, 32°S Indian Ocean	7.0	Toole and Warren (1993)
Heat budget, closed box 14°S–165°E–10°N	0–8.0	Wijffels et al. (1992)
Geostrophy/current meter, Australia – Bali, JADE, August, 1989	18.6±7	Fieux et al. (1994)
Geostrophy/current meter, Australia – Bali, JADE, February 1992	–2.6±7	Fieux et al. (1996)
Geostrophy, XBT time series, peak-to-trough, top 400 m	5.0	Meyers (1996)
Mooring, Timor Strait	3.4–5.3	Molcard et al. (1996)
Tidal-induced residual transport, Seram-east Banda	3.8	Hatayama et al. (1996)
Mooring, Makassar Strait, 1997	8.0±2	Susanto and Gordon (2005)
Current meters, Molucca Sea	7.0	Luick and Cresswell (2000)
Current meters, Halmahera Sea	1.5	Cresswell and Luick (2001)
Current meter, Lifamatola	1.5	van Aken et al. (1988)
Heat, freshwater, oxygen and silica budget, Lifamatola	>3.0	Talley and Sprintall (2005)
XBT, 20-year mean of geostrophic transport, reference to 750 m	8.9±1.7	Wijffels et al. (2008)

Method	Transport (Sv)	Author(s)
OGCM, 1250–2800 m deep	21.0–24.0	
3700 m deep	16.0	Hirst and Godfrey (1993)
OGCM	9–10.5	Inoue and Welsh (1993)
OGCM	20.0+3	Miyama et al. (1995)
OGCM	9.5	Masumoto and Yamagata (1996)

In Table 2, the various estimates of the ITF transport show a number of shortcomings, i.e. they lack temporal and spatial coherence since the data used cover different time periods and depths that lead to ambiguity in the ITF transport estimates. Even the transformation of the thermohaline can be misinterpreted. The aim of the INSTANT program was to resolve these shortcomings and derive a more accurate estimate of the ITF transport. The INSTANT program (Sprintall et al. 2004) is a coordinated endeavour by Indonesia, Australia, France, The Netherlands and United States, deploying 11 moorings in the major straits of the interior Indonesian seas to simultaneously and contemporarily measure the throughflow currents over three years (see Figure 3). The field measurements were completed in 2006. According to Gordon et al. (2006), who reported on the first year and a half years of data, the estimated Makassar Strait transport is about the same as that observed in 1997 during the Arlindo program at 8–9 Sv (Table 2) (Susanto & Gordon 2005). The complete three-year mooring data of INSTANT indicate an average increase of the ITF transport by 25% (Gordon et al. 2008). The ITF across the major straits reveals strong variability with a wide range of temporal scales, marked by month-long periods of significant imbalance between the inflow and export, implying substantial convergence and divergence within the interior seas.

The most recent report of the final results of the INSTANT mooring analysis confirms a 20–30% of increase in the total ITF transport (Gordon et al. 2008). This transport is derived under a relatively weak El Niño – Southern Oscillation (ENSO), suggesting that the mean of the previous individual estimates of the ITF transport could be underestimated. Under a strong ENSO condition, the ITF transport is likely much higher than previously thought. From the water-mass point of view concerned here, INSTANT is a program for measuring transport of the ITF within the

Indonesian straits. The program cannot address water mass content, the Indo-Pacific source waters or their route to the Indonesian seas. One of the aims of PACSWIN is to resolve the water-mass properties of various sources with a significant Pacific origin entering the Indonesian seas, their *en route* pathways, transformation, mixing, variability and constituents in the Indonesian seas.

Another ITF pathway for waters below the lower thermocline of South Pacific origin, the intermediate and deep waters, is derived through the eastern route, via the Maluku (or Molucca Sea) and Halmahera Seas into the Banda Sea (Cresswell & Luick 2001). Since the western side of the Indonesian seas is connected with the eastern Indian Ocean by a sill depth of over 1500 m, an influence of Indian water seems inevitable, especially during La Niña (Molcard et al. 1996). This Indian Ocean reverse flow was observed in recent moorings (Gordon et al. 2006). With careful examination of deep topographic barriers, Gordon (2001) and Gordon et al. (2003) proposed three possible paths of Pacific water into the Indonesian Seas (Figure 3).

Many ocean general circulation models (OGCMs) and coupled atmospheric and oceanic models (AGCM or CGCM) have simulated a closing down of the Indonesian straits resulting in a significant global and regional impact on climate. This kind of experiment treats the ITF as a pipe flow regardless of the horizontal and vertical resolution. Various model results show a positive heat transport of 0.6–0.9 PW (1 PW = 10^{15} Watt) contributed by the ITF to the Indian Ocean (Godfrey 1996; Hirst and Godfrey 1993; Schneider & Barnett 1997; Schneider 1998; Gordon 2001; Vranes et al. 2002). Modelling of the water mass of the Indonesian seas and the ITF is challenging because of the complicated boundary conditions. To resolve the ITF, sufficient resolution of the narrow straits in OGCMs is essential.

The consistent positive heat flux from the Pacific to the Indian Ocean illustrated in ocean models and other previous studies may be misleading, as argued recently by Gordon (2005) who points out that the positive heat flux can only be referred to be the value at the exit point when the water flows out of the Indian Ocean from south of Africa, while the net heat flux is actually negative. Gordon clarifies the concept by stating:

> But what of simpler questions? Does the ITF warm or cool the Indian Ocean? Does the ITF alter the meridional overturning circulation of the Indian (and Pacific Ocean, too)? And, is the ITF important to the climate system? My short answers are, respectively: cools; yes; I think so.

The ITF's heat flux contributes to cooling rather than warming in the eastern Indian Ocean as can be seen in Figure 1 of Gordon (2005) of the upper thermocline temperature and salinity across the Pacific and Indian oceans. You and Tomczak (1993) showed a low temperature and salinity AAMW tongue in the ambient Indian Ocean water in the upper 500 m (see their Figures 2 and 3). Gordon (2005) pointed out that strong tidal-induced mixing by the IMM allows various inflow waters to be mixed into a unique Indonesian seas water, i.e. AAMW. Some of those inflow waters have not been well accounted. For example, the inflow of cool, fresh water (due to precipitation and run-off) from the marginal seas, the South China Sea and the Bay of Bengal via the shallow Gulf of Thailand and the Sunda Shelf was regarded as being too small. Another major source, usually neglected, is the NPIW. The NPIW shoals substantially to 300 m in the Indonesian seas and freshens as well as cools the main thermocline in the Banda Sea, Timor Sea and eastern Indian Ocean. NPIW contributes significantly for about 2.7 Sv of transport to the Indonesian seas (You 2003).

4. Scientific issues and proposal for investigation

The proposed PACSWIN program consists of various components covering broad scientific issues and monitoring techniques. To carry out the ocean climate monitoring in the region, three priority monitoring infrastructures have been defined, i.e. submarine cables, floats and moorings, although other monitoring components are also of interest.

4.1 Priority monitoring programs

4.1.1 Moorings

The objective of the INSTANT mooring program was to better quantify the ITF volume transport and to contribute to a future long term monitoring strategy (Gordon et al. 2006). The 11 moorings (see Figure 3 for INSTANT mooring locations) were first deployed from August 2003 to January 2004 and finally recovered at the end of 2006 with mid-term recovery /redeployment in summer 2005. The first ever contemporary measurement of the ITF from the first half phase of INSTANT shows a similar and slightly larger transport in the western route in the Makassar Strait at 8–9 Sv than the Arlindo but a much larger transport in the eastern route across the Lifamatola Strait at 2.9 Sv compared to 1.5 Sv observed previously (van Aken et al. 1988). As mentioned above, the final results of the three-year INSTANT moorings have indicated about 20–30% increase of the ITF transport under a weak ENSO period (Gordon et al. 2008). With respect to resolving source water mass transport, INSTANT had only one mooring at the Lifamatola Strait. One mooring is insufficient in such a wide and deep strait. There was no mooring deployed to record the inflow of South Pacific water through the Halmahera Sea. Thus, the INSTANT mooring results suggest that a continuing and improved mooring program should be pursued as part of PACSWIN. After INSTANT, there has been only one mooring funded by NOAA to continue the measurement of the ITF at the Makassar Strait (Gordon, personal communication, 2008).

As shown in Figure 5, PACSWIN mooring component retains the INSTANT 11 moorings plus five additional moorings, one more at the Lifamatola Strait, two new moorings between Obi and Lifamatola for measuring the South Pacific source and two new moorings at the entrance for measuring the low latitude western boundary transport. The strategy for new moorings is to better address the South Pacific source water via the Halmahera Sea and the mixed source of North and South Pacific through the Maluku Sea. Since there is a deep strait between Halmahera island and Lifamatola (over 2000 m deep), moorings are placed in a deep inlet between Obi and Lifamatola rather than between Halmahera and Sorong. Such a mooring distribution would not only provide comparable transport with INSTANT moorings but better resolve water mass transport in the eastern route for the South Pacific source through the Halmahera Sea and mixed source through Maluku.

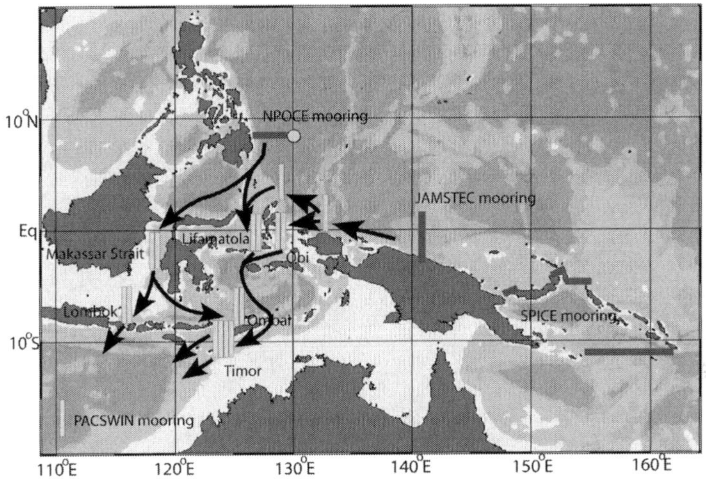

Figure 5 The proposed PACSWIN moorings component and its bridging with NPOCE, JAMSTEC and SPICEmooring programs.

4.1.2 Submarine cables

A submarine cable is a cost-effective tool to measure ocean current transport. The ITF flowing through a strait or channel above a cable acts as a conductor cutting the vertical lines of the Earth's magnetic fields and thus generating a horizontal electric field that can be measured as motion induced voltage (MIV) from the cable. The theoretical development of electromagnetic fields for oceanographic application was first advanced by Stommel (1948). For monitoring of the ITF in a particular strait, only a voltage recorder and a computer are needed for net transport of the ITF except for calibration costs. Note that the cable base and ground facility are very expensive. Cables can satisfy both present and future needs for implementing a long-term ocean climate monitoring of the ITF and a global network in generally supplementing the Argo array. Indeed, the technique complements the Argo program, which leaves most straits and marginal seas and water deeper than 2000 m unobserved, as, for example, within the Indonesian seas.

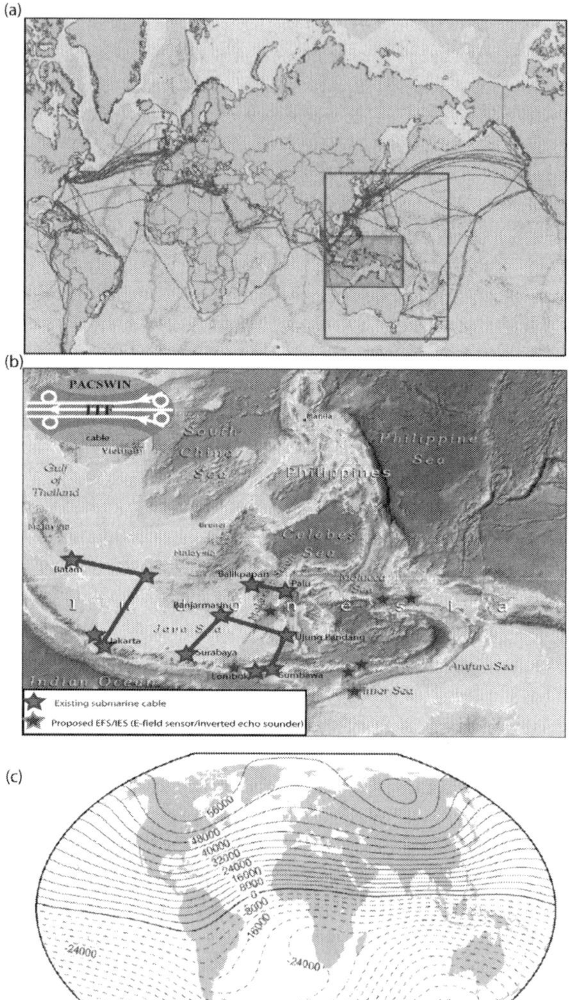

Figure 6 Maps showing a) the global submarine cable route (by courtesy of Alcatel-Lucent), b) the existing submarine cables in the Indonesian seas (solid line with black star) and proposed EFS/IES

(E-field sensor/inverted echo sounder) for the straits without cables (grey star) for undertaking the same job (Sanford, personnel communication, 2007) and c) the global vertical magnetic component (www.geomag.bgs.ac.uk/earthmag.html). As seen, the magnetic equator (zero line) lies well north of the Indonesian seas existing cables at Mindanao. This means that submarine cable measurement of the ITF should be a valid exercise.

A submarine cable has been utilised to monitor the Florida Current transport since the 1980s and others have been used elsewhere, including in the region of the Kuroshio between Taiwan, Luzon and Okinawa as well as at the Izu Islands; the Tsushima Current in the Korea Strait; transports between Gran Canaria and Tenerife; in the Baltic Sea and in channels on the Portugal continental shelf. However, so far, the application of a submarine cable to current measurement has been carried out for only a minor portion of the existing cable networks (see Figure 6a) and, as yet, the technique has not been given enough attention by the ocean climate community.

The western Pacific includes many of the marginal seas and straits of the world's oceans. Thus, submarine cables could be a useful tool for the region, especially in the Indonesian seas. It is very costly to maintain a set of moorings for monitoring the ITF continuously for timescales longer than a decade and, indeed, there is no guarantee of funding for such long-term monitoring. However, a submarine cable can satisfy the requirement for monitoring the ITF continuously for a century or longer. Several submarine cables cross the ITF passages (see Figure 6b). In particular, two cable routes cross the Makassar Strait, a main ITF pathway. PACSWIN has chosen submarine cables as one of its priority monitoring infrastructures along with moorings and floats. The cable routes in the Indonesian seas avoid the magnetic equator which lies at Mindanao (see the zero line in Figure 6c).

A first submarine cable workshop was endorsed by CLIVAR on 23 April, 2008 and held in Taipei on 9–10 September 2009. The proposed workshop will provide an opportunity to update the community and to review and promote the technique. It will enable exchanges to take place among users and the opportunity to share the technique with non-users. Since meaningful current transport measurements from cables depend upon accurate determination of parameters such as tides, sediment condition, local magnetic field and intercomparison of real current measurements, the workshop will standardise the methodologies to be used and intercomparison technology. Another task of the workshop is to design the PACSWIN cable component using the existing Indonesian seas cables (Figure 6b) for cost-effective long-term monitoring of the ITF.

On the other hand, a new technique has been developed recently to combine inverted echo sounder (IES) with E-field sensor (EFS) (Tom Sanford, personal communication, 2007) and will be available for those straits without existing cables. The workshop will provide technical support and design for the PACSWIN cable component and also set a road map for building a global network and open-access data bank.

4.1.3 Floats

Argo

Argo started in 1999 and by November 2007 had reached its target of 3000 floats. Currently, Argo has 3325 floats as of 15 March 2009 (see Figure 2). Argo floats are useful in PACSWIN for tracking the North and South Pacific source waters and the pathways on the Pacific side entrance of the Indonesian seas. In the eastern Indian Ocean, Argo will help tracing the components of the ITF as transformed Pacific source waters spread in space, depth and time. The existing Argo floats are ready to use for the purpose of PACSWIN in a broad equatorial current system. However, Argo has limitations for inshore currents within 100 km from the coast since Argo floats spend 5–10 hours on the surface transmitting data and may be swept up on shore or caught in strong near-shore currents. Another problem may be relevant to marine jurisdiction permission and EEZ (Exclusive Economic Zone) sovereign rights. In this latter context, Argo floats for Lagrangian tracing of Pacific source waters may be released offshore while inshore measurement is implemented by other means such as gliders and hydrographic survey sections applying a CTD.

Gliders

Autonomous underwater gliders (Davis et al. 2002) operate in a depth range of 200 to 1500 m over time periods from a few weeks to several months, and measure current velocity and water properties.

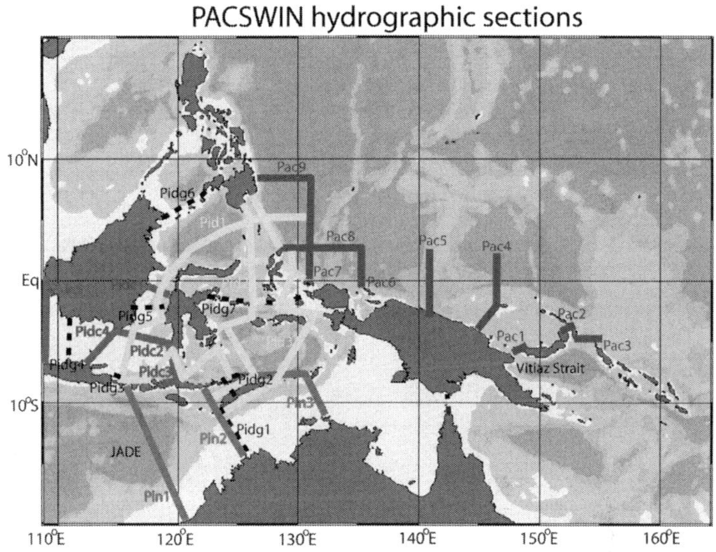

Note:1. Pac1-9: hydrographic sections for the Pacific inflow and Pin1-3: Indian outflow waters
2. Pid1-3: hydrographic sections within the Indonesian seas linked to inflow and outflow sections
3. Dashed line: gliders missions (Pidg1-7)
4. Pidc1-5: intercalibration section for submarine cable

Figure 7 The PACSWIN hydrographic sections including the survey sections at the Pacific side entrance and Indian side outflow regions.

Several glider sections are planned for the main straits of the Indonesian seas and north-west Australia (see dashed line in Figure 7). A glider's (Figure 8) position is located by GPS at the surface where data are transmitted to satellite and instructions are provided for the glider's continued mission. This new type of instrument has been successfully used around the world's oceans. As mentioned above, gliders may also be used to measure the inshore currents along the Pacific survey sections Pac1 to Pac9. However, since glider speeds are about 25 cm s^{-1}, they are perhaps too slow to overcome the strong, low latitude western boundary currents such as those currents through the Vitiaz Strait. In sections with strong boundary or throughflow currents, other measurement tools such as ADCP, LADCP and current meters have to be used.

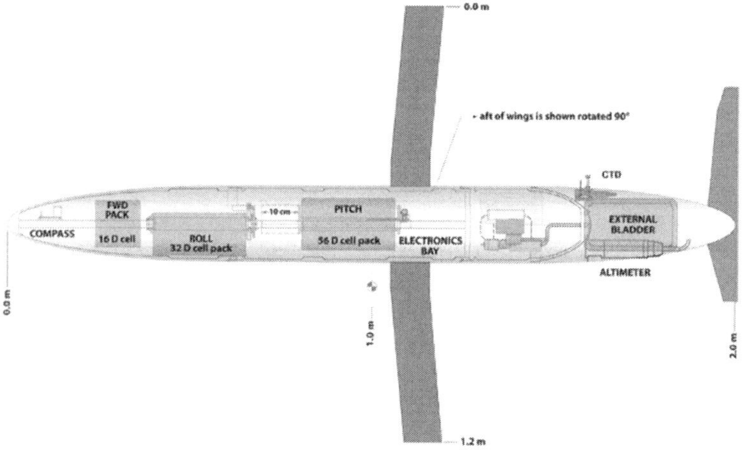

Figure 8 Schematic display of the main components of the Spray glider. The 6061 aluminium pressure case is made in three pieces and houses batteries, hydraulic system, compass, attitude sensor, GPS receiver, Iridium transceiver and the microprocessor controller. An aft flooded bay houses the external bladders that expand to increase buoyancy. This bay also houses the sensor suite selected for each mission. Sensors that have been used include: Precision Measurement Engineering CTD; Sea Bird 41 CP CTD with seawater pump; Sea Point Optical Backscatter Sensor; Sea Point Chlorophyll Fluorometer; Tritech PA200 acoustic altimeter for bottom avoidance; and Sontek Argonaut-SGP 750-kHz Acoustic Doppler Current Profiler (see spray.ucsd.edu or Davis et al. 2002 for details).

Lagrangian tracking technique of Pacific source waters

The application of Lagrangian techniques is one way to fill the ARGO gap (Figure 2) and to overcome the spatial problem of Eulerian observations. Surface drifters with deep drogues are one possibility for near-surface current observations. Unfortunately, they often become the subject of vandalism especially in regions in reach of archipelagos. Sub-surface floats can also be used as roving current meters. They are much more protected against severe weather conditions and human interference. This is particularly important for the Indonesian seas.

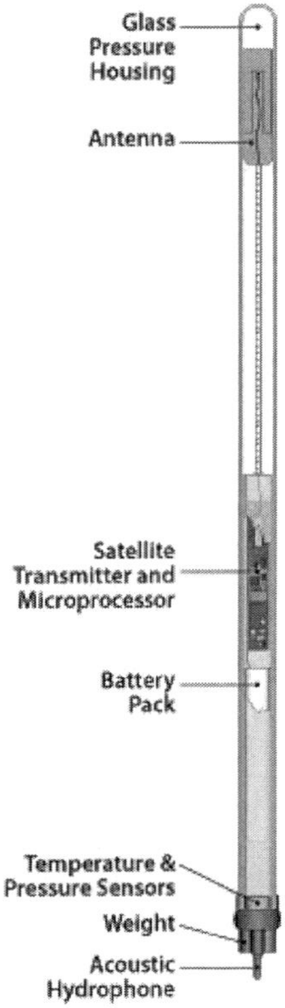

Figure 9 A typical RAFOS device structure. While freely drifting at a pre-ballasted pressure level, this eddy resolving subsurface drifter is tracked acoustically with daily resolution. The glass housing is about 1.5 m long. After a pre-programmed time, the instrument drops a weight and returns to the surface. Since its upper half is then exposed above the sea surface, it can upload recorded data via satellite transmission.

RAFOS floats (Rossby et al. 1986) are acoustically tracked devices (Figure 9), which can be used to track the source water pathway in the western equatorial Pacific. Since the core of the ITF is supposed to lie in the main thermocline, sub-surface floats appear more suitable than surface drifters. Once an array of RAFOS sound sources has been installed, an unlimited number of RAFOS floats can be deployed and be part of the system (cf. a radio station on land with any number of receivers). RAFOS floats are about three times less expensive than Autonomous Profiling EXplorer (APEX) etc. floats. This brings up the issue of the number of sound sources to be moored. In principle, two moorings are sufficient for an ocean basin, three are better. Since the western tropical Pacific is divided into several basins, the number of sound sources depends critically on the suspected region of the source waters of the ITF. A RAFOS sound array can be also used for Eulerian observations at strategic locations (mooring of opportunity). The RAFOS technology has been successfully applied to track the AAIW paths in the Bismarck Sea by Zenk et al. (2005). Lagrangian observations with eddy resolving RAFOS floats and extended model simulations at the source region of this ITF branch have identified a region of enhanced mixing at an intermediate level of 27.2–27.3 σ_θ (~580–800 m depth) in the equatorial western Pacific. The apparently increased eddy kinetic energy (EKE) may play a major role in the preconditioning process of this ITF source, augmented directly from the South Pacific. It reflects strikingly the convergence zone of the New Guinea Coastal Undercurrent (NGCUC) compared with the Equatorial Intermediate Current (EIC). Analogous particle tracking experiments with monthly resolution in this inter-hemispheric source region demonstrate great seasonal differences. Towards the Maluku and Sulawesi seas, an enhanced outflow from the south Pacific seems to develop in late summer (Figure 10). The pathway through the Halmahera Sea is probably of secondary importance (Zenk et al. 2005).

Up to 200 RAFOS floats would be released at the eastern entrance of the Indonesian seas on the Pacific side during the PACSWIN operation (Figure 11), which should be adequate to track the various Pacific source waters at different density levels.

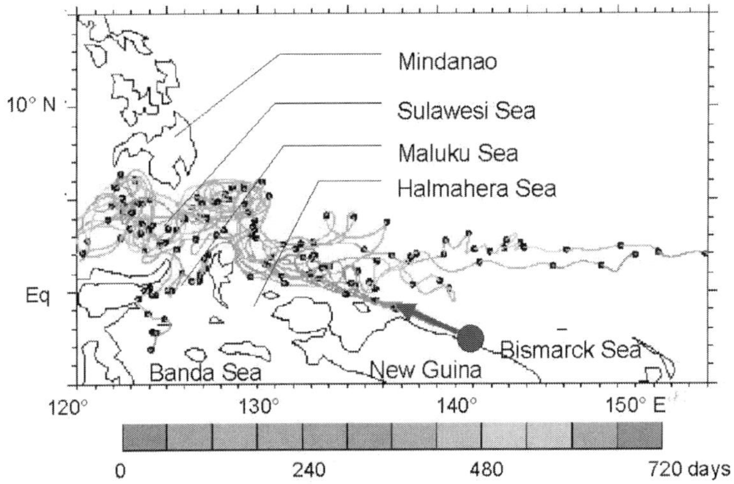

Figure 10 Map showing trajectories from a cluster of model floats all started at the same location in the Bismarck Sea (a big dot). These model markers were injected in the New Guinea Coastal Undercurrent at the level of the Antarctic Intermediate Water. The grey colour code applied to the drift paths (see grey colour bar) indicates the elapsed drift time from the start. Observed RAFOS tracks were reasonably reproduced by the model path ways shown (Zenk et al. 2005).

4.2 Other monitoring programs

4.2.1 Hydrographic survey

The last systematic hydrographic survey in the Indonesian seas was conducted 15 years ago during the ARLINDO program in 1993–94. However, the data measured were mainly CTD/CFC but not nutrients (Ilahude & Gordon 1996; Gordon & Fine 1996). Nutrients and other biochemical tracers are very much needed for PACSWIN in order to identify various Pacific source waters. In the past, nutrients were poorly sampled in the Indonesian seas.

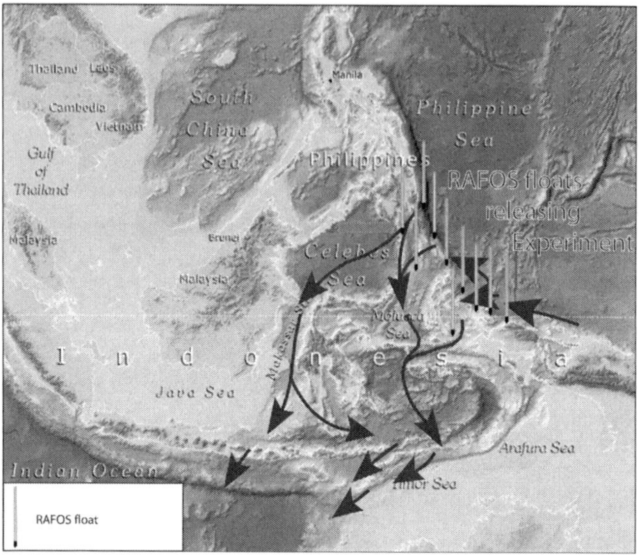

Figure 11 PACSWIN RAFOS component floats releasing experiment. About 200 floats will be released at the Pacific entrance to track the various source waters of Pacific origin in the ITF.

The hydrographic survey sections are designed for several parts (Figure 7). The Pacific side consists of Pac1 to Pac9 for tracing the North and South Pacific source waters via the Vitiaz Strait, St George channel and New Ireland to Solomon Islands and east of Mindanao. In the Indonesian seas, the survey sections follow roughly the pathways of the North Pacific source in the western route (Pid1), the South Pacific source in the eastern route (Pid3) and the mixed North and South Pacific sources in the middle route (Pid2). These sections align mostly with the Indonesian sea lane (Figure 12). Other sections in the Indonesian seas are the repeat survey along the submarine cables for intercalibration (Pidc1–4) and a glider mission at critical straits/channels (Pidg1–7 in dashed line). In the outflow region in the eastern Indian Ocean, three sections west and east of Timor and at JADE (Pin1–3) are perpendicular to the ITF. These sections can further trace the transformed Pacific source waters to the eastern Indian Ocean through the Indonesian seas.

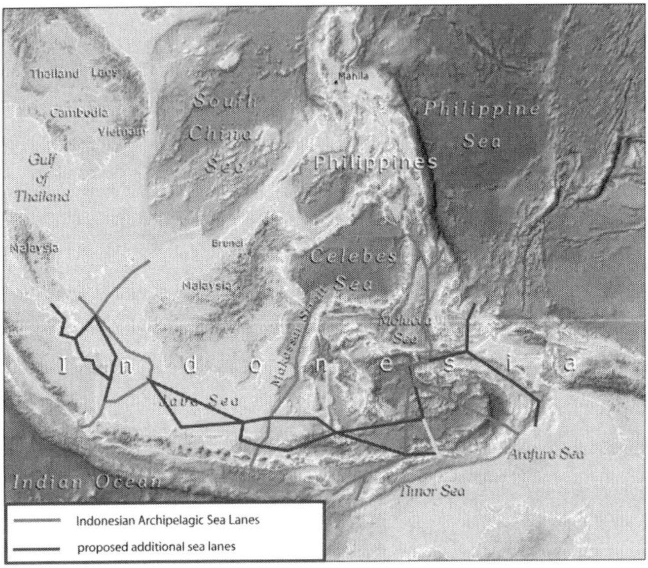

Figure 12 The present Indonesian archipelagic sea lanes (grey line) and proposed sea lanes in the near future (black line).

Except for traditional surveys by CTD and ship mounted ADCP, various nutrients, CFC and biochemical tracers will be sampled with Rosette bottle samplers such as dissolved oxygen, phosphate, silicate, nitrate, CFC11, CFC12 etc. These surveys will be conducted in each season for the first year and repeated for another year if possible.

4.2.2 ADCP/LADCP: using commercial freight ships for long-term measuring of the ITF and variability

It has long been practice for oceanographers to use commercial freight ships installed with a downward-looking acoustic Doppler current profiler (ADCP) mounted to the hull to routinely measure currents. One of the success stories is the Oleander project started in 1992 with weekly round trips between Elisabeth, New Jersey and Hamilton, Bermuda (Flagg et al. 1998) (Figure 13). To facilitate the data collection and dissemination, an autonomous data collection system called AutoADCP has been developed to allow wireless internet data connection as soon as a freight vessel arrives ashore. This important upgrade makes the monthly trips to the ship to collect the data unnecessary. The time period of data availability is shortened from over a month to a week or so. The new data processing system handles different data formats.

Figure 13 The Oleander ADCP project between the port of Elisabeth, New Jersey and Hamilton, Bermuda since 1992.

Using the technique in the Indonesian seas and for ITF monitoring strategy is effective because there are many regular container and freight ships navigating through the Indonesian archipelago, along and across the ITF. The Indonesian archipelagic sea lanes can be used by international freight ships with ADCP installed to measure the ITF regularly. On the other hand, Indonesia is the largest archipelagic country in the world. There are numerous passenger shipping/ferry lines in the Indonesian archipelago (Figure 14a), which are potentially large resources for installing ADCPs for long term climate monitoring.

In Figure 14b, only a portion of the Indonesia domestic container shipping lines from Jakarta and Surabaya are presented. The adjacent Darwin link with Asian ports is a perfect example for using the Oleander project model. We haven't shown other numerous international freight shipping lines, especially for the western Pacific and Asian countries. It can be anticipated that an international network of AutoADCP can be established to support the simultaneous monitoring of the ITF. ADCP measures currents down to below 400 m, which covers a depth range of the thermocline and NPIW in the Indonesian seas. Since the major source waters in the ITF are derived from the North Pacific, the depth counts for at least 80% of the total ITF transport. In conclusion, the commercial ship ADCP is considered a cost-effective long-term means for monitoring the ITF. This new technique is under consideration by SCOR/IAPSO working group (known as SCOR WG#133 or OceanScope) for exploring ways of developing a new paradigm for working with the merchant marine shipping industries (Rossby, personal communication, 2009).

4.2.3 Expendable Bathythermographs (XBT)

The above subsection calls for the traditional Expendable Bathythermographs (XBTs) to be considered by PACSWIN. The XBTs are devices to measure the temperature of the upper ocean using moving vessels including the aforementioned merchant vessels. XBT can measure temperature down to 800 m. Combined with the ADCP/LADCP, one can obtain a time series profile of velocity and temperature and the transport and heat flux in the upper ocean.

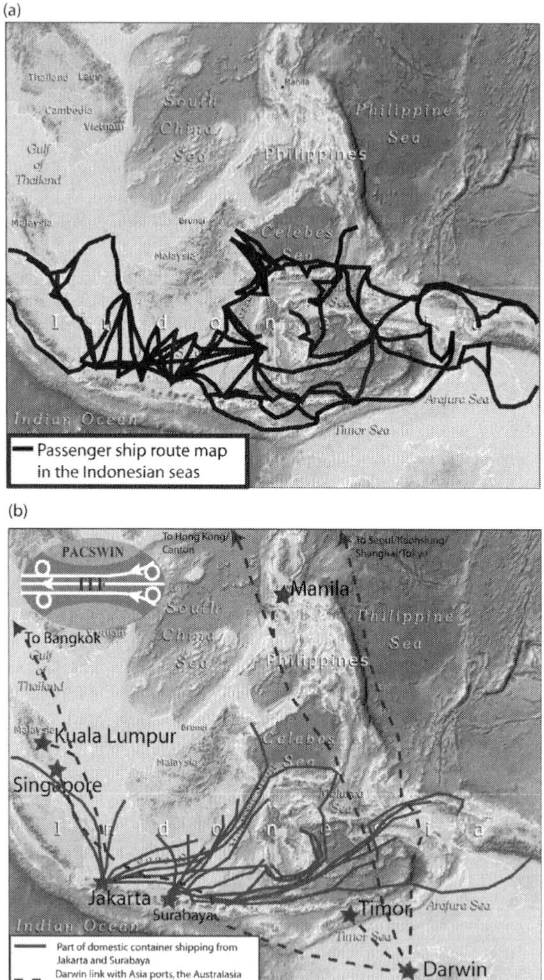

Figure 14 Maps for a) the passenger ship routes in the Indonesian seas and b) part of the Indonesian domestic container shipping lines from Jakarta and Surabaya (grey line) and the Darwin link with Asia ports, the Australia express for cargo, iron ore, oil and gas (dashed line).

Since TOGA in 1983, CSIRO initiated an XBT network in the Indonesian seas and adjacent region. After 1997, the project was jointly supported by CSIRO, the Joint Australian Facility for Ocean Observations (JAFOOS) and Australia's Bureau of Meteorology (BOM). Currently, there are three frequent XBT lines (12–18 times per year) across the ITF from Australia to the Indonesian shelf: one from Perth to Java (IX1), a second one that is a quasi zonal line from Java to the Banda Sea in the east (PX2) and a third one from Port Hedland to the Banda Sea, Maluku Sea and then east of Mindanao (IX22). These XBT lines have been running for 25 years and have provided the only long time series of information on ITF transport and other water-mass characteristics. Recently, Wijffels et al. (2008) estimated a 25-year mean of ITF transport from these XBT lines above 750 m of 8.9±1.7 Sv. The XBT program satisfies the target of PACSWIN pursuing a cost-effective long-term climate monitoring of the ITF. PACSWIN will use vessel opportunities to continue the Indonesian seas' XBT sampling.

4.2.4 Satellite altimetry and scatterometry

With their superior spatial coverage, satellite observations (e.g. altimetry and scatterometry) are complementary elements to the in situ elements of the PACSWIN program. Dominant variability of the ITF and related marginal seas are primarily driven by wind forcing, which can be monitored by scatterometers (e.g. SeaWinds on QuikSCAT and ASCAT on Metop). Scatterometer-derived wind measurements can be used to estimate near-surface Ekman transport as well as to diagnose signals in the thermocline or deeper layers in response to Ekman pumping. The latter not only generates local variability in the ITF region but can cause perturbations in the Pacific and Indian oceans that propagate into the ITF region to affect property transports. Oceanic responses to local and remote Ekman pumping have signatures in the variation of sea surface height (SSH), which can be measured by altimeters (e.g. JASON-1, ENVISAT, GFO). Altimetry data will be used to determine the Indonesian seas' sea-level and ocean-bottom-pressure (OBP). Song (2006) and Qu and Song (2009) recently proposed a method that allows the use of satellite-observed sea level and OBP for estimating the transport through a strait. Although the method has its limitations due to the simplified formula and the satellite data resolution, it may provide a useful upper bound of strait transport in the Indonesian seas. Therefore, satellite altimetry and scatterometry provide important sources of information to help diagnose the physics of circulations in the ITF region, for instance, to evaluate the relative role of local, Pacific and Indian Ocean forcing. Moreover, satellite observations can help interpret in situ observations and constrain ocean model and assimilation systems.

4.2.5 Process monitoring components

Other monitoring components of PACSWIN are process studies including 1) marginal sea effects, 2) biochemical tracers and 3) tides and tidal mixing, which are briefly described here but will be detailed in an operation plan. These components require some special measurements and monitoring. For example, the marginal sea input of mass and property through the Sunda shelf is monitored through the Kalimantan Strait section by hydrographic surveys, gliders and cables. Monitoring of biochemical tracers is through hydrographic survey and ship cruises from Rosette bottle samplers. Monitoring of tides and tidal mixing is realized through tidal gauges, tidal stations and the high resolution of CTD surveys. These components have an important impact on the ITF and associated variability in transport, heat and fresh water fluxes and ecosystem. They resolve some concerns of processes relevant to the ITF. The South China Sea (SCS) and Celebes Sea are known to have a great impact on the ITF, particularly during the winter monsoon. Qu et al. (2006) estimated that up to 0.2 PW of heat and 0.1 Sv of freshwater are transported into the Indonesian seas from the SCS. Gordon (2005) conjectured up to 1 Sv of total marginal sea input to the ITF during the boreal winter monsoon. Though the volume transport of the SCS throughflow is small in comparison with the ITF, its impact on the Pacific-to-Indian Ocean heat transport is hypothesised to be large (Qu et al. 2005). Tozuka et al. (2007) recently conducted a numerical experiment and provided a better insight into this hypothesis. The results demonstrated that the SCS throughflow reduces the ITF heat transport by as much as 47% in the model, thereby playing a potentially important role in regulating the SST pattern in the tropical Indo-Pacific Ocean.

A project called South China Sea – Indonesian Seas Transport/Exchange (SITE) has been carried out, a collaboration between the LDEO-Columbia University, BRKP-Indonesia and FIO-China. Five moorings (three moorings at the Kalimantan Strait and two moorings in Sunda Strait) were deployed in December 2007 and November 2008 to measure the flow between the SCS and Java Sea and Indian Ocean (Susanto, personal communication, 2009). Measurement of the SCS throughflow via the Kalimantan Strait and Celebes and Sulu Seas is considered in the hydrographic survey sections and glider mission (Figure 7).

The Arlindo CFC survey showed stratification of CFCs in the vertical structure of the Indonesian seas' water column, consistent with the thermohaline stratification of the water mass (Gordon & Fine 1996). Spatial and vertical fluxes of biochemical tracers are crucial to the study of the Indonesian seas' ecosystem. On the other hand, without understanding the physical process of the ITF system, many ecosystem phenomena are difficult to interpret. This is because the Indonesian seas ecosystem is a consequence of many environmental conditions such as ITF, tides/tidal mixing, upwelling/downwelling, heat/fresh water fluxes, the monsoon and ENSO. The large marine ecosystem is considered a Class II, moderately high (150–300 grams of carbon m^2/year) productivity ecosystem based on SeaWiFS global productivity estimates (NOAA 2008). The Indonesian seas' ecosystem is vulnerable to environmental change. Human activity and pollution as well as climate change are threatening the Indonesian seas' ecosystem. This topic, as a component of PACSWIN, could be better addressed under high priority monitoring components.

Tides and tidalmixing play a crucial role in the Indonesian Mixing Master (Gordon 2005). Various source waters from the Pacific and marginal seas are mixed and blended by tidal forcing into a single Australasian Mediterranean Water (AAMW) (You & Tomczak 1993) feeding the Indian ocean thermocline and intermediate water masses. Tides and tidal mixing in the Indonesian seas are most complicated. Tides from the Pacific and Indian oceans meet in the Indonesian seas, interact with the intricate bottom topography, numerous islands, deep and shallow straits, shore lines and shallow shelves inducing barotropic and baroclinic tides as well as small-scale internal tides (Robertson & Ffield 2008). The monitoring component of PACSWIN on tides and tidal mixing will rely on existing tidal gauges, the aforementioned mooring observations and high resolution CTD and XBT data along the hydrographic sections (Figure 7).

4.3 Scientific studies

4.3.1 Lagrangian characterisation of archipelago strait dynamics

In addition to the observational phase for tracing the Pacific source waters with Lagrangian techniques proposed in section 4.1.3.3, here we propose a comprehensive study of archipelago strait dynamics for PACSWIN using combined Lagrangian and other observational tools and laboratory experiments. We aim (1) to collect and analyse novel and exploratory Lagrangian velocity and stratification measurements to characterise the time- and space-dependent mesoscale variability in and around two specific Indonesian straits (Figure 15) and (2) to synthesise these observations with idealised laboratory simulations to elucidate the underlying dynamics and to facilitate assessment of more complex numerical simulations and predictions. Our goal is to determine not only what makes each strait unique but also to identify fundamental processes and dynamical regimes with relevance to other locations.

The proposed study focuses on the meso- and submesoscale circulation within two structurally different but adjacent Indonesian straits, Makassar and Lombok, as well as the region between these straits and the areas immediately up- and downstream of each. The major study will focus on three main topical areas. These are, in order of emphasis:

1) Exploratory measurements of horizontal flow patterns, including detailed descriptions of meso- and submesoscale variability and investigation of the responsible dynamical processes
2) Measurements of the temporal and spatial variation of upper-ocean stratification and horizontal velocity shear using temporal sampling resolution appropriate to the detection of internal waves and solitons

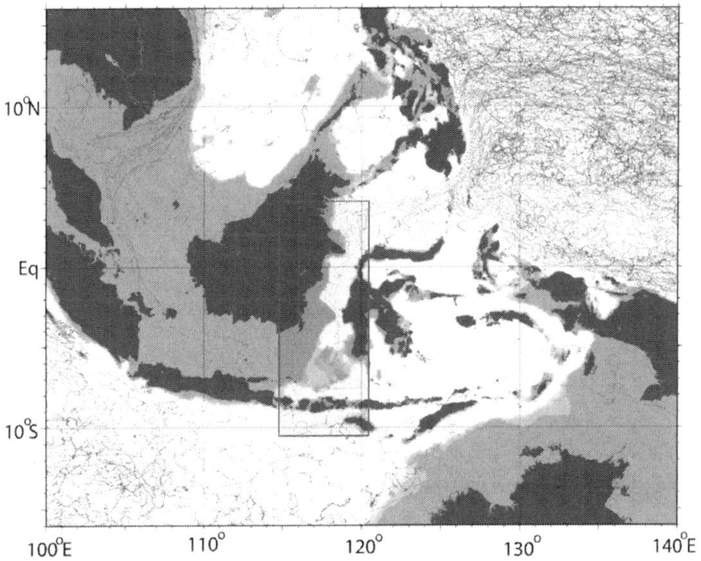

Figure 15 Display of all existing surface drifter trajectories in Indonesian waters retrieved from the NOAA global drifter archive. The proposed study area encompassing the Makassar and Lombok Straits is highlighted in the square. Water depths less than 800 m are shaded.

3) Estimates of low-frequency acoustic transmission loss and relationships to mesoscale variability, coherent features and topography.

In addition to providing unparalleled horizontal and temporal sampling coverage, Lagrangian measurements are uniquely capable of addressing fundamental questions regarding intrinsic time and length scales, the influence of topography on both the mesoscale and low-frequency circulation, and the relationship of coherent structures (eddies, filaments, etc) to upstream dynamic conditions. Float and drifter trajectories, either singly or in clusters, provide access to estimates of horizontal or isopycnal dispersion and may be directly compared with and/or assimilated into numerical ocean models. We propose to focus our measurements on the time-varying mesoscale velocity field, including explicit resolution of the semidiurnal tide. A subset of the proposed instrumentation will also permit measurements in the vertical dimension, including upper-ocean stratification and burst sampling of isopycnal displacements.

The proposed instrumentation to be employed includes a total of 270 satellite-tracked surface drifters, acoustically tracked RAFOS floats (both isopycnal and isobaric) (Rossby et al. 1986) and acoustically-tracked bobbing profiling floats (bobbers). Lagrangian measurements will be obtained from at least three vertical levels in order to roughly characterise vertical shear and its time dependence. For example, Murray and Arief (1988) found significant vertical shear north of Lombok Strait, with surface, thermocline and deep velocities all exhibiting substantially different character and intensity. A subset of the drifting instrumentation will be deployed as a function of time and/or in response to environmental cues using several Submerged Autonomous Launch Platforms (SALPs) installed on moorings in the upstream portion of the straits.

We propose to deploy 75 standard 15-m drogued GPS/Argos surface drifters from Indonesian vessels during the experiment. While the specific deployment plan will depend on the activities of our collaborators, we anticipate deploying a majority of these drifters upstream of sills, in sidewall boundary layers and/or in coherent features identified during synoptic CTD or towfish surveys. An additional 75 GPS/Argos drifters specifically designed for subsurface deployment are proposed to be launched from three SALP moorings located upstream of each of the two constrictions in Makassar Strait and north of Lombok Strait. Each of the SALP moorings will carry an acoustically linked upward-looking ADCP so that drifter launches can be keyed to specific flow regimes and/or seasons without dependence on ship availability. All of the drifters will report hourly positions and surface temperatures over a lifetime of 9–12 months.

An array of 10 acoustic source moorings must be installed to permit tracking of 100 RAFOS floats and 20 profiling bobbers. Only acoustic tracking will allow resolution of the full range of time and space scales expected to be important in and near the narrow straits. The majority (70) of the floats will be isopycnal (i.e. constant density) floats ballasted to remain in the centre of the main thermocline near 13°C (approximately 250 m). An additional 30 isobaric RAFOS floats are proposed for measuring the horizontal circulation at a depth of 800 m (shallower than the controlling sill in the Makassar Strait). To resolve the semidiurnal tide, acoustic positions will be obtained eight times per day. Based on recent work in the Luzon Strait (e.g. Ramp et al. 2004) we expect internal waves to be generated by the interaction of strong tidal or mesoscale velocities with topography. The isopycnal floats will measure temperature and pressure hourly and provide estimates of thermocline excursions associated with internal waves and solitons. To improve temporal resolution, we will burst sample T & P at approximately one minute intervals several times per day. The bobbers will alternate between acoustically tracked isopycnal drift (using active ballasting) and rapid 0–500 profiles of temperature and salinity. These floats will yield unique records of temporal and spatial variability including the Lagrangian evolution of potential vorticity.

In parallel with the observational program, we will conduct idealised laboratory modelling in order to identify the fundamental dynamics governing inertial flow through archipelago straits. The laboratory results will bring dynamical insight on a wide range of straits, including but not limited to those addressed by this proposal. In fact, the laboratory modelling will provide a dynamically consistent means to transfer the understanding gained during this field program to the multitude of other strategically relevant archipelago straits in the western Pacific and globally. The laboratory experiments will be highly quantitative, using both particle tracking and particle image velocimetry. Both stationary and rotating tank experiments will be performed. For this effort, three nondimensional parameters will be varied: the ratio of the channel width, W, to the Rossby radius of deformation, R_d, the Froude number, Fr, and the Reynolds number, Re. The external parameters controlling the ratio W/R_d will be varied to examine three regimes: $W \ll R_d$, $W \approx O(R_d)$ and $W \gg R_d$, and, by varying the velocity of the inertial flow through the strait, we will span different values of Fr and Re. We hypothesise the occurrence of at least three regimes: laminar (low Re and Fr), hydraulically controlled (for $W \ll R_d$, $Fr > 1$, and low Re) and mesoscale eddies (for $W \ll R_d$ and large Re).

Laboratory models, while highly complementary with numerical simulations, are particularly desirable in this instance because they use *real* fluids, with known viscosities and diffusivities. There is a true no-slip boundary condition at the bottom, and dynamics are fully nonlinear with no approximations involved. Fully turbulent flows with large Reynolds numbers are possible, with all scales of motion (within the tank) fully resolved. Direct comparison with in situ observations will be possible using synthetic floats in the laboratory. We anticipate that such comparisons will both aid in the interpretation of the field results and help guide observational strategies.

4.3.2 Other topical studies

Other major topical studies related to the ITF include 1) low latitude western boundary currents, 2) the equatorial current system, 3) ENSO and the 4) monsoons. The low latitude western boundary currents (LLWBCs) are defined as those western boundary flows that are directly connected to the strong wind-driven zonal flows of the tropics and an important component of the equatorial heat and mass budgets, contributing to fraction of the cross-equatorial transport of mass, heat and salt. Thus they are an important element of the global climate system (Lukas et al. 1996). Since the major sources of North and South Pacific origin are carried by the equatorward LLWBCs, the Mindanao Current and New Guinea Coastal Undercurrent (NGCUC), it is particularly important to study these currents, transport and water-mass properties from the hydrographic survey along the Pacific sections Pac1 to Pac9. In addition to physical properties such as temperature and salinity, nutrients and biochemical tracers will also be used to identify the source waters. Analysis of water-mass property in the LLWBCs' water will distinguish different sources. Hu et al. (1992) discovered the Mindanao Undercurrent, which may imply the pathway of the cross equatorial South Pacific source waters. However, its northward intrusion in the western North Pacific is an issue of debate. The open question as to whether or not the South Pacific sources cross the equator northward may help to identify the pathway of South Pacific-originating water along the middle route (Figure 7).

The LLWBCs are connected with the equatorial current system as seen in Figure 1. The North and South Pacific sources brought to the far western equatorial Pacific by LLWBCs are regulated by two prominent eddies, the Mindanao and the Halmahera Eddy, which draw the source waters eastward to the equatorial interior in the North Equatorial Counter Current. It is unknown how the eddy and equatorial current system controls the source waters of North and South Pacific origin, partly effectively entering the Indonesian seas and partly removed away from the entrance into the equatorial interior. In the eastern Indian Ocean, the blended Indonesian water (AAMW) is mainly contained in the westward South Equatorial Current (Figure 1). However, it is unclear how the southward Sumatra-Java Current interplays with the outflow.

It is well understood that the ITF contains an ENSO signal showing an interannual variability with a phase of the ENSO cycle. During El Niño, ITF transport is usually smaller and larger during La Niña. One of the obvious reasons is that sea level during El Niño is lower than that during La Niña affecting the Indo-Pacific geopotential levels. Gordon et al. (2008) observed a strong ENSO suppression on the Makassar Strait throughflow resulting in a 2004–06 transport and temperature of 27% and 1°C higher than 1997 at 11.6±3.3 Sv and 15.6°C. During INSTANT, ENSO is weak from a transition of the El Niño phase to the La Niña phase. However, the simultaneously measured ITF transport by main strait moorings show a three year mean 20–30% higher than the previous averaged value based on individual estimates from various means (Gordon et al. 2008, unpublished manuscript). For a La Niña suppressed phase, the total ITF transport could be expected to be much higher. Nevertheless, the INSTANT result urges an ENSO monitoring component and the PACSWIN program in general.

The dominant seasonal forcing in the Indonesian seas and adjacent region are the monsoons since the region lies in the Intertropical Convergence Zone (ITCZ). The Indonesian seas are covered by the Asia-Australia monsoon with a much stronger winter monsoon than summer monsoon. The winter (north-west) monsoon can force the surface currents of the ITF to reverse (Gordon 2005). The imbalance of the monsoon forcing induces intraseasonal variability of the upper ITF transport. The most significant impact of monsoon forcing on the ITF is a winter monsoon that drives cool/fresh water out of the marginal seas into the Indonesian seas and intensifies mixing. In the boreal summer, the cool/fresh Indonesian seas water flows back to the South China Sea and Sulu Sea. The total cool/fresh water transport from the marginal seas contributing to the ITF is conjectured as roughly 1 Sv (Gordon 2005). Monitoring of the monsoon as one of the PACSWIN components is important not only to the ITF volume transport but also for heat and fresh water fluxes. The PACSWIN observational programs outlined above should satisfy the need for monsoon monitoring.

4.4 Modelling of the ITF and its role in global climate change

There have been many attempts to estimate the magnitude of the volume and heat transport associated with the ITF and its pathways within the Indonesian archipelago (see reviews by Godfrey 1996; Gordon 2001). Results demonstrate a wide spread of the values even for the mean magnitude, with significantly large amplitude variations at the seasonal and interannual time-scales. The difficulty in simulating the ITF lies in the complex geometry and bottom topography in the region and the strong mixing generated by tidal currents. For example, with a 1/6 degree resolution global Parallel Ocean Program (POP) model, Gordon and McClean (1999) found that, due to the sill depth being almost 400 m shallower in the model than observed in the Makassar Strait, only the upper 250 m (sill depth in POP) current was correctly simulated. Using a recent higher-resolution ocean general circulation model (OGCM), called OFES, Masumoto et al. (2008) indicate that magnitude of the ITF is reasonably simulated with a mean transport of about 11 Sv. However, they also demonstrated that the water mass properties within the Indonesian archipelago are quite different from those observed due to the lack of a tidal mixing parameterisation in the OFES. A recent modelling effort by Koch-Larrouy et al. (2008a, 2008b) successfully reproduced the water mass properties by including the effect of tidal mixing in the Indonesian archipelago, confirming the importance of mixing in this particular region. However, detailed volume, heat and salt transports through each major ITF passages and their relationships to the Pacific inflows and the Indian outflows are not yet well understood. For example, the simulated large positive heat flux of the ITF into the Indian Ocean, contrary to observation, is probably due to an unresolved mixing of the IMM mechanism and the omission of some Pacific source waters such as NPIW in numerical models.

Influences of the ITF on the global climate system cannot be correctly addressed without considering air-sea interaction. Only when ocean and atmosphere feedbacks are accounted for in a coupled general circulation model (CGCM) can the full global impact of the ITF on the climate system be assessed. In general, the ITF tends to warm the upper Indian Ocean and deepen the thermocline there due to the warm water flowing from the western tropical Pacific to the eastern Indian Ocean, while the opposite situation occurs in the western tropical Pacific where water gets cooler due to the loss of heat to the Indian Ocean. By blocking the ITF in a CGCM, an ENSO-like response is produced over the equatorial Pacific and there is a dipole mode signature over the Indian Ocean (Hirst & Godfrey 1993). They also demonstrated that there was a significant difference in the surface heat flux distribution between the open and closed ITF cases. With the opening of the Indonesian straits, the net surface heat flux from the Pacific to the Indian Ocean is doubled in contrast to an ocean-only simulation (Wajsowicz & Schneider 2001). An open ITF in the CGCM also shows an accelerated equatorial undercurrent in the Pacific Ocean and a decreased Indian Ocean response to a change of wind stress and meridional and zonal thermocline gradients (Schneider 1998). Using the Geophysical Fluid Dynamics Laboratory (GFDL) CGCM, Song et al. (2007) restate a mean condition by closing the ITF – that is, SST in the eastern Pacific (Indian) is warm (cool), the near-equatorial Pacific (Indian) thermocline flattens (shoals), the Indo-Pacific warm pool precipitation shifts eastwards, the tropical Pacific trade wind relaxes and anomalous easterlies occur over the equatorial Indian Ocean. When their CGCM permits interannual diagnostics, interannual variability intensifies in the equatorial Pacific and the eastern Indian Ocean and ENSO becomes more energetic and frequent.

Considering the proposed intensive observations of PACSWIN, that will provide an invaluable data set in and around the Indonesian archipelago, and the recent capability of high-resolution OGCMs and CGCMs, now is the time to revisit the evaluation of the detailed transport and property distributions, the ITF pathways and its impact on the regional, basin-scale and global ocean circulations and climate.

The PACSWIN model components comprise studies utilising ocean-only models, atmosphere-ocean coupled models, fully coupled climate models and palaeoclimate models. In addition to outputs from prognostic models, results from ocean state estimations with a data assimilation scheme will be used.

For the ocean component, current high-resolution models can reach a horizontal resolution of 0.1 degree or higher, such as POP and OFES, which is crucial for resolving the narrow Indonesian straits and the complicated bottom topography. A locally enhanced high-resolution model nested in a global model with less high resolution is also utilised for studying the global climate impact on the ITF. Such a model has been developed by Australia's CSIRO and BOM under the Bluelink program. The model has a regional resolution of 10 km, resolving the mean and eddy flow fields but, outside of the Australia-Indonesian seas' region, the model has a coarse resolution. The INSTANT moorings provide a benchmark for various modelling validation tests of the PACSWIN modelling components. With recent advances in ocean observations, including Argo, surface drifting buoys and tropical mooring arrays, and especially the improved measurement of the ITF from INSTANT, high-resolution data assimilated models would provide a reasonably good simulation of the three-year mooring results.

The initial task of PACSWIN modelling will be to use high-resolution model outputs and to assess and validation the PACSWIN observational projects on the pathways and transports of the ITF and circulation in the western tropical Pacific Ocean, as well as the water mass distributions associated with the ITF system. The focus will be on the three-year mean condition, mean seasonal variations, interannual variations associated with El Niño – Southern Oscillation and Indian Ocean Dipole mode, and intra-seasonal variability in and around the Indonesian archipelago. High-resolution, eddy resolving models can reproduce meso-scale eddies and ocean frontal structures generated by internal instabilities in the ocean. The roles of such an oceanic instability on the large-scale ITF system will be investigated. All models will be further adjusted and improved as PACSWIN develops in the coming decade.

For coupled model and climate model components, the impacts of the ITF on the regional, basin-scale and global ocean circulation and climate systems will be one of the main targets to be studied. At the same time, the impacts of many climate variations in remote areas on the ITF will be investigated.

At the completion of the PACSWIN program (in 5–10 years), our understanding of the mean condition of the ITF and its variability will be enhanced significantly, and the target models should attain optimal performance in simulating the ITF and associated ocean and climate conditions.

5. *Programmatic development*

5.1 An international coordinated program

A successful PACSWIN requires a truly strong international effort. The survey plans, instruments and technologies must be well coordinated among the many participating countries and international organisations, especially those in the Asia-Pacific region. A time span of about two years for pilot studies and between three and five years for operation is currently suggested. Some long-term monitoring projects such as submarine cables will be needed for as long as possible.

PACSWIN has set three priority monitoring projects: submarine cables, floats and moorings. An international submarine cable workshop on cable voltage measurement to monitor transport of ITF has been planned. PACSWIN will integrate with other two proposed ocean programs, SPICE and NPOCE, to constitute a major ocean climate monitoring network in the Indo and western Pacific for at least the coming decade.

Sponsorship will be sought for the PACSWIN program from individual participating national governments and funding agencies including governmental and intergovernmental sectors and other international organisations such as CLIVAR, IOC, SCOR and APEC as well as the private sector. Support from the Indonesian scientific community and government and collaboration with the international community will be crucial to ensure the success of this international program. A memorandum for joint funding will be submitted to participating national funding agencies.

5.2 CLIVAR endorsement

CLIVAR (www.clivar.org) is one of the World Climate Research Programme (WCRP) (wcrp.wmo.int) projects, which addresses Climate Variability and Predictability with a particular focus on the roles of ocean-atmosphere interactions in climate and of biochemical cycles. PACSWIN in many ways meets the interests of CLIVAR, for example, the First PACSWIN Submarine Cable workshop in Taipei on 9–10 September 2009 endorsed by CLIVAR on 23 April 2008. The CLIVAR endorsement will not only allow PACSWIN to become an officially recognised program but help make it an international coordinated endeavour to satisfy the need for accurate ocean climate information from the region.

A roadmap towards CLIVAR endorsement has been drawn up. PACSWIN was proposed at the CLIVAR Pacific panel meetings in Guang Zhou, China, on 29–30 November 2007 and in Perth, Australia, on 27–28 March 2009. PACSWIN was also proposed to the CLIVAR Indian Ocean panel meeting in Bali, Indonesia on 12–14 May 2008. The progress of PACSWIN was reported to Pacific Panel Meeting 5th session in Perth. A further progress report of PACSWIN along with this document will be submitted to CLIVAR Pacific and Indian panel sessions for review and feedback.

6. Towards a PACSWIN implementation plan

This is the initial report on the PACSWIN scientific description aimed at providing a state-of-the-art ocean climate monitoring program in the Indonesian seas and adjacent regions. The chapter brings an up-to-date view of the scientific proposal from individual program component chairs/co-chairs for future operation. To develop PACSWIN into an international coordinated program, the proposed program components will be agreed upon through the participating country chairs and the wider community in their own countries. Each country chair leads at least one component of PACSWIN. Each component will involve a number of participating countries and scientists. To make PACSWIN operational, participating countries and scientists need to apply for funding from their own government funding agencies. Therefore, PACSWIN will be achieved by coordinated international efforts including personal, technical and funding resources. Detailed operation and activities will appear in future reports.

References

Church JA (2007). A change in circulation? *Science*, 37, 908–09.

Cresswell GR & Luick JL (2001). Current measurements in the Halmahera Sea. *J Geophys Res*, 106, 13945–152.

Cunningham SA, Kanzow T, Rayner D, Baringer MO, Johns WE, Marotzke J et al. (2007). Temporal variability of the Atlantic meridional overturning circulation at 26.5°N. *Science*, 37, 935–38.

Davis RE, Eriksen CE & Jones CP (2002). Autonomous buoyancy-driven underwater gliders. In G Griffiths (Ed), *The technology and applications of Autonomous Underwater Vehicles*, London: Taylor and Francis.

Ffield A & Gordon AL (1992). Vertical mixing in the Indonesian thermocline. *J Phys Oceanogr*, 22, 184–95.

Fieux M, Andrié C, Delecluse P, Ilahude AG, Kartavtseff A, Mantisi F et al. (1994). Measurements within the Pacific–Indian Ocean Throughflow region. *Deep-Sea Res, Part 1*, 41, 1091–130.

Fieux M, Andrié C, Charriaud E, Ilahude AG, Kartavtseff A, Mantisi F et al. (1996). Hydrological and chlorofluoromethane measurements of the Indonesian throughflow entering the Indian Ocean. *J Geophys Res*, 101, 12,433–54.

Fine RA (1985). Direct evidence using tritium data for the throughflow from the Pacific to the Indian Ocean. *Nature*, 315, 478–80.

Flagg CN, Schwartze G et al. (1998). Operating an acoustic Doppler current profiler aboard a container vessel. *J Atm Ocean Tech*, 15, 257–71.

Fu L-L (1986). Mass, heat and freshwater fluxes in the South Indian Ocean. *J Phys Oceanogr*, 16, 1683–93.

Ganachaud A, Kessler W, Wijffels S, Ridgway K, Cai W, Bowen M et al. (2007). *Southwest Pacific Ocean circulation and climate experiment. Part I: scientific background.* International CLIVAR Project Office, CLIVAR Publication Series No. 111, NOAA OAR Special Report. NOAA/OAR/PMEL, Seattle, WA, 30 pp.
Godfrey JS (1989). A Sverdrup model of the depth-integrated flow for the world ocean allowing for island circulations. *Geophys Astrophy Flu Dyn*, 45, 89–112.
Godfrey JS (1996). The effect of the Indonesian Throughflow on ocean circulation and heat exchange with the atmosphere: a review. *J Geophys Res*, 101, 12,217–38.
Godfrey JS & Golding TJ (1981). The Sverdrup relation in the Indian Ocean, and the effect of Pacific–Indian Ocean Throughflow on Indian Ocean circulation and on the East Australian Current. *J Phys Oceanogr*, 11, 771–79.
Gordon AL (1986). Interocean exchange of thermocline water. *J Geophys Res*, 91, 5037–46.
Gordon AL & Fine RA (1996). Pathways of water between the Pacific and Indian Oceans in the Indonesian seas. *Nature*, 379, 146–49.
Gordon AL & Susanto RD (1999). Makassar Strait transport: initial estimate based on Arlindo results. *Marine Tech Soci*, 32, 34–45.
Gordon AL & McClean JL (1999). Thermohaline stratification of the Indonesian seas: model and observations. *J Phys Oceanogr*, 29, 198–216.
Gordon AL (2001). Interocean exchange. In G Siedler, J Church & J Gould (Eds). *Ocean circulation and climate*, (pp303–14). London, UK and San Diego, California: Academic Press.
Gordon AL, Giulivi CF & Ilahude AG (2003). Deep topographic barriers within the Indonesian seas. *Deep-Sea Res II*, 50, 2205–28.
Gordon AL (2005). Oceanography of the Indonesian seas and their throughflow. *Oceanography*, 18, 14–27.

Gordon AL, Soesilo I, Brodjonegoro I, Ffield A, Jaya I, Molcard R et al. (2006). The first 1.5 years of INSTANT data reveal the complexities of the Indonesian Throughflow. *CLIVAR Exch,* 11, 10–11.
Gordon A, Susanto RD, Ffield A, Huber BA, Pranowo W & Wirasantosa S (2008). Makassar Strait throughflow, 2004 to 2006. *Geophys Res Letts,* 35, L24605, doi: 10.1029/2008GL036372.
Gordon A, Wijffels J, Sprintall J, Susanto D, Molcard R, van Aken H et al. (2008). The Indonesian Throughflow, 3-year INSTANT composite view. An unpublished manuscript, 18pp.
Hatayama T, Awaji T & Akimoto K (1996). Tidal currents in the Indonesian Seas and their effect on transports and mixing. *J Geophys Res,* 101, 12,353–73.
Hautala SL, Reid JL & Bray N (1996). The distribution and mixing of Pacific water masses in the Indonesian seas. *J Geophys Res,* 101, 12,375–89.
Hirst AC & Godfrey JS (1993). The role of the Indonesian Throughflow in a global GCM. *J Phys Oceanogr,* 23, 1057–86.
Hu DX, Cui MC, Qu TD & Li YX (1992). A subsurface northward current of Mindanao identified by dynamic calculation, In K Takanao (Ed), *Oceanography of Asian marginal seas.* Vol. 54, (pp359–65). New York: Elsevier.
Ilahude AG & Gordon AL (1996). Thermocline stratification within the Indonesian seas. *J Geophys Res,* 101, 12,401–09.
Inoue M & Welsh SE (1993). Modelling seasonal variability in the wind-driven upper-layer circulation in the Indo–Pacific region. *J Phys Oceanogr,* 23, 1411–36.
Kanzow T, Cunningham SA, Rayner D, Hirschi JJ-M, Johns WE, Baringer MO et al. (2007). Observed flow compensation associated with the MOC at 26.5N in the Atlantic. *Science,* 37, 938–41.
Koch-Larrouy A, Madec G, Iudicone D, Atmadipoera A & Molcard R (2008a). Physical processes contributing to the water mass transformation of the Indonesian Throughflow. *Ocean Dynamics,* 58, 275–88.

Koch-Larrouy A, Madec G, Blanke B & Molcard R (2008b). Water mass transformation along the Indonesian Throughflow in an OGCM. *Ocean Dynamics*, 58, 289–309.

Luick JL & Cresswell GR (2000). Current measurements in the Maluku Sea. *J Geophys Res,* 106, 13,953–58.

Lukas R, Yamagata T & McCreary JP (1996). Pacific low-latitude western boundary currents and the Indonesian throughflow. *J Geophys Res,* 101(C5), 12,209–16.

Masumoto Y, & Yamagata T (1996). Seasonal variations of the Indonesian throughflow in a general ocean circulation model. *J Geophys Res,* 101 (C5), 12,287–93.

Masumoto Y, Morioka Y & Sasaki H (2008). High-resolution Indian Ocean simulations: recent advances and issues from OFES. In MW Hecht & H Hasumi (Eds), *Ocean modelling in an eddying regime.* Geophysical Monograph 177, (pp199–212). Washington DC: American Geophysical Union.

McCreary JP, Miyama T, Furue R, Jensen, T, Kang H-W, Bang B & Qu T (2007). Interactions between the Indonesian Throughflow and circulations in the Indian and Pacific Oceans. *Progress in Oceanography*, 75(1), 70–114.

Meyers G (1996). Variation of Indonesian throughflow and the El Niño – Southern Oscillation. *J Geophys Res,* 101, 12,255–64.

Miyama T, Awaji T & Akitomo K (1995). Study of seasonal transport variations in the Indonesian Sea. *J Geophys Res,* 100, 20,517–41.

Molcard R, Fieux M & Ilahude AG (1996). The Indo-Pacific throughflow in the Timor Passage. *J Geophys Res,* 101, 12,411–20.

Murray SP & Arief D (1988). Throughflow into the Indian Ocean through the Lombok Strait, January 1985–January 1986. *Nature, 333*, 444–47.

National Oceanic and Atmospheric Administration (Content source); M McGinley (Ed) (2008). Indonesian Sea large marine ecosystem. In CJ Cleveland (Ed), *Encyclopedia of Earth.* Washington, DC: Environmental Information Coalition, National Council for Science and the Environment.

Pariwono JI, Ilahude AG & Hutomo M (2005). Progress in oceanography of the Indonesian seas: a historical perspective. *Oceanogr,* 18, 42–49.

Piola AR & Gordon AL (1984). Pacific and Indian-Ocean upper-layer salinity budget. *J Phys Oceanogr,* 14, 747–53.

Qu T & Song TY (2009). Mindoro Strait and Sibutu passage transports estimated from satellite data. *Geophys Res Lett,* 36, L09601, doi: 10.1029/2009GL037314.

Qu T, Du Y & Sasaki H (2006). South China Sea Throughflow: a heat and freshwater conveyor. *Geophys Res Lett,* 33, L23617, doi: 10.1029/2006GL028350.

Qu T, Du Y, Meyers G, Ishida A & Wang D (2005). Connecting the tropical Pacific with Indian Ocean through South China Sea. *Geophys Res Lett,* 32, L24609, doi: 10.1029/2005GL024698.

Ramp SR, Tang TY, Duda TF, Lynch JF, Liu AK, Chiu C-S et al. (2004). Internal solitons in the northeastern South China Sea. Part I: source and deep water propagation, *IEEE J Oceanic Eng,* 29, 1157–81.

Robertson R & Ffield A (2008). Baroclinic tides in the Indonesian Seas: tidal fields and comparisons to observations. *J Geophys Res,* 113, C07031, doi: 10.1029/2007JC004677.

Roemmich D & The Argo Steering Team (2009). Argo: the challenge of continuing 10 years of progress. *Oceanography,* 22(3), 26–35.

Rossby HT, Dorson D & Fontaine J (1986). The RAFOS system. *J Atmos Oceanic Tech,* 3, 672–79.

Saji NH, Goswami BN, Vinayachandran PN & Yamagata T (1999). A dipole mode in the tropical Indian Ocean. *Nature,* 401, 360–63.

Schneider N & Barnett T (1997). The competition of freshwater and radiation in forcing the ocean during El Niño. *J Climate,* 8, 980–92.

Schneider N (1998). The Indonesian Throughflow and the global climate system. *J Climate,* 11, 676–89.

Song Q, Vecchi GA & Rosati A (2007). The role of the Indonesian Throughflow in the Indo-Pacific climate variability in the GFDL coupled climate model. *J Climate,* 20, 2434–51.

Song YT (2006). Estimation of interbasin transport using ocean bottom pressure: Theory and model for Asian marginal seas. *J Geophys Res,* 111, C11S19, doi: 10.1029/2005JC003189.
Sprintall J, Wijffels S, Gordon AL, Ffield A, Molcard R, Susanto D et al. (2004). INSTANT: a new international array to measure the Indonesian Throughflow. *EOS Trans,* 85(39), 369.
Stommel H (1948). The theory of the electric field in deep ocean currents. *J Marine Res*, 7, 386–92.
Susanto RD & Gordon AL (2005). Velocity and transport of the Makassar Strait throughflow. *J Geophys Res*, 110, doi:10.1029/2004JC002425.
Talley LD & Sprintall J (2005). Deep expression of the Indonesian throughflow: Indonesian Intermediate water in the South Equatorial Current. *J Geophys Res*, 110, doi:10.1029/2004JC002826.
The Argo Science Team (1998). On the design and implementation of Argo: a global array of profiling floats. An unpublished manuscript, 32 pp.
Toole JM, Zou E & Millard RC (1988). On the circulation of the upper waters in the western equatorial Pacific Ocean. *Deep-Sea Res*, 35, 1451–82.
Toole J & Warren B (1993). A hydrographic section across the subtropical South Indian Ocean. *Deep-Sea Res,* 40, 1973–2019.
Tozuka T, Qu T & Yamagata T (2007). Dramatic impact of the South China Sea on the Indonesian Throughflow. *Geophys Res Lett*, 34, L12612, doi: 10.1029/2007GL030420.
van Aken HM, Punjanan J & Saimima S (1988). Physical aspects of the flushing of the East Indonesian basins. *Neth J Sea Res,* 22, 315–39.
van Riel PM (1932). The Snellius Expedition, Conseil Perm. Internat. p. l'Explor De la Mer, *Journal du Conseil,* 7, 212–17.
Vranes K, Gordon AG & Ffield A (2002). The heat transport of the Indonesian Throughflow and implications for the Indian Ocean heat budget. In F Schott (Ed), *Physical oceanography of the Indian Ocean during the WOCE period. Deep-Sea Res II*, 49, 1391–410.

Wajsowicz RC & Schneider EK (2001). The Indonesian Throughflow's effect on global climate determined from the COLA coupled climate system. *J Climate,* 14, 3029–42.
Wijffels SE, Schmitt RW, Bryden HL, & Stigebrandt A (1992). Transport of freshwater by the oceans. *J Phys Oceanogr,* 22, 155–62.
Wijffels SE, Meyers G & Godfrey JS (2008). A 20-yr average of the Indonesian Throughflow: regional currents and interbasin exchange. *J Phys Oceanogr*, 38, 1965–78.
Wunsch C, Hu D & Grant B (1983). Mass, heat, salt and nutrient fluxes in the South Pacific Ocean. *J Phys Oceanogr,* 13, 725–53.
Wyrtki K (1961). Scientific results of marine investigations of the South China Sea and the Gulf of Thailand, 1959–1961, Vol. 2, *Physical oceanography of Southeast Asian waters*. NAGA Rep. 2, Scripps Inst of Oceanogr, Univ of Calif, San Diego, La Jolla, 195 pp.
You Y & Tomczak M (1993).Thermocline circulation and ventilation in the Indian Ocean derived from water mass analysis. *Deep-Sea Res I*, 40, 13–56.
You Y (1998). Intermediate water circulation and ventilation of the Indian Ocean derived from water-mass contributions. *J Marine Res*, 56, 1029–67.
You Y (2003).The pathway and circulation of North Pacific Intermediate Water. *Geophys Res Lett,* 30(24), doi: 10.1029/2003GL018561.
Zenk W, Siedler G, Ishida A, Holfort J, Kashino Y, Kuroda Y et al. (2005). Pathways and variability of the Antarctic Intermediate water in the western equatorial Pacific Ocean. *Prog Oceanogr,* 67, 245–81.

Appendix 1: Acronyms

AAMW	Australasian Mediterranean Water
AC	Agulhas Current
ADCP	Acoustic Doppler Current Profiler
AGCM	atmosphere general circulation model
APEC	Asia-Pacific Economic Cooperation
APEX	Autonomous Profiling EXplorer
Argo	Array for Real-time Geostrophic Oceanography
ARLINDO	Arus Lintas Indonesia
ASCAT	advanced scatterometer
BOM	Bureau of Meteorology
CC	California Current
CGCM	coupled atmosphere and ocean general circulation model
CLIVAR	Climate Variability and Predictability Experiment
CSIRO	Commonwealth Scientific and Industrial Research Organization
CTD	conductivity-temperature-depth
EAC	East Australia Current
EEZ	Exclusive Economic Zone
EFS	E-field sensor
EIE	Equatorial Intermediate Current
EKE	eddy kinetic energy
ENSO	El Niño – Southern Oscillation
ENVISAT	Environmental Satellite
GAIW	Gulf Alaska Intermediate Water
GCOS	Global Climate Observing System
GFDL	Geophysical Fluid Dynamics Laboratory
GFO	GEOSAT (Geophysical/Geodetic Satellite) Follow-On
GODAE	Global Ocean Data Assimilation Experiment
GOOS	Global Ocean Observing System
GPS	Global Positioning System
HE	Halmahera Eddy

iAAIW	Indian Ocean Antarctic Intermediate Water
IAPSO	International Association for the Physical Sciences of the Oceans
ICW	Indian Central Water
IES	inverted echo sounder
IIW	Indonesian Intermediate Water
IMM	Indonesian Mix-Master
INSTANT	International Nusantara Stratification and Transport
IOC	Intergovernmental Oceanographic Commission
ITCZ	Intertropical Convergence Zone
ITF	Indonesian Throughflow
JADE	Java Australia Dynamic Experiment
JAFOOS	Joint Australian Facility for Ocean Observations
JASON	Joint Altimetry Satellite Oceanography Network
KS	Kuroshio
LADCP	Lowered Acoustic Doppler Current Profiler
LC	Leeuwin Current
lCDW	lower Circumpolar Deep Water
LLWBCs	low latitude western boundary currents
MC	Mindanao Current
ME	Mindanao Eddy
MOC	meridional overturning circulation
NGCUC	New Guinea Coastal Undercurrent
NICW	North Indian Central Water
NOAA	National Oceanic and Atmospheric Administration
NPCW	North Pacific Central Water
NPIW	North Pacific Intermediate Water
NPOCE	Northwestern Pacific Ocean Circulation Experiment
NPSW	North Pacific Subtropical Water
OGCM	ocean general circulation model
OIW	Okhotsk Intermediate Water
pAAIW	Pacific Ocean Antarctic Intermediate Water
PACSWIN	Indonesian Throughflow: PACific Source Water INvestigation

PC	Peru Current
PGW	Persian Gulf Water
POP	Parallel Ocean Program model
QuikSCAT	Quick Scatterometer
RAFOS	Sound Fixing and Ranging (SOFAR) spelled backwards
RSW	Red Sea Water
SALPs	Submerged Autonomous Launch Platforms
SC	Somali Current
SCOR	Scientific Committee on Oceanic Research
SCS	South China Sea
SeaWiFS	Sea-viewing Wide Field-of-view Sensor
SEC	South Equatorial Current
SISW	South Indian Subtropical Water
S-JC	Sumatra-Java Current
SPCW	South Pacific Central Water
SPICE	Southwest Pacific Ocean Circulation and Climate Experiment
SPSW	South Pacific Subtropical Water
SSH	sea level height
TOGA	Tropical Ocean Global Atmosphere
uCDW	upper Circumpolar Deep Water
WAC	West Australia Current
WCRP	World Climate Research Programme
WOCE	World Ocean Circulation Experiment
XBT	eXpendable BathyThermograph

Appendix 2: Available resources and technique for PACSWIN from participating countries and institutes

University of Sydney

National facility: R/V Southern Surveyor

National facility: Gliders, moorings

Australian Centre for Field Robotics

University of Rhode Island, www.gso.uri.edu/

Research vessel: 185 foot R/V Endeavor

RAFOS floats and sound sources (Tom Rossby)

The Inverted Echo Sounder (IES) (Randy Watts)

BPPT (Agency for the Assessment and Application of Technology

Research vessels: R/V Baruna Jaya I to IV

JAMSTEC (Japan Agency for Marine-Earth Science and Technology)

Research vessels: R/V Natsushima, R/V Kaiyo, R/V Yokosuka, R/V Mirai, R/V Kairei, R/V Hakuho Maru, R/V Tansei Maru

Submersible Shinkai 6500

Deep-Sea Drilling Vessel: Chikyu

Other Vehicle: Deep-Sea cruising AUV Urashima; Deep-Sea Survey System: Deep Tow; 3000m remotely operated vehicle: Hyper-Dolphin; 7000m Class remotely operated vehicle: Kaiko 7000II

Observation facility: Doppler Radar, Radiosonde, TRITON Buoy, CTD, ADCP, Argo, POPS

Earth Simulator

Scripps Institute of Oceanography

Research vessels: R/V Roger Revelle, R/V Melville, R/V New Horizon, R/V Robert G Sproul

Floating Instrument Platform (FLIP)

Underwater plane

Weather buoy

3D Sea Scan

Gliders, Argo, moorings

Institute of Oceanography, Chinese Academy of Sciences

Research vessel: R/V Kexue-I, R/V Kexue-III, R/V Jinxing-II

Flow Cytometer, Marine Gravimeter, towed oceanographic vehicle, marine CTD winch

Other facility: network server SGI3400, stable isotope MS system, modern gas chromatography-mass spectrometry, ICP-MS, X-ray diffractometer, 500 M nuclear magnetic resonance spectrometer

LDEO, Columbia University

Research vessel: R/V Marcus G. Langseth

OTIC, CICAR, Environment Tracer Group, ICPMS, LCSN, MARGINS

Woods Hole Oceanographic Institution

Research vessels: R/V Atlantis, R/V Knorr, R/V Oceanus, R/V Tioga

National Deep Submergence Facility: HOV Alvin, ROV Jason/Medea, AUV ABE/Sentry

Observing systems: moorings, buoys, floats, drifters

Instruments: Sensors/samplers, Seafloor mapping system, Lighting and camera system, Communication tools, MISO

Underwater vehicles: HROV Nereus, AUVs, Towed vehicles

Marine Science Institute, University of the Philippines

Research vessel: MV DA-BFAR

Specialise on tropical marine sciences

Institute of Oceanography, National Taiwan University

Research vessel: R/V Ocean Researcher I, R/V Ocean Researcher II, R/V Ocean Researcher III

CTD, ADCP

Korea Ocean Research and Development Institute (KORDI)

Research vessels: R/V Onnuri, R/V Eardo

Observational equipment: wave meter, CTD, ADCP, Gravity meter, Precise depth recorder

Institut Français de Recherche pour l'Exploitation de la Mer (IFREMER)

Research vessels: R/V Pourquoi pas? R/V L'Atalante, R/V Thalassa, R/V Le Suroît, R/V L'Europe, R/V Gwen Drez, R/V Thalia

Underwater systems: Nautile, Victor 6000, AUV Aster x, SAR, Scampi, MIMOSA, ADELIE, VEMO+, POSIDONIA

Acoustic facilities: Multibeam echo-sounders – Simrad EM12/Dual 13khz, Simrad EM1000, Simrad EM300 30 khz; Doppler current metersomètres Doppler – RDI 300 khz, RDI 150 khz, RDI 75 khz; Towed sonar – Edge TechDF10; Sediment sounders – Elics 2 khz, EDO 3.5 khz; Pingers – Genisea, Suber; Fishing sounders and sonars; Trawling – Scanmar 42 khz, TMS Pacha 2000, Geonet; Acoustic environment – Censea Orca, Sabrina

Seismic survey methods used by the Ifremer fleet. (English Version – pdf format): Multitrace seismic survey (SMT), High Resolution Seismic Survey (HR), Rapid digital seismic survey (SISRAP)

PENFELD penetrometer

Other facilities: Gravimeters - KSS30, Lockheed BGM5; Magnetometers – Baringer M244; Thermosalinometer – Seabird 21, SIS CTD 1000; Fluorometer; Sea water temperature – TQP, TPP; Bathythermographs – Sippican MK12; Weather unit – Vaisala Milos 5, Anderaa, Informedic (NOAA/METEOR), Batos Meteo France

International Pacific Research Center, School of Ocean and Earth Science and Technology, University of Hawaii

Research vessels: R/V Kai'imikai-O-Kanaloa, R/V Kilo Moana

two research submersibles, the Pisces IV and Pisces V (2000 meters), and the ROV RCV-150 (914 m)

ADCP/LADCP

RSMAS/MAC, University of Miami

Research vessel: R/V FG Walton Smith

Mobile instruments: Air Sea Interaction Buoys, 94-GHz Doppler Cloud Radar, X-Band Cloud Radar, 915 MHz Wind Profiler, Marine-Atmosphere Emitted Radiance Interferometer (M-AERI)

State-of-the-art instruments including a salt-water wave tank, an isotopic mass spectrometer, five-tank Conditioning and Spawning Systems, multi-tank Aplysia Culture Laboratory, Controlled Corals Climate Tanks, Applied Biosystems 3730xl DNA Analyser, Evolution 3 automated liquid-handling system

University of Tokyo

Research vessels: R/V Keikaku, R/V Hakuho

School of Oceanography, University of Washington

Research vessels: R/V Thompson, R/V Barnes

Argo program, SeaGlider system, OOI/RSN

California Institute of Technology, JPL

Various satellite data and products such as altimetry and scatterometry

9

The Australian Integrated Marine Observing System (IMOS) and the regional implementation in New South Wales

Moninya Roughan, Iain Suthers and Gary Meyers

*Abstract**

The marine community in Australia has recently commenced the establishment of marine research-infrastructure to systematically observe and monitor the ocean around Australia. The Integrated Marine Observing System (IMOS) was established as part of the Australian Government's National Collaborative Research Infra-structure Strategy (NCRIS) initially for a five year period to June 2011, with an extension through to June 2013 announced in 2009. The IMOS strategic research-goal is to assemble and provide free, open and timely access to streams of data that support research on: the role of the oceans in the climate system, and the impact of major boundary currents on continental shelf environments, ecosystems and biodiversity . Here we describe the implementation of IMOS by the NSW node as an example, and present preliminary data from the NSW-IMOS observing system.

1. Overview

In recent years Australia has commenced the establishment of marine research-infrastructure in Australia to systematically service Australia's significant requirements and responsibilities for one of the largest marine

* Chapter originally submitted in February 2009.

jurisdictions of any nation on earth. At over 14 million km^2 Australia's Exclusive Economic Zone (EEZ) is nearly twice the surface area of the Australian continent. It extends from the tropics to high latitudes in Antarctic waters and much of it is unexplored.

The surrounding Pacific, Southern and Indian oceans strongly affect the continental climate-system at all time scales, from seasons to decades. The major ocean currents on its eastern, western, northern and southern boundaries, (best known of these being the East Australian Current and the Leeuwin Current in the west), affect regional climatic conditions and marine ecosystems . There is evidence that these currents are changing on decadal time scales and have already impacted marine ecosystems, but the data are sparse and neither the currents nor ecosystems have been monitored in a systematic way. Research on marine climate impacts is an open book in Australia, and the pages are nearly blank, because long term data has been missing.

The Integrated Marine Observing System (IMOS) was established as part of the Australian Government's National Collaborative Research Infrastructure Strategy (NCRIS) with a \$A50M allocation. Co-investments from universities, state and commonwealth agencies make nearly \$A100M available to develop IMOS during the period 2007 to 2011. In May 2009 a \$A52M extension through to June 2013 was announced. It is coordinated and financially managed by the IMOS Office, located at the University of Tasmania. The system is comprised of a distributed set of oceanographic instruments, which provide streams of in situ data resulting in information services which contribute to meeting the needs of marine research in Australia's open ocean and coastal waters. The in situ data when combined with satellite data, enables the modelling required to explain the role of the oceans in seasonal prediction and climate change. Sustaining the project will allow identification and management of climate change in the coastal marine environment. It will also provide an observational nexus to better understand and predict the fundamental connections between coastal biological processes and regional/oceanic phenomena that influence biodiversity. As an NCRIS project IMOS was primarily designed to support research; however, the data streams are also critical for societal, environmental and economic applications. Some of these include: management of marine natural resources and their associated ecosystems, support and management of coastal and offshore industries, safety at sea, marine tourism and defence.

The IMOS strategic research-goal is to assemble and provide free, open and timely access to streams of data that support research on:

- the role of the oceans in the climate system, and
- the impact of major boundary currents on continental shelf environments, ecosystems and biodiversity .

Given the extent and challenge of addressing the broad range of marine issues in the Australian EEZ, IMOS is considered only the beginning of the observing system that Australia needs. The cost of an adequate observing system is great considering the length of coastline and the complexity of the system. Nevertheless, staged enhancements are being implemented. The return from investing in ocean observations around Australia was estimated through an economic analysis undertaken in 2006 by the Australian Academy of Technological Sciences and Engineering and the Western Australian Global Ocean Observing System Inc. (*Economics of Australia's sustained ocean observation system, benefits and rationale for public funding*). That study, based on only a limited set of benefiting industries concluded that the cost benefit to the Australian economy of investing in ocean observations was better than 1:20. While this study was a preliminary cost-benefit analysis and has some conceptual and calculation limitations, it shows that the return on investment is considerable and investment in an ocean observing system is worthwhile.

The scientific rationale and details of the need for observations were identified by five regional nodes covering the Great Barrier Reef, New South Wales (south-eastern Australia), Southern Australia, Western Australia and the Bluewater and Climate Node (Figure 1). Each node has 50 to 100 members and design of the observing system was determined by consensus.

Governance of IMOS is led by an Advisory Board with an independent Chair. The IMOS Office established at the University of Tasmania coordinates and manages all of the investments as a national system, with advice from the Board and from a Scientific Steering Committee made up of node leaders.

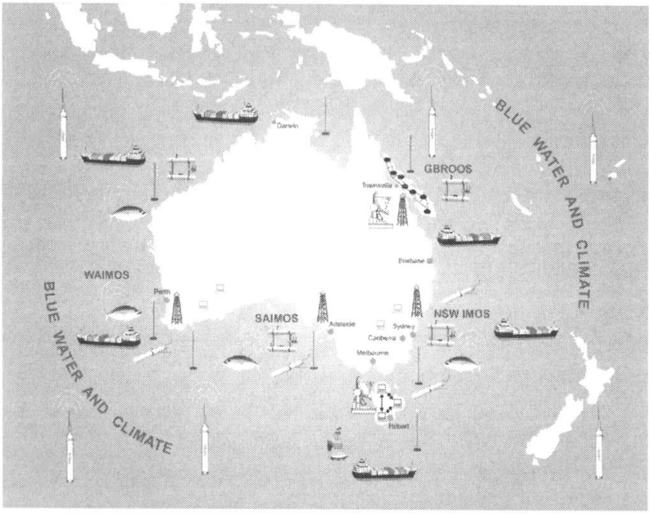

Figure 1 The IMOS data facilities and their regional node implementation

Nine national facilities are building the infrastructure illustrated in Figure 1. The observing facilities include three for ***bluewater and climate observations*** (Argo Australia, Enhanced Measurements from Ships of Opportunity and Southern Ocean Time Series), three facilities for ***coastal currents and water properties*** (oceanographic and acoustic Moorings, Ocean Gliders and High Frequency coastal Radar) and three for ***coastal ecosystems*** (Acoustic Tagging and Tracking, Autonomous Underwater Vehicle and a biophysical sensor network on the Great Barrier Reef). The operators of the facilities are the major players in marine research in Australia.

Data management is an essential, and in many ways the most important part of the observing system. A satellite remote sensing facility assembles satellite data for the region and the electronic Marine Information Infrastructure (eMII facility) provides access to all IMOS data, enhanced data products and web services via imosmest.emii.org.au, in a searchable and interoperable framework. An IMOS development officer has been appointed to aid the data uptake in the community. The value from this infrastructure investment lies in the nationally coordinated deployment of a wide range of equipment within a region aimed at deriving critical data sets that serve multiple applications. Additional information on IMOS is available at www.imos.org.au. Implementation of IMOS facilities began in 2007. About 70% of the planned infrastructure is operating as of June 2009 (Figure 2).

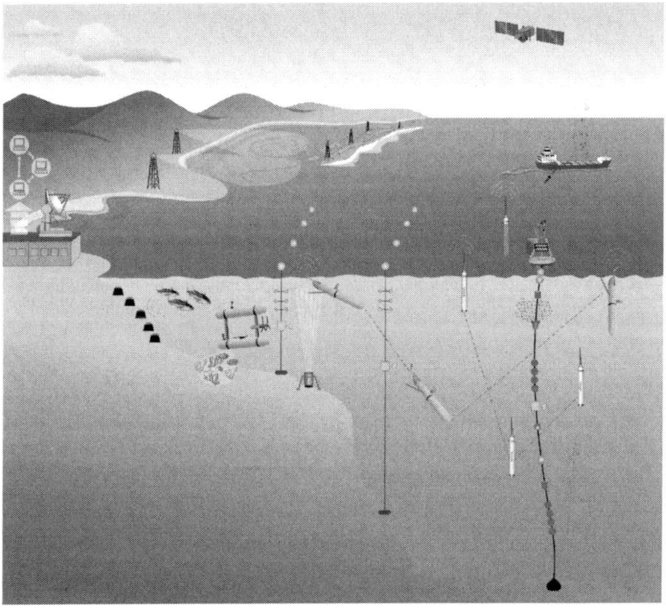

Figure 2 A regional observing system using all of the IMOS facilities.

2. The socio-economic context of NSW-IMOS

Most of Australia's urban population is located on the eastern and southern seaboards, making the associated coastal waters among the nation's most exploited and often stressed environments. More than 80% of Australia's residents live within 50 km of the coast and more than half the nation lives in and between Brisbane and Melbourne. The major pressures include urbanisation, water quality, freshwater availability, and beach erosion. The coastline of NSW alone has over 450 coastal discharge sites along the NSW coast, the largest three being off Sydney amounting to nearly 1000 $ML.d^{-1}$. Desalination plants are planned or being constructed for the Gold Coast, Sydney (Kurnell) and Melbourne (at Wonthaggi on the South Gippsland coast).

The largest marine industry in NSW is marine tourism, the second largest of all states contributing 22% of the national marine industry ($27 billion in value added during 2002–03, The Allen Report 2004). The value of the marine industry (i.e. all recreational and light commercial vessels) in NSW is valued at over $2 billion p.a. and employs over 11,000 – both figures are almost equivalent to all other states combined (mostly Victoria and Queensland, www.bia.org.au/data.html). Over a third of the national marine industry employment (36%) is in NSW – and mostly in marine tourism. These figures are more remarkable considering that our estuaries, while numerous (>130) are small and we have the nation's narrowest continental shelf (only 20–60 km, Figure 2, Chapman et al. 1982).

Most of the marine technology industry is related to recreational fishing, which faces significant problems of declining catch rates. Nearly 20% of the Australian population fish at least once a year (17% in NSW), and recreational fishers harvest up to 27,000 t of finfish with an annual value of around $2 billion (Henry & Lyle 2003). Recreational fishing effort increases annually, sometimes up to 7.5% on the previous year (Gwynne 1994) and this is especially important around Australia's largest cities on the eastern seaboard. In NSW the recreational catch is about 30% of the commercial catch, but for six major species the recreational catch is actually greater than the commercial. In NSW the recreational fishing fee bought out commercial fishing licenses in 25 estuaries in 2001, now described as recreational fishing havens. Target species that are especially relevant to recreational fishing, tourism and IMOS are bream, flathead, mulloway, prawns, Australian salmon, grey nurse shark, bull shark and white shark. The Recreational Fishing Trust is driving research on fish attraction devices (FADS, moorings up to 20 km offshore), re-stocking projects, and on estuarine artificial reefs. Large artificial reefs are proposed for the continental shelf off Newcastle,

Sydney and Port Kembla with the main issue being whether these structures are fish-attractors or fish-producers.

Lord Howe Island and the Solitary Islands were declared state marine parks in 1998/1999. Four other NSW marine parks have been declared over the past six years, which extend out to the 3 nm limit of state waters: Cape Byron (22,000 ha), Batemans Bay (85,000 ha), Jervis Bay (21,000 ha) and the largest, Port Stephens-Great Lakes (98,000 ha). The Batemans Bay and Port Stephens parks have been the most controversial with the recreational fishing community, and have challenged in state parliament the science on which they are based. The NSW Marine Parks Authority (MPA) through NSW-DECCW aims to establish and manage a system of multiple-use marine parks designed to conserve marine biodiversity, maintain ecological processes and provide for ecologically-sustainable use, public appreciation, education, understanding and enjoyment of the marine environment. Key issues remaining are larval connectivity amongst areas, and the degree of "spill-over".

For 2005–06, ports of the East Marine Planning region (mostly Newcastle, Sydney (Sydney Harbour and Botany Bay), Brisbane, Port Kembla) accounted for 42% of the nation's exports and 51% of national imports by tonnage (Anon 2007). These ports accounted for 18% of freight loaded and 67% unloaded by all Australian ports. The busiest sea lanes are through the Coral Sea.

In summary the main, socio-economic issues of relevance to NSW-IMOS are the warming and strengthening of the EAC , the impact of storm surge and beach erosion, a need for marine park planning, and management of invasive species, marine tourism and fisheries. These socio-economic issues tie in with the major scientific goals of IMOS where we are striving to obtain an understanding of the East Australian Current and its role in the climate system, and its influence on continental shelf environments, ecosystems and biodiversity.

3. *The implementation of IMOS in NSW*

Here we describe the implementation of IMOS by the NSW node as an example. IMOS is implemented in the other coastal nodes in a similar regionally relevant fashion.

The Sydney Institute of Marine Science (SIMS) is host to NSW-IMOS , one of five national nodes of IMOS. Through its industry partners (NSW Department of Environment, Climate Change and Water, Sydney Water Corporation, Manly Hydraulics Laboratory, Oceanographic Field Services, and NSW Department of Primary Industries), NSW-IMOS inherits a wealth of historical oceanographic observations off Port Hacking back to 1942; nearly three decades of wave climate observations at seven locations along the NSW coast; the Ocean Reference Station's 17 years of oceanographic data off Sydney, as well as an existing array of 70 acoustic telemetry listening posts for tagged sharks along the NSW coast.

The East Australian Current (EAC) influences the climate and marine economies of nearly half of the Australian population and yet we do not understand some of its most fundamental behaviour (dynamics of current strength and separation from the coast; ecological effects of cross-shelf flows and larval trajectories). We do know the EAC sporadically stimulates upwelling and phytoplankton blooms; that it transports phytoplankton and the larvae of fish from southern Queensland to Bass Strait and Tasmania. The EAC extends into Tasmanian waters, particularly during the past decade where the Maria Island reference station has recorded an increase of over 2°C per century – the fastest recorded rise of any regional sea in the world (Ridgway 2007). Climate change scenarios consistently indicate further warming of the Tasman Sea. Ocean forecasts consistently reveal large relative error due to mesoscale variability in the Tasman Sea.

The goal is to examine the physical and ecological interactions of the East Australian Current and its eddy field with coastal waters, to assess the synergistic impacts of urbanisation and climate change. Our specific aims during this IMOS program are to:

1) quantify the seasonal and annual variation in EAC inflow upstream of the EAC's characteristic separation off mid-NSW

2) make sustained observations of the coastal separation of the EAC and the EAC eddy field and biological consequences
3) determine the biological response to oceanographic and climate effects (eddies, upwelling, rainfall, dust storms), from fish movements, to phytoplankton communities, to benthic habitats.

These aims contribute to the major scientific goals of IMOS by adding to the understanding of the East Australian Current and its role in the climate system, and its influence on continental shelf environments, ecosystems and biodiversity.

NSW-IMOS will address these aims by (Figure 3):

1) establishing a national reference transect of **oceanographic moorings** off Sydney, and further moorings off Coffs Harbour in the north, and Jervis Bay and Eden to the south, supported by a **high frequency coastal radar** off Coffs Harbour
2) monthly sampling near the oceanographic moorings (for calibration and especially chlorophyll and zooplankton), supported by **autonomous ocean gliders**
3) deploying two cross-shelf transects of **acoustic receivers** ("listening posts") from the shore to the shelf break off Sydney and off Coffs Harbour, and using an **Autonomous Underwater Vehicle** (AUV).

The expected outputs from NSW-IMOS are numerous and broad in scope. They include knowledge of the latitudinal gradient in EAC effects and climate impacts; the availability of near-real-time in situ observations that could be used to evaluate or initialise ocean models, such as Bluelink (www.cmar.csiro.au/bluelink/); evidence-based prediction of the biophysical response to climate impacts on beaches, coastal lowlands; evidence-based planning for marine parks; estimates of larval connectivity along the coast of southeastern Australia, amongst ports, harbours as well as among marine parks; predictions of fish landings based on rainfall and oceanographic variation; extension of NSW-IMOS data products to high schools, the public and the media (especially over the internet of temperature and current fields, shark and glider tracks). Other outputs are the postgraduate research theses and associated publications resulting from IMOS activities. Details of the

NSW-IMOS Science and Implementation Plan, and plans for the other IMOS Nodes are at imos.org.au/nodes.html.

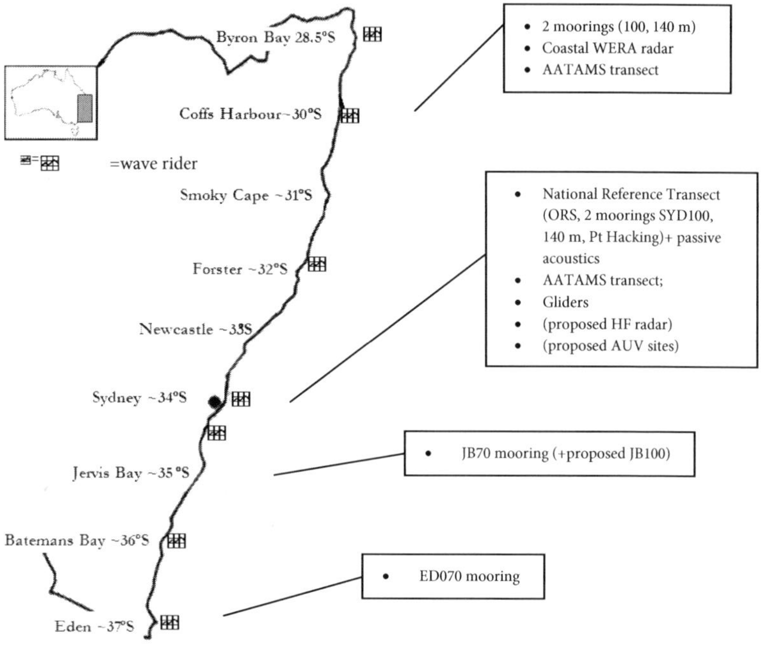

Figure 3 Schematic diagram showing the installed and proposed infrastructure along the NSW coastline.

4. Preliminary results from NSW-IMOS

IMOS data collection commenced in NSW during 2008, with the deployment of the two new moorings off Sydney (SYD100 and SYD140), a mooring off Jervis Bay (JB070), the extension of the hydrographic sampling line off Port Hacking, and the two successful glider deployments (and recovery!). Unfortunately the Jervis Bay mooring operated by UNSW@ADFA was lost after four months of operation in 2008. These activities complemented the previously existing instrumentation and data collection along the NSW coastline (Figure 3). For full details of the NSW-IMOS instrumentation program refer to Roughan et al. 2010.

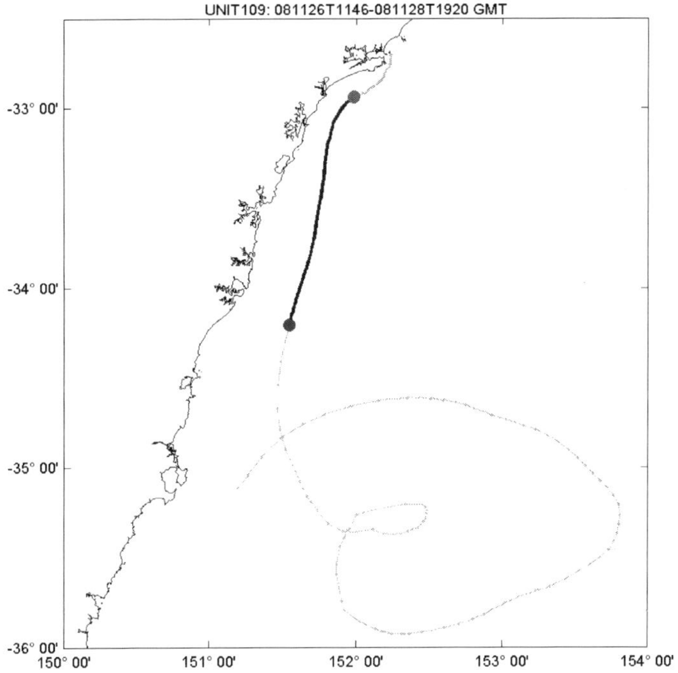

Figure 4: Pathway of the first NSW-IMOS glider deployment November–December 2008.

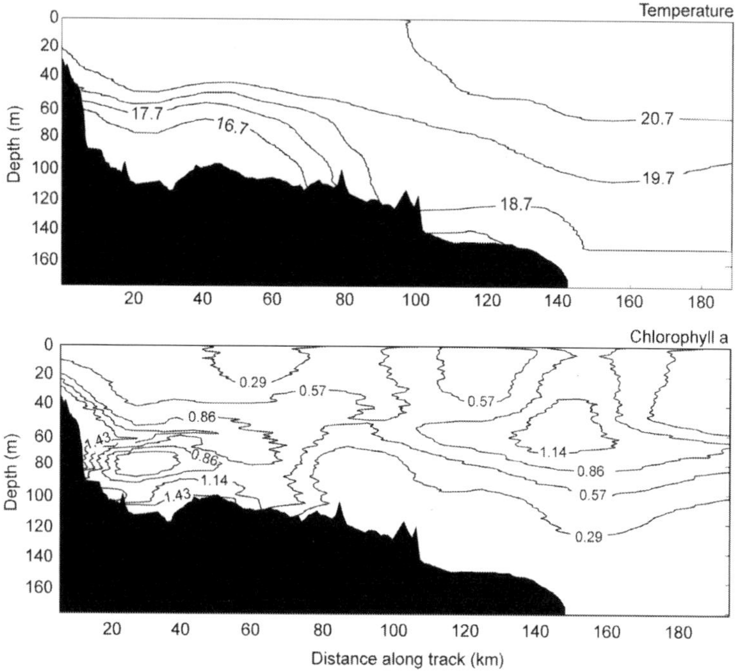

Figure 5 Data from the first NSW-IMOS glider deployment along the transect shown in Figure 4. Figure courtesy ANFOG.

Data from the first of the NSW-IMOS glider deployments (Figures 4 and 5) show two key features of the NSW shelf. The first is a cold, low salinity plume at the start (northern end) of the transect, associated with high turbidity and high chlorophyll a concentrations. This end of the transect captured the known high chlorophyll plume in the Stockton Bight, associated with the separation of the EAC (Roughan & Middleton 2002). Typically upwelling is associated with the separation of the EAC which slowly comes to the surface downstream of the separation point often in the Stockton Bight area (where the coastline bends westward). Towards the southern end of the transect (140 km along track) the glider crossed the shelf break. Evidence of this is again in the chlorophyll transect, where the sub-surface chlorophyll maximum occurs at the shelf edge. Concomitant with this glider deployment, the IMOS Remote sensing facility obtained satellite imagery of sea surface temperature, and the NSW-IMOS mooring sub-facility obtained vertical velocity profiles at the shelf edge off Sydney. This data show the presence of a large anti-cyclonic eddy encroaching onto the continental shelf, with velocities in excess of 130 cm/s in 140 m of water.

Further glider deployments will give insight into the nature of warm and cold core eddies along the NSW shelf as well as the EAC driven upwelling. Co-incident mooring data at Coffs Harbour, Sydney, Jervis Bay and Eden will give the temporal evolution of these events as well as a baseline from which to monitor changes in the EAC and its impact on the continental shelf. Surface radar will enhance the picture with large spatial coverage at the surface. Together these data sets give an integrated view of the coastal ocean along one of the most populated coastlines of Australia.

References

The Allen Report (2004). The economic contribution of Australia's marine industries. 1995–96 to 2002–03. Report to the National Oceans Office. Available at: www.environment.gov.au/coasts/mbp/publications/general/marine-economic.html.

Anonymous (2007). *East Marine Planning Region. Non-fisheries resource use activities in the east marine planning region.* Sinclair Knight Mertz report to DEWR. Available at: www.environment.gov.au/coasts/mbp/publications/east/pubs/non-fisheries.pdf.

Chapman DM, Geary M, Roy PS & Thom BG (1982). *Coastal evolution and coastal erosion in New South Wales: a report prepared for the Coastal Council of New South Wales.* Coastal Council of New South Wales.

Gwynne HL (1994). Management of recreational fishing in Queensland. In DA Hancock (Ed), *Recreational fishing: what's the catch*. pp220–223. Canberra: Australian Society for Fish Biology.
Henry GW & Lyle JM (2003). *The national recreational and indigenous fishing survey*. Canberra: Department of Agriculture, Fisheries and Forestry.
Ridgway KR (2007). Long-term trend and decadal variability of the southward penetration of the East Australian Current, *Geophys Res Lett*, 34, L13613, doi:10.1029/2007GL030393.
Roughan M & Middleton JH (2002). A comparison of observed upwelling mechanisms off the east coast of Australia. *Continental Shelf Research*, 22, 2551–2572.
Roughan M, Suthers IS, Morris BD, Baird ME & Pritchard T. 'NSW-IMOS An integrated Marine Observing System for South Eastern Australia: the East Australian Current and its interaction with coastal environments'. *Deep Sea Research* 2010 (Submitted 22 Oct 2009).

10

South Pacific Ocean dynamics: potential for enhancing sea level and climate forecasts

Neil J Holbrook

*Abstract**

Long ocean Rossby waves are critically important to changes in South Pacific Ocean climate, western boundary circulation and sea level. The southwest Pacific boundary converts impinging Rossby wave energy into Kelvin wave transfers to the equatorial zone, contributing to the timing of equatorial wave dynamics and ENSO climate variability. This complex boundary also hosts the poleward flowing East Australian Current (EAC) which transports heat to the mid-latitudes and moderates and modulates the regional climate of the southwest Pacific. This study uses wind stress forced upper ocean dynamics in the South Pacific to investigate interannual to decadal changes in sea level at the New South Wales (NSW) coast and the relationship with summer rainfall totals across NSW. It is found that there is a highly significant relationship between modelled first baroclinic-mode ocean dynamics, the baroclinic component of EAC intensity at 34°S, and observed sea level at Fort Denison. There is also a strong relationship with NSW summer rainfall, but only during the negative phase of the Interdecadal Pacific Oscillation. This chapter concludes with a discussion of the large scale ocean observing and monitoring of extra-tropical Rossby waves and the potential for enhancing future ENSO climate forecasts.

* Chapter originally submitted in January 2009.

1. Introduction

The South Pacific Ocean plays a very important role in the Earth's climate system. For example, the western tropical Pacific is critical in contributing to the timing of El Niño – Southern Oscillation (ENSO) variations, as it includes the complex boundary system for reflection of long (westward-propagating) Rossby waves, the generation of coastal Kelvin waves which propagate equatorward, and the transformation into equatorial Kelvin waves which propagate eastwards (e.g. Clarke 1991; du Penhoat & Cane 1991; Boulanger & Menkes 1999). In the subtropical southwest Pacific, the East Australian Current (EAC) represents the western boundary current of the South Pacific (subtropical) Gyre and results from bifurcation of the South Equatorial Current at the Queensland Plateau around 18°S (Church 1987; Ridgway & Dunn 2003). The EAC transports heat polewards (Boland & Church 1981) and tends to separate from the east Australian coast at around 32°S (Godfrey et al. 1980) to form the Tasman Front. However, the large EAC variations pulse much farther polewards than the mean separation point, with EAC meanders and shed eddies often extending to Tasmania. The timing and intensity of EAC changes have been shown to influence both regional climate (e.g. Sprintall et al. 1995) and severe weather events along the east Australian coast (Hopkins & Holland 1997). By identifying the relevant forcing mechanisms and timing that lead to EAC variations in position and intensity, our ability to forecast weather and climate changes across eastern Australia may be enhanced.

Satellite altimeter derived sea surface heights, available since the early 1990s, combined with in situ subsurface temperature and salinity measurements from World Ocean Circulation Experiment (WOCE) hydrographic data, a repeat XBT survey, and data from the global Argo array, confirm that the South Pacific subtropical gyre circulation has been intensifying over the past decade or so (Willis et al. 2004; Qiu & Chen 2006; Roemmich et al. 2007). In addition, historical observations show evidence for a multi-decadal upper ocean warming of the southwest Tasman Sea east of Tasmania (Holbrook & Bindoff 1997; Ridgway 2007). This suggests that the reported intensification ("spin-up") of the South Pacific Gyre circulation

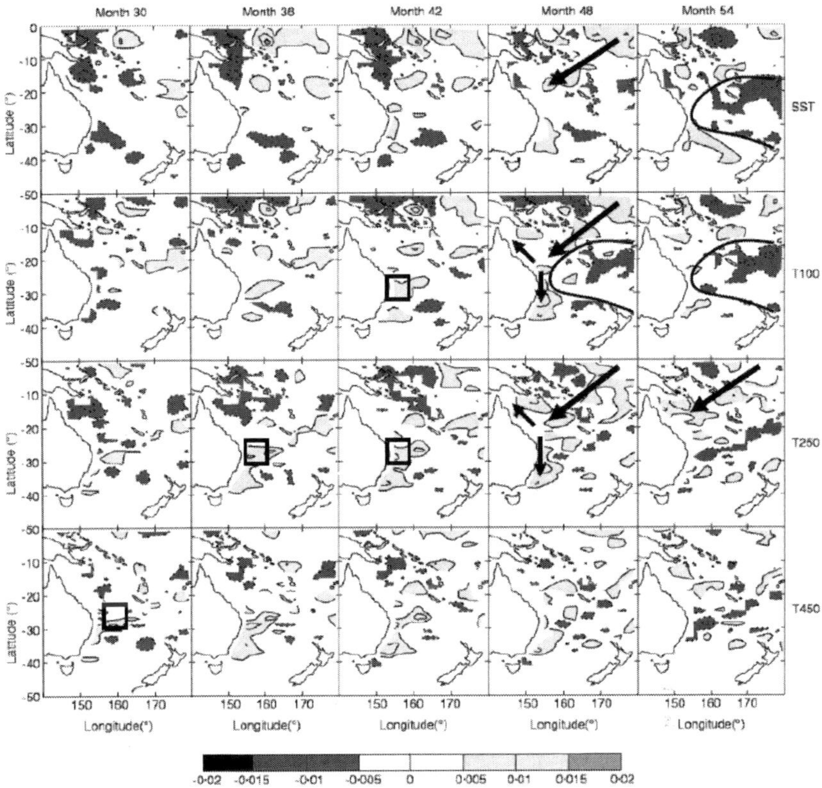

Figure 1 A sample of the mode 2 spatial lag sequence (mode 2 explains 25% of the 3–7-year Butterworth filtered data variance) from an EEOF analysis of 1973–1988 normalised upper ocean temperature anomalies from the Digital Atlas of Southwest Pacific Upper Ocean Temperatures (DASPOT). The complete lag sequence is six-monthly over 72 months (six years) – this sample represents two years in the sequence. Data correlations in the EEOF analysis were taken across four vertical levels – sea surface, 100 m, 250 m and 450 m depth (modified from Holbrook et al. 2005b; Figure 4).

since the early 1990s may actually be part of a multi-decadal, or longer-term, increase in the gyre intensity and southward extension of the EAC . Indirect evidence from corals suggests that this may also be part of a 300-year trend in this EAC behaviour (Thresher et al. 2004).

Since the late 1970s, climate change in the Southern Hemisphere has been marked by a strengthening of the circumpolar westerly winds extending from the stratosphere to the troposphere and Earth's surface (e.g. Hurrell & van Loon 1994; Thompson & Solomon 2002; Gillett & Thompson 2003). Modelling studies suggest that these positive changes in the Southern Annular Mode (SAM) may result from greenhouse warming (e.g. Kushner et al. 2001) and ozone depletion (e.g. Shindell & Schmidt 2004). More recent modelling studies, using a linear Sverdrup ocean model forced with wind stress trends since 1978, suggest that Antarctic ozone depletion and changes in the SAM may be responsible for the southward shift and decadal spin-up of the "super-gyre" since the 1990s, including a strengthening and southward penetration of the EAC through the Tasman Sea (Cai 2006), which would warm the region. Projected warming experiments using the CSIRO Mark-3 climate model have also been performed (Cai et al. 2005) while analogous multi-decadal changes in the circulation were identified in an earlier ocean general circulation model simulation forced by changes in the latitude of the Southern Hemisphere subpolar westerly winds (Oke & England 2004). These studies improve our understanding of 20th century southwest Pacific Ocean climate changes.

Baroclinic Rossby waves play a critical role in South Pacific Ocean climate changes, including changes in western boundary circulation, from seasonal to interannual to multi-decadal time scales (e.g. Holbrook & Bindoff 1999; Qiu & Chen 2006; McGregor et al. 2007, 2008; Kessler & Gourdeau 2007). Taking account of at least first order baroclinic effects is essential in order to successfully model the timing of both ENSO and decadal scale climate changes in this region. The relatively slow westward propagation speeds of baroclinic Rossby waves – which decrease polewards (it takes about two years for Rossby waves to cross the Pacific basin at 10°S and close to a decade at 30°S) – make Rossby waves an ideal mechanism for research and monitoring of the ocean's "memory". Moreover, these waves provide deterministic information with relatively long lead times (months to years and longer) that have the potential to be translated into enhanced forecasts of the timing and amplitude of interannual-to-decadal climate changes over Australia.

This chapter explores and critically appraises recent research undertaken into Pacific Ocean Rossby waves as a mechanism to understand interannual-to-decadal time scale climate changes. The chapter also investigates the connection between first-order baroclinic ocean dynamics, modelled EAC baroclinic component intensity changes, and observed sea level changes at Fort Denison, Sydney. Finally, large scale ocean observing and monitoring of extra-tropical Rossby waves are discussed in terms of potential for enhancing future ENSO climate predictability and improving Australian climate risk management.

2. Historical observations

Recent analysis of 3–7-year Butterworth filtered upper ocean temperature observations from the Digital Atlas of Southwest Pacific upper Ocean Temperatures (DASPOT; objectively mapped historical expendable bathythermograph (XBT) data), using an extended empirical orthogonal function (EEOF) analysis technique, demonstrates that ENSO time scale variability in the southwest Pacific Ocean is dominated by two propagating modes (Holbrook et al. 2005a,b). The first mode EEOF dynamically describes vertical movements of the near-equatorial thermocline, and sea level variations, consistent with the delayed oscillator mechanism. This mechanism consists of westward-propagating long Rossby waves entering the region from the east (equatorward of 10°S) being reflected at the western boundary as eastward-propagating Kelvin waves along the equator. The second mode EEOF is more interesting as it shows the ENSO time-scale connection between the low and mid-latitudes through the large-scale subtropical gyre circulation (see Figure 1). It is consistent with variations in the subtropical gyre circulation strength, including the EAC and its separation, and continuous with the Tasman Front. The upward propagating anomalies over the time scale of a year observed in the western Tasman Sea

(27°–33°S, 157°–161°E) appear to be characteristic of Rossby waves entering this zone. Hence, the historical observations suggest that Rossby waves may be modulating the subtropical gyre intensity, including the EAC and its extension, on ENSO time scales.

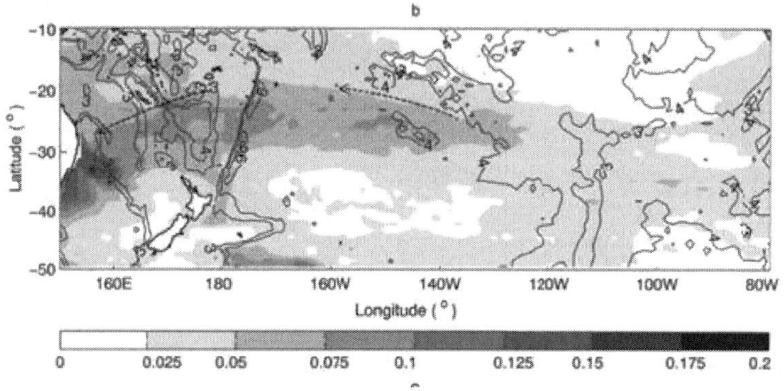

Figure 2 Standard deviation of westward filtered sea level anomalies (m) from European Remote Sensing and TOPEX/Poseidon satellite altimeter observations. The dashed arrows show the westward propagation pathway through the middle of the South Pacific Gyre. Bottom topography is contoured in thousands of metres depth (Source: Figure 2b in Maharaj et al. 2005).

3. Satellite observations

Satellite altimeter observations provide strong evidence for the continuous and uninterrupted westward propagation of Rossby waves throughout the South Pacific Gyre interior (Maharaj et al. 2005; see Figure 2) – although we acknowledge the complex contribution from nonlinear eddy-eddy interactions (Qiu et al. 2008). The important feature to note in Figure 2 is the large scale variability of the westward propagating anomalies identified throughout the interior of the western half of the South Pacific subtropical gyre. The South Equatorial Current impinges on the western South Pacific boundary at around 18°S where it bifurcates to flow southwards as the East

Australian Current and northwards as the Hiri Current. This bifurcation point is well to the north of the strong westward variability identified in the westward filtered satellite altimeter data.

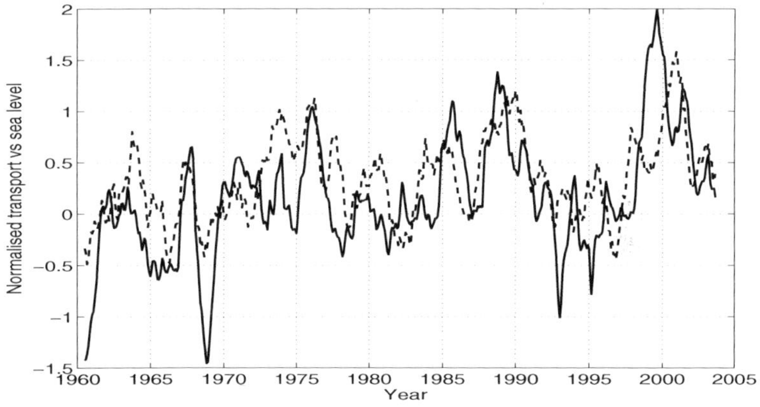

Figure 3 Time series of the modelled monthly EAC baroclinic transport anomaly across 34°S [bold line] plotted against the observed sea level anomaly at Fort Denison, Sydney [dashed line]. An annual running mean has been applied. Sea level lags gyre intensity by nine-month [$r = 0.54$, significant at 99% confidence level after accounting for serial correlation]. The series are normalised and shifted by the nine-month lag for visual comparison.

This highlights two important points: (1) there is large scale, relatively uninterrupted westward propagation throughout the interior of the South Pacific Gyre ; and (2) the largest variability in the westward propagation is observed in the western Tasman Sea region. This western Tasman Sea region includes the EAC, the EAC's mean separation point that forms the Tasman Front, and the EAC extension which consists of variable pulses, meanders and eddy shedding towards Tasmania. It is this region where Rossby waves would be expected to modulate the western boundary flow, affecting the EAC meanders and eddy shedding, and analogous to similar processes observed and modelled in the Kuroshio region (e.g. Jacobs et al. 1994; Qiu &

Chen 2005). This region is mostly well represented by the first baroclinic mode, although it is noteworthy that the second mode also appears to play a significant role in the southwest Pacific between about 28–34°S (Maharaj et al. 2007; see their Figure 3b).

4. Modelling the mechanisms

4.1 Baroclinic ocean model

The baroclinic ocean model , developed by the author of this chapter for the MATLAB environment, is a 1½-layer model of the stratified ocean (e.g. Philander 1990, pp106–108) resolved on a 1° x 1° spatial grid (the Arakawa *C*-grid of Mesinger and Arakawa 1976). The upper and lower ocean density layers are separated by an interface, which represents the pycnocline. Since ocean water density is a function of temperature, salinity and pressure, the pycnocline represents the depths over which the largest vertical density gradient occurs, taking account of all three factors. The reduced gravity, g', reflects the density difference between the upper and lower layers. Following McGregor et al. (2009) we use the geographically-varying reduced-gravity, g', calculated from the geographically-varying first baroclinic mode gravity wave speed of Chelton et al. (1998), assuming a mean depth thermocline H=300m.

Church et al. (2005) have shown that there is a clear relationship [their Fig.1b,c] between global mean sea level (full ocean depth) and global ocean heat content integrated over the upper 300 m. This highlights the importance of the ocean's vertical structure (or "baroclinicity") to sea level, with the upper 300 m being paramount. For our relatively simple 1½-layer baroclinic model ocean used in a large-scale context, we consider the thermocline as a surrogate for the pycnocline. With this in mind, the baroclinic model is configured with a mean depth thermocline of 300m. Changes in the thermocline depth provide the climate base metric to define changes in the upper ocean heat content (or warm water volume).

Motion in the upper layer is driven by the applied wind stresses (per unit density), τ (m^2 s^{-2}), which are anomalies from the long term monthly means (i.e. seasonal cycle removed). The associated response of the ocean is displayed by the vertical displacement of the pycnocline [or thermocline], η (m), and the horizontal components of the flow velocity. The ocean dynamics are described by the linear reduced-gravity form of the shallow-water equations detailed below (Equations 1–3):

$$u_t - fv + g'\eta_x = \tau^x / H \quad (1)$$

$$v_t + fu + g'\eta_y = \tau^y / H \quad (2)$$

$$g'\eta_t + c_1^2 (u_x + v_y) = 0 \quad (3)$$

where u and v are the eastward and northward components of velocity respectively (m s^{-1}) and t is time (s).

The relatively simple model dynamical equations describe the generation and propagation of baroclinic planetary waves in response to large scale wind stress forcing at the surface and oceanic adjustment from horizontal pressure gradients. These planetary waves include not only first baroclinic mode westward propagating Rossby waves, but also boundary Kelvin waves including Kelvin waves along the equatorial boundary. With a mean depth thermocline of 300m, the first baroclinic mode Kelvin wave speed c = 3 m s^{-1}. The first baroclinic mode Rossby wave speed (m s^{-1}) is $C_R = \beta(c_1^2/f^2)$, where f (s^{-1}) is the Coriolis parameter and β (m^{-1} s^{-1}) is the derivative of f northwards. The Fischer's (1965) numerical scheme is utilised for model time stepping, with a time-step of two hours sufficient for a 1° resolution domain. Rossby wave speeds decrease polewards, resulting in Pacific basin transit times of <2 yr at 10°S, ~8 yr at 25°S, and more than 20 yr at 40°S.

4.2 Modelling sea level variations

Climate change is arguably the most significant problem facing our planet. The recent Fourth Assessment Report (AR4) of the Intergovernmental Panel on Climate Change (IPCC) (IPCC 2007) indicates that global sea level rose at an average rate of 1.8 [1.3–2.3] mm per year during the period 1961–2003. Based on oceanic instrumental records since 1955, the global ocean heat content also increased during the past 50 years. Continuous sea level time series extending back before 1950 are sparse and based on palaeo-climatic records and a small number of tide-gauge records. One recent global sea-level reconstruction using instrumental records and empirical orthogonal functions facilitates studies back to 1870 (Church & White 2006). Detecting climate and sea level change above natural variability is an ongoing challenge – model hindcasts help us to understand the important mechanisms that aid this detection.

We explore the potential for the baroclinic ocean model to predict the timing and nature of sea level changes in response to large-scale changes in wind stress. Following recent studies which suggest the increased intensity of the South Pacific Gyre and EAC is caused by strengthened westerlies in the Southern Hemisphere (e.g. Cai et al. 2005; Cai 2006; Qiu & Chen 2006; Roemmich et al. 2007), we use the baroclinic model to investigate the fundamental large-scale ocean dynamic mechanisms that connect the Southern Hemisphere wind stresses with EAC transport anomalies above the thermocline. To test this, in the South Pacific Gyre context, the baroclinic model was forced with National Center for Environmental Prediction/National Center for Atmospheric Research (NCEP/NCAR) wind stresses (Kalnay et al. 1996) across the Southern Hemisphere from 1960–2004. To extend this, we also compare the modelled EAC transport anomalies against the observed Fort Denison sea level changes over the 45-year period.

Modelled southward transport anomaly variations above the thermocline were calculated from the baroclinic model output near to the east Australian coast and across the EAC at 34°S. This model simulation of the upper ocean transport anomalies was compared against observed sea level variations from the long-record and well established Australian east coast tide-gauge site at Fort Denison, Sydney (see Figure 3). The sea level time series is provided by the Permanent Service for Mean Sea Level (PSMSL) [www.pol.ac.uk/psmsl/] (Woodworth & Player 2003). The results show that the relatively simple linear baroclinic model simulation of the upper ocean EAC transport anomalies across 34°S represents an impressive hindcast surrogate of the interannual-to-decadal time scale sea level changes that have been observed at Fort Denison, Sydney since 1960.

While the Fort Denison tide gauge site is contained well within the semi-enclosed Sydney Harbour, and would seem to be well away from direct influences of the EAC, the observed sea level variations are surprisingly well represented by the modelled EAC transport variations. Aside from the direct relationship between the Southern Hemisphere winds and barotropic circulation around the South Pacific Gyre (e.g. Cai 2006), this study demonstrates the significant link that also exists between Fort Denison sea level, the baroclinic component of EAC intensity and large-scale variations in the Southern Hemisphere winds on interannual to multi-decadal time scales. This is useful since the model-determined EAC baroclinic transport and projected sea level variations represent the Pacific Ocean's western boundary response to large-scale climate changes expressed in the Southern Hemisphere winds. Hence this provides scope for model experimental design to investigate and diagnose the role of large-scale climate modes, including ENSO and SAM, on EAC transport and sea level changes. It also provides scope to investigate the role of multi-decadal and long time scale wind stress changes on sea level rise, including human-induced climate change forcing.

4.3 Interannual-to-decadal climate variability and predictability

ENSO is the dominant mode of interannual climate variability observed globally (e.g. Ropelewski & Halpert 1987, 1989; Philander 1990). Substantial advances in our understanding since the late 1960s have demonstrated that ENSO is a natural mode of climate variability that fundamentally exists because of the strong coupling between the tropical ocean and atmosphere (Bjerknes 1969; Zebiak & Cane 1987). Despite this close coupling between the ocean and atmosphere in the tropics, ENSO also results in large-scale teleconnections through the mid- to high latitudes (e.g. Glantz et al. 1991).

Our fundamental understanding of the mechanisms that underpin ENSO can be attributed to several leading paradigms. These are: (1) the delayed action oscillator theory (Suarez & Schopf 1988; Battisti & Hirst 1989); (2) the advective-reflective oscillator theory (Picaut et al. 1997); (3) the western Pacific oscillator theory (Weisberg & Wang 1997); (4) the recharge-discharge oscillator theory (Jin 1997); and (5) the unified oscillator theory (Wang 2001). Rather than detailing the differences between each of these theories here, it is sufficient to point out the strong connection between theories in terms of their collective requirement for coupling the tropical *upper* ocean with the atmosphere as the important mechanism setting the timing of ENSO climate variability.

Aside from Australia's climate being strongly influenced by ENSO (while acknowledging the important signals that exist regionally also from other large scale climate modes, in particular the Indian Ocean Dipole and Southern Annular Mode), it also changes on decadal and longer time scales (Power et al. 1999a,b; Timbal et al. 2006; CSIRO-BoM 2007). But ENSO itself varies on (inter-)decadal time scales, and its impact on Australia appears to be modulated. Specifically, previous research has shown that Australia's rainfall, temperature and wheat crop yield are highly and significantly correlated (at the 95% confidence level) with the Southern Oscillation index (SOI) when the Interdecadal Pacific Oscillation index (IPOI) is less than –0.5, i.e. when the background state of the tropical Pacific Ocean is in its cool phase (Power et al. 1999a). Conversely, there appears to be no relationship between Australia's rainfall, temperature or wheat crop yield and the SOI during the IPOI warm phase (Power et al. 1999a).

This asymmetric relationship between interdecadal climate variability and ENSO raises questions about interannual-to-decadal climate predictability. For example, "Are individual ENSO events predictable in advance?", and "Are multi-year El Niño events (e.g. 1991–1995; Trenberth & Hoar 1996) due to random (stochastic) fluctuations of the climate system or are they intrinsically and deterministically predictable?" With regards to the second question, one could think of a simple experiment where a coin is tossed 14 times, say, and the experiment repeated. In the first series of tosses, the experiment results in the following series (where H is a head and T is a tail): HTHTHTHTHTHTHT. The resultant series of seven Hs and seven Ts are recorded alternatively in a symmetric manner, perhaps naively suggesting that every H should be followed by a T, and vice versa. Of course, the following series is equally likely: HTTHHHHTTTHHTT, which also comprises of seven Hs and seven Ts, but here we can draw the analogy of a background "red noise" (low-frequency) spectrum where we find repeat Hs and Ts in the series. The mechanism of a stochastically generated low-frequency event incidence is one paradigm that one might consider to be the reason for, for example, the increased prevalence of El Niño events between 1976/77 and the end of the 20th century. Alternatively, the climate system may also contain intrinsically deterministic physically-based mechanisms and metrics that, with increased knowledge and effective monitoring, enable us to predict (or enhance probabilistic forecasts of) El Niño event frequency and/or intensity in advance. It is this paradigm that relates deterministic physically-based climate mechanisms with our knowledge of the background climate state that excites interest here in future climate change monitoring.

There are two established paradigms that link Pacific (inter-)decadal climate variability and ENSO event changes. These are: (1) interdecadal variability of ENSO is generated within the tropical Pacific by the physical mechanisms that operate as part of the interannual ENSO cycle (e.g. Tziperman et al. 1994; Newman et al. 2003; Timmermann et al. 2003; Rodgers et al. 2004; Power et al. 2006); and (2) low-frequency extra-tropical ocean exchanges with the tropical Pacific generate SST variations on interdecadal time scales, i.e., physical mechanisms not directly operating within the interannual ENSO cycle (e.g. Gu & Philander 1997; Kleeman et al. 1999; Capotondi & Alexander 2001; Giese et al. 2002; Galanti & Tziperman 2003; Luo et al. 2003; Wang et al. 2003; McGregor et al. 2004; Tourre et al. 2005). The first paradigm is concerned with pure randomness (i.e. toss of a coin analogy) and/or non-linearities in (the higher frequency) ENSO processes. In contrast, the second paradigm considers the longer adjustment times of the off-equatorial Pacific Ocean (from latitudes greater than +/−10°) as an important mechanism which helps to determine the magnitude (the timing is set by mechanisms utilised in the ENSO system) of the equatorial region climate variability. One such mechanism here is the subtropical cell which impacts ENSO by subtropical-tropical exchanges through the thermocline, with recent analyses of the AR4 models showing increased western boundary pathways under global warming (Luo et al. 2009).

An excellent example of the second paradigm with a focus on the extra-tropical Rossby wave mechanism is provided in the study by Jacobs et al. (1994). Using a combination of wind stress forced model sea surface height (SSH) anomalies and SSH observations from satellite altimeter and derived from in situ data, it was shown that the northward re-routing of the Kuroshio Extension in 1993 was the result of a decadal-scale transit of a mid-latitude North Pacific Rossby wave following the 1982/83 El Niño event (see Figure 3 of Jacobs et al. 1994). This Rossby wave mechanism is the hypothesis applied here to the South Pacific. In more recent modelling studies it has been shown that extra-tropical Rossby waves generated polewards of 12.5°S, and which are able to propagate to the western boundary, will reflect and transform into equatorward propagating coastal Kelvin waves. They then propagate eastwards as equatorial Kelvin waves and affect significant changes to the NINO3 region thermocline depth and SSTs more than two years later (McGregor et al. 2007, 2008). This >2-yr lead time information may be crucial for enhancing ENSO forecasts.

The implications for climate predictability are further considered in terms of the following question: "Can understanding long time-scale changes in the equatorial background state improve ENSO forecasts?" This question was considered in a recent stochastically forced coupled modelling study by McGregor et al. (2009) using the same dynamic baroclinic ocean model together with a linear SST anomaly (SSTA) equation dependent on the thermocline depth displacement and longitude (Kleeman 1993), coupled with a relatively simple steady-state, two-pressure level (250 mb and 750 mb) atmosphere (Kleeman 1991). The important result of this study is Rossby-Kelvin wave induced exchanges between the extra-tropics and the tropics can act to modulate modelled ENSO behaviour (see Figure 4). The utility of this result is demonstrated through an interesting thought experiment. First, consider the likelihood of an ENSO event occurring at least a year ahead. This is represented as a probability density function (PDF) of our *a priori* knowledge of ENSO event likelihood (this PDF is based on the modelled NINO3 region SSTA) (see Figure 4, solid line). This knowledge is based on historical experience, and contains very little in the way of deterministic skill. In practical terms, there is a 90% chance of SSTA falling between +/−1.6°C, whereas there is an equally likely chance of an El Niño exceeding +1.6°C or a La Niña exceeding −1.6°C. Next, consider instead that we take account of information regarding the extra-tropically induced background state of the NINO3 thermocline depth – this might include observations of the IPO phase and its contribution to the background state. The results of McGregor et al. (2009) show that an extra-tropically induced deeper (shallower) equatorial region thermocline acts to shift the PDF of ENSO variability towards more El Niño (La Niña) like conditions. Thus, the forecast of an extra-tropical Rossby-Kelvin wave induced equatorial thermocline deepening of +0.4m (and corresponding increase in SST) can provide deterministic prediction skill of increased El Niño event likelihood and/or intensity up to >2-yr in advance. Conversely, the prediction of a 0.4m decrease (-0.4m) in the NINO3 region thermocline depth (and corresponding decrease in SST) leads one to choose the left hand PDF and therefore an increased La Niña event likelihood and/or intensity.

4.4 Human-induced climate change: the Rossby wave speed-up

Recent observational analyses, based on satellite altimeter derived sea surface heights during the 1990s, showed that long Rossby wave speeds through the low to mid-latitudes travel up to twice as fast as the standard theory predicts (Chelton & Schlax 1996; Killworth et al. 1997; Chelton et al. 1998). Closer investigation of the standard theory shows that westward propagating planetary wave speeds are predicted to increase with increasing density stratification (Gill 1982). Historical observations show that the human-induced upper ocean warming over the last 40 years has tended to increase the density contrast between the surface and the deep ocean (Barnett et al. 2005; Delworth et al. 2005; Pierce et al. 2006). A recent modelling analysis of the low latitude North Pacific (to 20°N) suggests that the observed human-induced changes in ocean stratification would be expected to cause a speed-up of oceanic planetary (Rossby) waves in a manner similar to the present-day mismatch between satellite observed Rossby wave speeds and predictions from the standard theory (Fyfe & Saenko 2007). This causal increase in Rossby wave speeds due to human-induced ocean stratification changes is a factor that will need to be considered in future mechanistic studies that aim to predict the timing of future climate events.

5. *Climate change monitoring potential*

The ocean and atmosphere clearly operate on different time scales. In simple "rule-of-thumb" terms, it has been said that atmospheric processes are defined by large space and short time scales, while ocean processes comprise of small space and long time scales. Intense atmospheric low pressure weather systems (atmospheric eddies) such as tropical cyclones, for example, have typical horizontal scales of around 1000 km and time scales of a few days to a couple of weeks. In contrast, oceanic eddies, which are the analogy for ocean weather events, have typical horizontal scales of 30–100 km and may last for several months or longer. Hence the ocean as the slow component of the coupled climate system provides excellent potential to act as an informing agent for deterministic climate prediction.

Planetary wave dynamics (Rossby-Kelvin waves) play a critical role in setting the timing of climate variations through the low to mid-latitudes globally. For example, ENSO represents the dominant interannual mode of climate variability across the globe – a climate mode intimately dependent on the atmosphere's coupling with, and response to, low-latitude ocean planetary wave dynamics in the Pacific Ocean. On decadal to multi-decadal time scales, Pacific region climate variability, identified in the recent literature as the Interdecadal Pacific Oscillation, also appears to contain a very important and significant dependence on the atmosphere's interaction with ocean planetary wave dynamics – as discussed above. To summarise: (1) there is an important and significant dependence of ENSO (interannual) and Pacific Ocean interdecadal time scale changes on planetary wave dynamics (Section 4.3); (2) extra-tropical oceanic Rossby waves provide potential for deterministic climate prediction of ENSO event likelihood at least two years in advance (Section 4.3); (3) based on previous analyses of observed and filtered historical upper ocean temperatures and satellite altimeter sea surface heights, there appears to be an important and significant ENSO time scale modulation of the East Australian Current intensity, and its extension, that is most likely due to westward propagating Rossby waves reaching the western boundary through the South Pacific Gyre interior (Section 2); and (4) using a relatively simple wind stress forced linear baroclinic ocean model, based primarily on planetary (Rossby & Kelvin) ocean wave dynamics, the modelled East Australian Current intensity changes at 34°S provides a highly significant and impressive hindcast surrogate of the observed monthly sea level changes recorded at the Fort Denison tide gauge in Sydney Harbour, New South Wales (Section 4.2).

Of particular climatic interest is the possible connection between large scale climate predictors and regional rainfall. Here we ask "Is there any connection between changes in the South Pacific Gyre circulation and New South Wales rainfall?" To provide a first-order answer to this question, New South Wales summer rainfall is plotted against the modelled EAC baroclinic transport anomalies at 34°S and observed sea level variations at the Fort Denison tide gauge shown earlier (Figure 5). Interestingly, Figure 5 shows a very strong relationship between NSW rainfall variations and variations in the modelled gyre circulation/observed sea level at 34°S between 1960 and 1976. However, between 1976 and the early part of the 21st century, there appears to be no relationship between the gyre/sea level and NSW rainfall. This result corroborates the previous research demonstrating the very strong and highly significant relationship between all-Australian rainfall and the SOI during the Interdecadal Pacific Oscillation (IPO) negative phase (i.e. prior to 1976/77) compared with the absence of any relationship between all-Australian rainfall and the SOI during the IPO positive phase (i.e. following 1976/77) (Power et al. 1999a). While this study demonstrates the existence of a strong and highly significant relationship between the modelled South Pacific Gyre circulation, based purely on linear first baroclinic-mode dynamics, and observed Fort Denison sea level, the further connection with NSW summer rainfall is only evident in our record during the IPO negative phase prior to the 1976/77 climate shift (e.g. Giese et al. 2002; Rakich et al. 2008). Hence, while there is a robust multi-decadal connection between modelled EAC intensity and Fort Denison sea level over the entire 45-year record, the disconnect between Fort Denison sea level and NSW summer rainfall changes following the 1976/77 climate shift is symptomatic of the phase change in IPO (as reported Power et al. 1999a). This disconnect amounts to there apparently being no relationship following the 1976/77 shift between year-to-year climate forced gyre variations, NSW summer rainfall and expected changes in Parramatta River discharge into Sydney Harbour which might otherwise also affect Fort Denison sea level. Alternatively, the 'predictive potential' of NSW summer rainfall might be possible by our increased understanding of wind forced Rossby wave dynamics on gyre intensity (and sea level), but only during the negative phase of the IPO.

The connection between climate variations on ENSO-to-interdecadal time scales and planetary wave dynamics in the South Pacific Ocean affords excellent potential for the value of large scale climate monitoring and observing systems. Internationally, the Southwest Pacific Ocean Circulation and Climate Experiment (SPICE: Ganachaud et al. 2007, 2008) aims to observe, simulate, and understand the role of the southwest Pacific Ocean circulation in (i) the large-scale, low-frequency modulation of climate from the low to mid-latitudes, and (ii) the generation of local climate signatures whose diagnosis will aid regional sustainable development. A recent major investment through the National Collaborative Research Infrastructure Scheme (NCRIS) to establish an Australian Integrated Marine Observing System (IMOS) makes an important contribution to Australia's future ocean climate prediction capability. In the context of South Pacific Ocean climate monitoring and prediction, it will be particularly important to take full advantage of the basin-wide coverage and real-time in situ temperature and salinity profiles provided by the significant introduction of Argo float technology into the world's oceans since the beginning of the 21st century together with repeat-track sea surface heights provided by the JASON satellite altimeter. The uncertainty surrounding future satellite altimeter programs, and the longevity of satellite altimeter observing systems, must be taken seriously and is nevertheless of concern for climate monitoring (and prediction) into the future.

The planned and funded deployment of IMOS moorings across the New South Wales shelf off Sydney (M Roughan et al., Chapter 8 in this book) will be necessary for future monitoring of the cross-shelf transport connections between East Australian Current intensity changes along the southwest Pacific shelf break and changes within the coastal zone. A measure of the utility of climate change monitoring in the South Pacific will be determined by its effectiveness in integrating the monitored (and predicted) large scale South Pacific Ocean dynamics and its connection with the coastal zone. This presents a great challenge for the future of an effective integrated observing system, and should be considered as part of the Australian Climate Change and Earth Observatory being planned and developed by the Centre for Australian Weather and Climate Research (Cleugh 2008).

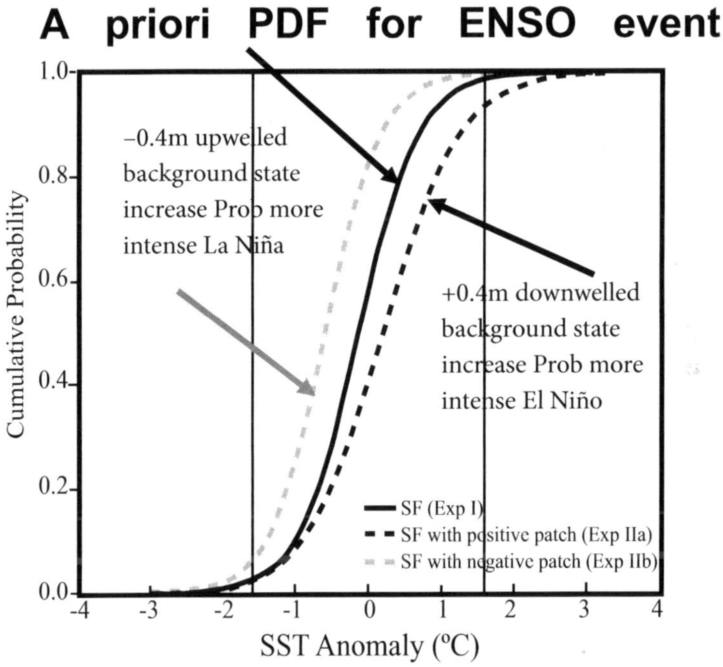

Figure 4 Cumulative probability distribution represented by a fitted PDF describing our a priori knowledge about ENSO event likelihood (solid line). Scenario of a wind stress forced extra-tropical Rossby wave that can potentially provide us with advance (lead) information regarding the downwelling (upwelling) of the Niño 3 region thermocline and the increased likelihood of a more intense El Niño (La Niña) event (modified from Figure 8 in McGregor et al. 2009a).

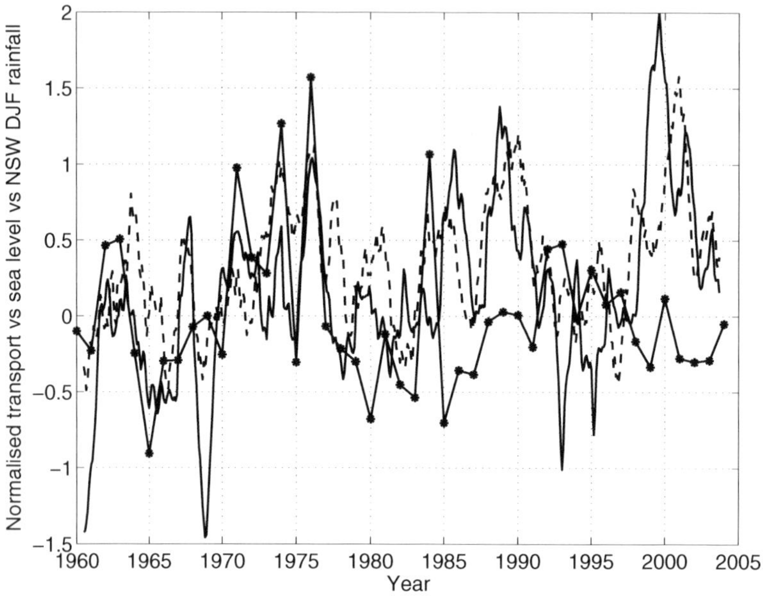

Figure 5 As in Figure 3 but here including the corresponding time series of total New South Wales summer rainfall variations (asterisked solid line).

6. Summary and conclusions

Historical observations in the southwest Pacific Ocean indicate that ENSO time scale variability is not only observed in the equatorial and tropical regions of the South Pacific but is also observed through the western extent of the South Pacific Gyre. Both observations and modelling studies suggest that Rossby waves play a role in modulating the South Pacific Gyre and East Australian Current intensity on interannual time scales. Using a relatively simple (intermediate-complexity) 1½-layer reduced-gravity model forced with observed wind stresses around the South Pacific Gyre, we find an impressive relationship between the modelled East Australian Current monthly baroclinic transport changes and observed sea level changes at Fort Denison. Despite the Fort Denison tide gauge site being located well within

Sydney Harbour, observed sea level at this site varies in manner that is significantly correlated with modelled linear baroclinic ocean dynamics – that is, the modelled intensity of the first-order baroclinic component of the EAC at 34°S appears to be well connected to interannual to decadal scale sea level variations within Sydney Harbour. Hence, EAC intensity and sea level are effectively modulated in a fundamental way by wind stress forced first baroclinic-mode ocean dynamics.

Finally, intermediate-complexity modelling investigations of extra-tropical oceanic Rossby waves in the South Pacific suggest that enhanced predictions of ENSO event intensity may be possible. With ongoing and future improvements in ocean observing and monitoring systems across the South Pacific, together with recent coupled ocean-atmosphere modelling research suggesting extra-tropical ocean Rossby waves can determine ENSO event intensity, there appears to be significant potential for enhancing ENSO event forecasts, up to two years or more in advance, in the future. Based on the relationship between first baroclinic-mode ocean dynamics, EAC intensity and sea level demonstrated in this chapter, there is also the potential for ocean monitoring of extra-tropical Rossby waves to provide useful advance information through the subtropics to mid-latitudes.

Acknowledgements

I am grateful to Shayne McGregor, Angela Maharaj, Scott Power and Paolo Cipollini for the many discussions over recent years that have ultimately helped motivate this research synthesis of the observations, modelled climate mechanisms and predictability. This provided important background to the new research presented here investigating the link between wind-stress forced modelled East Australian Current intensity and observed sea level at Fort Denison, Sydney. I want to extend my thanks to Clinton Rakich at the Australian Bureau of Meteorology for providing the New South Wales summer rainfall data. I wish to finally also thank the two anonymous reviewers for their constructive comments that have helped to improve the text. This chapter is a contribution to the Southwest Pacific Ocean Circulation and Climate Experiment (SPICE) and the Eastern Seaboard Climate Change Initiative (ESCCI).

References

Barnett TP, Pierce DW, AchutaRao KM, Gleckler PJ, Santer BD, Gregory JM & Washington WM (2005). Penetration of human-induced warming in the world's oceans. *Science,* 309, 284–87.

Battisti DS & Hirst AC (1989). Interannual variability in the tropical atmosphere-ocean system: influence of the basic state and ocean geometry. *J Atmos Sci,* 46, 1687–712.

Bjerknes J (1969). Atmospheric teleconnections from the equatorial Pacific. *Mon Wea Rev,* 97, 163–72.

Boland FM & Church JA (1981). The east Australian current 1978. *Deep-Sea Res,* 28, 937–57.

Boulanger J-P & Menkes C (1999). Long equatorial wave reflection in the Pacific Ocean from TOPEX/POSEIDON data during the 1992–1998 period. *Clim Dyn,* 15, 205–25.

Cai W (2006). Antarctic ozone depletion causes an intensification of the Southern Ocean super-gyre circulation. *Geophys Res Lett,* 33, L03712, doi: 10.1029/2005GL024911.

Cai W, Shi G, Cowan T, Bi D & Ribbe J (2005). The response of the Southern Annular Mode, the East Australian Current, and the southern mid-latitude ocean circulation to global warming. *Geophys Res Lett,* 32, L23706, doi: 10.1029/2005GL024701.

Capotondi A & Alexander MA (2001). Rossby waves in the tropical Pacific and their role in decadal thermocline variability. *J Phys Oceanogr,* 31, 3496–515.

Chelton DB & Schlax MG (1996). Global observations of oceanic Rossby waves. *Science,* 272, 234–38.

Chelton DB, de Szoeke RA, Schlax MG, El Naggar K & Siwertz N (1998). Geographical variability of the first baroclinic Rossby radius of deformation. *J Phys Oceanogr,* 28, 433–60.

Church JA (1987). East Australian Current adjacent to the Great Barrier Reef. *Aust J Mar Freshw Res,* 38, 671–83.
Church JA & White NJ (2006). A 20th century acceleration in global sea-level rise. *Geophys Res Lett,* 33, L01602, doi: 10.1029/2005GL024826.
Church JA, White NJ & Arblaster JM (2005). Significant decadal scale impact of volcanic eruptions on sea level and ocean heat content. *Nature,* 438, 74–77.
Clarke AJ (1991). On the reflection and transmission of low-frequency energy at the irregular western Pacific Ocean boundary. *J Geophys Res,* 96 (Supplement), 3289–305.
CSIRO-BoM (2007). *Climate Change in Australia.* Technical Report, CSIRO, 148 pp.
Cleugh HA (2008). *An Australian climate change and Earth observatory.* Presented at the First National Climate Change Monitoring Symposium, Webster Theatre, Veterinary Science Conference Centre, University of Sydney, 12 December 2008.
Du Penhoat Y & Cane MA (1991). Effect of low-latitude western boundary gaps on the reflection of equatorial motions. *J Geophys Res,* 96, 3307–22.
Delworth TL, Ramaswamy V & Stenchikov GL (2005). The impact of aerosols on simulated ocean temperature and heat content in the 20th century. *Geophys Res Lett,* 32, L24709, doi: 10.1029/2005GL024457.
Fischer G (1965). On a finite-difference scheme for solving the primitive equations for a barotropic fluid with application to the boundary problem. *Tellus,* 17, 405–12.
Fyfe JC & Saenko OA (2007). Anthropogenic speed-up of oceanic planetary waves. *Geophys Res Lett,* 34, L10706, doi: 10.1029/2007GL029859.
Galanti E & Tziperman E (2003). A midlatitudes-ENSO teleconnection mechanism via baroclinically unstable long Rossby waves. *J Phys Oceanogr,* 33, 1877–88.
Ganachaud A, Kessler W, Wijffels S, Ridgway K, Cai W, Holbrook N et al. (2007). Southwest Pacific Ocean Circulation and Climate Experiment

(SPICE) – Part I. Scientific Background. International CLIVAR Project Office, CLIVAR Publication Series No. 111, NOAA OAR Special Report, NOAA/OAR/PMEL, Seattle, WA, 37 pp.

Ganachaud A, Brassington G, Kessler W, Mechoso CR, Wijffels S, Ridgway K et al. (2008). Southwest Pacific Ocean Circulation and Climate Experiment (SPICE) – Part II. Implementation Plan. International CLIVAR Project Office, CLIVAR Publication Series.

Giese BS, Urizar SC & Fuckar NS (2002). Southern Hemisphere origins of the 1976 climate shift. *Geophys Res Lett,* 29, 1014, doi: 10.1029/2001GL013268.

Gill AE (1982). *Atmosphere-ocean dynamics.* New York: Academic Press.

Gille ST (2002). Warming of the Southern Ocean since the 1950s. *Science,* 295, 1275–77.

Gillett NP & Thompson DWJ (2003). Simulation of recent Southern Hemisphere climate change. *Science,* 302, 273–75.

Glantz MH, Katz RW & Nicholls N (1991). *Teleconnections linking worldwide climate anomalies: scientific basis and societal impact.* Cambridge, UK and New York: Cambridge University Press.

Godfrey JS, Cresswell GR, Golding TJ, Pearce AF & Boyd R (1980). The separation of the East Australian Current. *J Phys Oceanogr,* 10, 430–40.

Gu D & Philander SGH (1997). Interdecadal climate fluctuations that depend on exchanges between the Tropics and extratropics. *Science,* 275, 805–07.

Holbrook NJ & Bindoff NL (1997). Interannual and decadal temperature variability in the southwest Pacific Ocean between 1955 and 1988. *J Climate,* 10, 1035–49.

Holbrook NJ & Bindoff NL (1999). Seasonal temperature variability in the upper southwest Pacific Ocean. *J Phys Oceanogr,* 29, 366–81.

Holbrook NJ, Chan PS-L & Venegas SA (2005a). Oscillatory and propagating modes of temperature variability at the 3–3.5- and 4–4.5-yr time scales in the upper southwest Pacific Ocean. *J Climate,* 18, 719–36.

Holbrook NJ, Chan PS-L & Venegas SA (2005b). Corrigendum. *J Climate,* 18, 1637–39.

Hopkins LC & Holland GJ (1997). Australian heavy-rain days and associated east coast cyclones: 1958–92. *J Climate,* 10, 621–35.

Hurrell JW & van Loon HA (1994). Modulation of the atmospheric annual cycle in the Southern Hemisphere. *Tellus Ser A,* 46, 325–38.

IPCC (2007). *Climate change 2007: the physical science basis – summary for policy makers*. Contribution of Working Group I to the Fourth Assessment Report of the Intergovernmental Panel on Climate Change. Available at: www.aaas.org/news/press_room/climate_change/media/4th_spm2feb07.pdf

Jacobs GA, Hurlburt HE, Kindle JC, Metzger EJ, Mitchell JL, Teague WJ & Wallcraft AJ (1994). Decade-scale trans-Pacific propagation and warming effects of an El Niño anomaly. *Nature,* 370, 360–63.

Jin F–F (1997). An equatorial ocean recharge paradigm for ENSO. Part I: conceptual model. *J Atmos Sci,* 54, 811–29.

Kalnay E, Kanamitsu M, Kistler R, Collins W, Deaven D, Gandin L, Iredell M, Saha S, White G, Woollen J, Zhu Y, Leetmaa A, Reynolds R, Chelliah M, Ebisuzaki W, Higgins W, Janowiak J, Mo KC, Ropelewski C, Wang J, Jenne R, and Joseph D (1996). NCEP/NCAR 40-year reanalysis project. *Bull Amer Meteor Soc,* 77, 437–41.

Karoly DJ (1989). Southern Hemisphere circulation features associated with El Niño – Southern Oscillation events. *J Climate,* 2, 1239–52.

Killworth PD, Chelton DB & de Szoeke RA (1997). The speed of observed and theoretical long extra-tropical planetary waves. *J Phys Oceanogr,* 27, 1946–66.

Kleeman R (1991). A simple model of the atmospheric response to ENSO SST anomalies. *J Atmos Sci,* 48, 3–18.

Kleeman R (1993). On the dependence of hindcast skill on ocean thermodynamics in a coupled ocean-atmosphere model. *J Climate,* 6, 2012–33.

Kleeman R, McCreary JJP & Klinger BA (1999). A mechanism for generating ENSO decadal variability. *Geophys Res Lett,* 26, 1743–46.

Kushner P, Held I & Delworth T (2001). Southern Hemisphere atmospheric circulation response to global warming. *J Climate,* 14, 2238–49.
Luo J-J, Masson S, Behera S, Delecluse P, Gualdi S, Navarra A & Yamagata T (2003). South Pacific origin of the decadal ENSO-like variation as simulated by a coupled GCM. *Geophys Res Lett,* 30, 2250, doi: 10.1029/2003GL018649.
Luo Y, Rothstein LM & Zhang R-H (2009). Response of Pacific subtropical-tropical thermocline water pathways and transports to global warming. *Geophys Res Lett,* 36, L04601, doi: 10.1029/2008GL036705.
Maharaj AM, Cipollini P & Holbrook NJ (2005). Observed variability of the South Pacific westward sea level anomaly signal in the presence of bottom topography. *Geophys Res Lett,* 32, L04611, doi: 10.1029/2004GL020966.
Maharaj AM, Cipollini P, Holbrook NJ, Killworth PD & Blundell JR (2007). An evaluation of the classical and extended Rossby wave theories in explaining spectral estimates of the first few baroclinic modes in the South Pacific Ocean. *Ocean Dyn,* 57(3), 173–87.
McGregor S, Holbrook NJ & Power SB (2004). On the dynamics of interdecadal thermocline depth and sea surface temperature variability in the low to mid-latitude Pacific Ocean. *Geophys Res Lett,* 31, L24201, doi: 10.1029/2004GL021241.
McGregor S, Holbrook NJ & Power SB (2007). Interdecadal sea surface temperature variability in the equatorial Pacific Ocean. Part I: the role of off-equatorial wind stresses and oceanic Rossby waves. *J Climate,* 20, 2643–58.
McGregor S, Holbrook NJ & Power SB (2008). Interdecadal sea surface temperature variability in the equatorial Pacific Ocean. Part II: the role of equatorial/off-equatorial wind stresses in a hybrid coupled model. *J Climate*, 21, 4242–56.

McGregor S, Holbrook NJ & Power SB (2009a). The response of a stochastically forced ENSO model to observed off-equatorial wind-stress forcing. *J Climate*, 22, 2512–25.
Mesinger F & Arakawa A (1976). Numerical methods used in atmospheric models. GARP Publication Series 17, 64 pp.
Newman M, Compo GP & Alexander MA (2003). ENSO-forced variability of the Pacific decadal oscillation. *J Climate,* 16(23), 3853–57.
Oke PR & England MH (2004). Oceanic response to changes in the latitude of the Southern Hemisphere subpolar westerly winds. *J Climate,* 17, 1040–54.
Picaut J, Masia F & du Penhoat Y (1997). An advective-reflective conceptual model for the oscillatory nature of the ENSO. *Science,* 277(5326), 663–66. doi: 10.1126/science.277.5326.663.
Philander SGH (1990). *El Niño, La Niña, and the Southern Oscillation.* International Geophysics Series, Vol. 46, Academic Press.
Pierce DW, Barnett TP, AchutaRao KM, Gleckler PJ, Gregory JM & Washington WM (2006). Anthropogenic warming of the oceans: Observations and model results. *J Climate,* 19, 1873–900.
Power S, Casey T, Folland C, Colman A & Mehta V (1999a). Inter-decadal modulation of the impact of ENSO on Australia. *Clim Dyn,* 15, 319–24.
Power S, Haylock M, Colman R & Wang X (2006). The predictability of interdecadal changes in ENSO and ENSO teleconnections. *J Climate,* 8, 2161–80.
Power S, Tseitkin F, Mehta V, Lavery B, Torok S & Holbrook N (1999b). Decadal climate variability in Australia during the twentieth century. *Int J Climatol,* 19, 169–84.
Qiu B & Chen S (2005). Variability of the Kuroshio Extension jet, recirculation gyre, and mesoscale eddies on decadal time scales. *J Phys Oceanogr,* 35, 2090–103.
Qiu B & Chen S (2006). Decadal variability in the large-scale sea surface height field of the South Pacific Ocean: Observations and causes. *J Phys Oceanogr,* 36, 1751–62.

Qiu B, Scott RB & Chen S (2008). Length scales of eddy generation and nonlinear evolution of the seasonally modulated South Pacific Subtropical Countercurrent. *J Phys Oceanogr*, 38, 1515–28.

Rakich CS, Holbrook NJ & Timbal B (2008). A pressure gradient metric capturing planetary-scale influences on eastern Australian rainfall. *Geophys Res Lett*, 35, L08713, doi: 10.1029/2007GL032970.

Ridgway KR (2007). Long-term trend and decadal variability of the southward penetration of the East Australian Current. *Geophys Res Lett*, 34, L13613, doi: 10.1029/2007GL030393.

Ridgway KR & Dunn JR (2003). Mesoscale structure of the mean east Australian current system and its relationship with topography. *Prog Oceanogr*, 56, 189–222.

Rodgers KB, Friederichs P & Latif M (2004). Tropical Pacific decadal variability and its relation to decadal modulations of ENSO. *J Climate*, 17, 3761–74.

Roemmich D, Gilson J, Davis R, Sutton P, Wijffels S & Riser S (2007). Decadal spinup of the South Pacific subtropical gyre. *J Phys Oceanogr*, 37, 162–73.

Ropelewski CF & Halpert MS (1987). Global and regional scale precipitation patterns associated with El Niño – Southern Oscillation. *Mon Wea Rev*, 115, 1606–26.

Ropelewski CF & Halpert MS (1989). Precipitation patterns associated with high index phase of Southern Oscillation. *J Climate*, 2, 268–84.

Roughan M, Suthers I & Meyers G (2010). The Australian Integrated Marine Observing System (IMOS) and the regional implementation in New South Wales. In Y You & A Henderson-Sellers (Eds), *Climate alert: climate change monitoring and strategy*, (pp299–312). Sydney: Sydney University Press.

Shindell DT & Schmidt GA (2004). Southern Hemisphere climate response to ozone changes and greenhouse gas increases. *Geophys Res Lett*, 31, L18209, doi: 10.1029/2004GL020724.

Sprintall J, Roemmich D, Stanton B & Bailey R (1995). Regional climate variability and ocean heat transport in the southwest Pacific Ocean. *J Geophys Res,* 100, 15865–71.
Suarez M & Schopf P (1988). A delayed action oscillator for ENSO. *J Atmos Sci,* 45, 3283–87.
Thompson DWJ & Solomon S (2002). Interpretation of recent Southern Hemisphere climate change. *Science,* 296, 895–99.
Thresher R, Rintoul SR, Koslow JA, Weidman C, Adkins J & Proctor C (2004). Oceanic evidence of climate change in southern Australia over the last three centuries. *Geophys Res Lett,* 31, L07212, doi: 10.1029/2003GL018869.
Timbal B, Arblaster J & Power S (2006). Attribution of late-twentieth-century rainfall decline in southwest Australia. *J Climate,* 19, 2046–62.
Timmermann A, Jin F-F & Abshagen J (2003). A nonlinear theory for El Niño bursting. *J Atmos Sci,* 60, 152–65.
Tourre YM, Cibot C, Terray L, White WB & Dewitte B (2005). Quasi-decadal and inter-decadal climate fluctuations in the Pacific Ocean from a CGCM. *Geophys Res Lett,* 32, L07710, doi: 10.1029/2004GL022087.
Trenberth KE & Hoar TJ (1996). The 1990–1995 El Niño – Southern Oscillation event: longest on record. *Geophys Res Lett,* 23(1), 57–60.
Tziperman E, Stone L, Cane MA & Jarosh H (1994). El Niño chaos: overlapping of resonances between the seasonal cycle and the Pacific ocean-atmosphere oscillator. *Science,* 264, 72–74.
Wang C (2001). A unified oscillator model for the El Niño – Southern Oscillation. *J Climate,* 14, 98–115.
Wang X, Jin F-F & Wang Y (2003). A tropical recharge mechanism for climate variability. Part II: A unified theory for decadal and ENSO modes. *J Climate,* 16, 3599–616.
Weisberg RH & Wang C (1997). A western Pacific oscillator paradigm for the El Niño – Southern Oscillation. *Geophys Res Lett,* 24(7), 779–82.
Willis JK, Roemmich D & Cornuelle B (2004). Interannual variability in the upper ocean heat content, temperature, and thermosteric

expansion on global scales. *J Geophys Res,* 109, C12036, doi: 10.1029/2003JC002260.
Woodworth PL & Player R (2003). The permanent service for mean sea level: an update to the 21st century. *J Coastal Res,* 19, 287–95.
Zebiak SE & Cane MA (1987). A model El Niño – Southern Oscillation. *Mon Wea Rev,* 115, 2262–78.

11

Future cyclone season characteristics from a thermodynamic model driven by coupled model climate simulations

Kendal McGuffie

*Abstract**

The characteristics of tropical cyclone seasons under enhanced greenhouse conditions (SRESB1, stabilisation at 550 ppmv) have been examined using results from five CMIP3 models. The tropical cyclone season in this study is characterised by the integrated depression of maximum potential intensity below the mean monthly sea level pressure for grid boxes characteristic of the region of peak tropical cyclone activity for each tropical cyclone basin. Results are analysed as anomalies from the long term mean. There is an overall tendency for more severe cyclone seasons (either increased length of the season or increased maximum potential intensity) although this varies between basins. The method does not take into account changes in dynamics, which can act to mitigate against tropical cyclone development, but concentrates on the enhancement of the thermodynamic environment in which storms develop. Analysis of the changes on a mean case and worst case basis for a 550 ppmv greenhouse scenario indicate changes in maximum potential intensity of tropical cyclones of roughly half a category and one category respectively. Increases in maximum potential intensity are found to occur differentially across basins and it seems likely that no simple global conclusion can be made on overall intensity from this study, but a

* Chapter originally submitted in February 2009.

consistent prediction for increases in months of most intense cyclone activity of around 1 category is found.

1. *Tropical cyclones*

1.1 Introduction

Tropical cyclones, typhoons and hurricanes, being all essentially the same atmospheric phenomenon – warm-cored cyclonic storms with an intense core of destructive winds; are the most damaging natural catastrophe to affect human society. High winds, flooding and storm surge cause considerable damage and loss of life in both developed and developing nations. The potential for changes, particularly increases, in either intensity or frequency of storms as the climate changes in response to enhanced CO_2 has been the subject of considerable research and debate (e.g. Henderson-Sellers et al. 1998; Emanuel 2005, 2007a; Landsea et al. 2006; Mann & Emanuel 2006). Various techniques have been employed to address the problem, but there is limited consensus on the answer, partly because of the difficulties of modelling atmospheric disturbances of this scale and partly because of the confounding array of metrics associated with tropical cyclone activity. There is no reason why numbers of storms, intensity of storms and size of storms should change in the same way and the difficulty in characterising future tropical cyclone activity is further increased by the differing nature of climate responses in the tropical cyclone basins around the world. For this reason, it is important to maintain a broad approach to the analysis of future tropical cyclone climates. This chapter aims to introduce and document a process that can provide a relatively straightforward uniform global analysis of the changes in the environment in which tropical cyclones develop as modelled by general circulation models in the World Climate Research Programme Coupled Model Intercomparison Project, Phase 3 (WCRP/CMIP3) database (Meehl et al. 2007).

1.2 Past assessments

Some time ago, following on from the IPCC second assessment reports, Henderson-Sellers et al. (1998) made a comprehensive survey of the tropical cyclone community in order to form a coherent picture of the state of knowledge of potential effects of greenhouse warming on tropical cyclones. On the basis of a comprehensive analysis of modelling and observational studies available at the time, Henderson-Sellers et al. (1998) concluded that the broad geographic areas of cyclogenesis were not likely to change significantly in an enhanced CO_2 climate, but that the changed thermodynamic environment would likely lead to enhanced storm intensity of around 10–20%. Since then, significant effort has been devoted in the tropical cyclone community to quantify both the effects of recent climate change (Mann & Emanuel 2006; Kossin 2008) and to understand the effect of past climate changes on tropical cyclone activity (notably hurricane activity in the North Atlantic, e.g. Woodruff et al. 2008) and to clarify the arguments presented regarding the likely effects of climate change (Curry et al. 2006). Liu and Fearn (2000) found millennium scale variability in cyclone occurrence at a location in the Gulf of Mexico, which they attribute to changes in the large-scale circulation of the Northern Hemisphere. Although such studies (e.g. Liu & Fearn 2000; Donnelly & Woodruff 2007; Woodruff et al. 2008) provide information on the larger population of storms that affect these regions, the quality of storm data from such palaeo studies is not enough to inform us about the possible changes suggested by Henderson-Sellers et al. (1998) and others. Kossin (2008) has studied changes in the North Atlantic and finds a tendency towards more common early and late storms.

The short homogeneous record of observations for tropical cyclones (for the most part limited to recent satellite data) and a wide variation in data quality and quantity over the different tropical cyclone basins (Figure 1) has meant that the debate on changes in cyclone activity has been intense. Landsea et al. (1999) found significant multi-decadal variability in the north Atlantic basin and this tends to make interpretation of recent changes difficult (see Curry et al. 2006). In addition, results presented for a particular basin can be interpreted as being applicable more generally: the term hurricane is often used to refer to tropical cyclones generally, but meteorologically speaking it refers only to the North Atlantic. The confusion of terminology, together with the difficulty in separating intensity, frequency

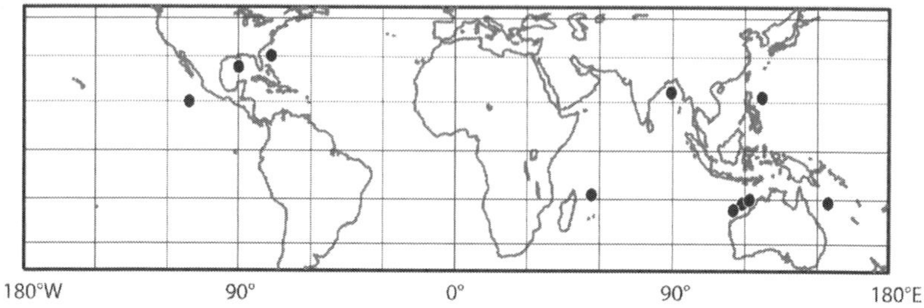

Figure 1 Locations of the points used to define the thermodynamic characteristics of the eight tropical cyclone basins analysed in this study. Only the central point in NW Australia is used in this study.

and storm size in the minds of the non-specialist reader means that conclusions relating to tropical cyclone changes in future climates are sometimes communicated poorly.

In spite of the caveats regarding the quality of the observation record in most basins, there seems to be enough evidence to indicate that changes in tropical cyclone characteristics in the North Atlantic, namely an increase in the frequency of the most intense storms, can at least in part be attributed to increases in ocean temperature that are occurring in this region (Hoyos et al. 2006; Holland & Webster 2007).

With regard to other tropical cyclone basins, Nicholls et al. (1998) found a slight trend in numbers of tropical cyclones affecting the northern parts of Australia, and Broadbridge and Hanstrum (1998) noted a relationship with the Southern Oscillation index (SOI) in north-western Australia. Tropical cyclone climatology of the eastern Australian region is strongly influenced by El Niño, with a maximum of activity near the Australian mainland during positive SOI, with the peak of cyclone activity being further east (away from the mainland of Australia) during negative SOI events. The usable record here and in other basins such as the southwest Indian Ocean is limited by the availability of satellite information. However, Landsea et al. (2006) have questioned even the recent satellite record by calling into question the consistency with which this record was created. In an alternative approach, analysing the power dissipation index of storms, an attempt to include size as well as lifetime, frequency and intensity, Emanuel (2005; 2007b) concluded that the destructiveness of hurricanes in the North Atlantic, measured by the total power dissipation over the life of the storm has increased over the last 30 years. Elsner et al. (2008) also found an increase in the intensity of the strongest storms correlated with SST. This approach has led to some convincing links with greenhouse induced warming in the North Atlantic (e.g. Mann & Emanuel 2006). Overall, despite considerable efforts to improve the record (e.g. Landsea et al. 2004; Webster et al. 2005; Kossin et al. 2007) it is clear that examination of the tropical cyclone record must be accompanied by interpretation of modelling studies.

1.3 Modelling studies

Model simulations of tropical cyclone vortices in climate change studies are limited. Global climate models do not as yet routinely offer the resolution required to simulate the detail of tropical cyclones, so other techniques must be used to analyse the likely changes in tropical cyclone activity. Despite the technical difficulties associated with high-resolution models, there have been attempts at high-resolution climate simulation of storms (e.g. Oouchi et al. 2006; Bengtsson et al. 2007). Some modellers have analysed changes in 'tropical cyclone like vortices' (Knutson et al. 1998) in lower resolution climate models and employed regional models (Knutson & Tuleya 1999) forced by low resolution models while others have applied downscaling techniques to look at storm frequency in greenhouse enhanced climates (e.g. Knutson & Tuleya 2004; Knutson et al. 2007, 2008; Emanuel et al. 2008). Emanuel et al. (2008) performed a range of different seeding studies, and found an increase in very intense storms with a tendency for reduction of events in the southern hemisphere.

In terms of the Australian region, Walsh and Ryan (2000) examined the changing climate of tropical cyclone activity around the northern coast using a regional climate model where they seeded a series of storms in the Australian region and examined the resulting storm intensities. They found slight increases in storm intensity associated with enhanced greenhouse conditions, but did not find the changes to be statistically significant.

Knutson et al. (1998) compared the simulated intensity changes of tropical cyclone like vortices in the North Pacific (TCLVs) in their model to the theoretical results of Holland (1997) and Emanuel (1988, 1999). They found an increase in intensity of 5–12% for the period studied (corresponding to a 7–20 hPa decrease in central pressure). Later, Knutson and Tuleya (2004) forced the GFDL model with results from a range of climate models. Focusing only on the North Atlantic, they found a likely increase in intensity of 'about half a category' that was not strongly dependent on the choice of climate model forcing the regional model. The high resolution approach of Oouchi et al. (2006) pointed to higher risks of more devastating storms in a global model with a 20 km mesh but reported an overall decrease in the numbers of storms in their model in enhanced CO_2 conditions.

1.4 Other strategies

Simpler models and diagnostic strategies offer the opportunity to analyse the results of complex climate models in a way that offers some insight into likely changes in tropical cyclones in an enhanced greenhouse climate. The nature of the problem calls for a wide range of approaches. The difficulty of providing an adequate assessment for all basins based on intensive modelling studies (Knutson and Tuleya (2004) tackle only the North Atlantic, while studies like Oouchi et al. (2006) and Bengtsson et al. (2007) are able to make broad conclusions for global or hemispheric level changes) leads to the multi-model thermodynamic approach used here, where the same assessment can be applied over multiple basins and models.

With such studies, it is possible to leverage the huge modelling effort that has been devoted to running and validating the CMIP3 project to gain information about tropical cyclone activity in a greenhouse warmed climate. This chapter examines the results of a number of models from the WCRP/CMIP3 archive (Meehl et al. 2007) using one model of tropical cyclone maximum potential intensity (Holland 1997). Section 2 discusses thermodynamic models of tropical cyclones and a technique is presented to derive a measure of season intensity for a tropical cyclone basin that can be used to compare the tropical cyclone environment for many basins and many models.

2. *Methodology*

2.1 Thermodynamic models

The development of a tropical cyclone is dependent on a number of factors. In addition to a suitably warm ocean surface, dynamic factors such as low-level cyclonic flow and low vertical wind shear, there must be a suitable thermodynamic environment in which the storm can develop. The nature of this thermodynamic environment determines how intense any storm can be and so imposes a maximum potential intensity (MPI) on the storm. Thermodynamic models of maximum potential intensity (e.g. Emanuel 1988; Holland 1997; Michaud 2001) have been shown to have value at a range of timescales for characterising the intensity of storms, should they occur, in a given thermodynamic environment and therefore have the potential to be used to characterise aspects of a future tropical cyclone climate. Thermodynamic models consider only the thermodynamic state of the atmosphere–ocean system and any analysis with these models cannot offer insight into the frequency or likelihood of such storms developing, only the maximum intensity to which any storm could develop. However, Emanuel (2007) hypothesised that, given a large enough sample, the average storm intensity is likely to increase by the same amount as the MPI. Some authors (e.g. Camargo et al. 2007) have developed this line of enquiry to formulate a cyclone genesis potential, where other factors such as low-level vorticity and tropospheric wind shear are combined to give a better relationship with cyclone frequency. MPI (Tonkin et al. 2000) and other indices such as the cyclone genesis index of Camargo et al. (2007), are related to the number of storms observed over a basin as well as the intensity reached by these storms. It is clear that given a supportive thermodynamic environment there is no dearth of convective disturbances that can develop into tropical cyclones. McDonald et al. (2005) concluded that such convective genesis parameters would be likely to provide useful information where models cannot resolve storms.

Schade (2000) identifies two theoretical frameworks for calculating the maximum potential intensity of tropical cyclones: static theories and dynamic theories. Dynamic theories such as that of Emanuel (1988), can be considered loosely as a 'heat engine' approach. In contrast, static theories derive the maximum potential intensity based on the thermodynamic structure of the environment within which the storm develops, along with some knowledge of the likely thermal structure of the eye and eye wall. Earliest appearances of this static approach are Miller (1957), though it has recently been enhanced by Holland (1997) and tested by Tonkin et al. (2000). Holland's (1997) development of this approach utilises a separate development of warming through ascent in the eye wall and subsequently through descent in the eye (which develops if the ascent produces a pressure drop beyond a defined threshold).

Differences between the MPI techniques have been examined by Tonkin et al. (2000), giving a thorough overview of the rationale for calculating the maximum potential intensity of tropical cyclones. Tonkin et al. (2000) made a detailed study of the performance of the thermodynamic techniques of Holland and Emanuel and summarised the performance of the models and provided some results for relevant station data and demonstrating that the concept works at a monthly timescale. Differences between the Emanuel (1988) and Holland (1997) models have also been examined by Schade (2000) and Camp and Montgomery (2001). Schade (2000) concluded that the differences in sensitivity of the models were due to the different formulation of the models and the differing roles the SST plays in the two models. Here we are applying Holland (1997) in a climatological situation to determine climatological statistics and to postulate possible climate changes as a result of increased CO_2. In such cases (climate model results and reanalyses), the use of Holland's (1997) method is appropriate since the ocean and atmosphere can be assumed to be in equilibrium.

2.2 Strategy

For this study, output from five coupled ocean–atmosphere models participating in CMIP3 and utilised in the Intergovernmental Panel on Climate Change Fourth Assessment Report (IPCC AR4) have been analysed for tropical cyclone season characteristics. The control runs are for pre-industrial conditions (PICNTRL) and the experiment consists of forcings appropriate to the climate of the 20th century (C20C) and for a future scenario of stabilisation at 550 ppmv of CO_2 equivalent (SRESB1). For each of the models, the oceanic grid point nearest the point of maximum tropical cyclone activity in each of the eight identified tropical cyclone basins is selected and the maximum potential intensity is calculated using the Holland (1997) method based on the monthly mean environment archived for the model for all cyclone basins indicated in the Global Guide to Tropical Cyclone Forcing (WMO 1993) (see Figure 1).

The representative point for each basin (Figure 1) is the point of maximum cyclone activity for each basin as charted in WMO (1993) and the locations of these points are also indicated in Table 1. These locations may, of course, not correspond with the optimal thermodynamic environment, but they can be considered typical. Notably, tropical cyclones in the North Atlantic region are split into two areas, one off the Eastern US coast and one in the Gulf of Mexico, a strategy somewhat at odds with other analyses, which tend to amalgamate these two areas. Australian region storms are split into SW Pacific and NW Australia, to reflect the presence of two quite different regimes. Section 3 describes how these results were further analysed.

Table 1 Locations of sample data points for cyclone season characteristics. Nearest 2.5° grid point reference to peak cyclone activity identified in Figure 1.10 of the Global Guide to Tropical Cyclone Forecasting (WMO 1993).

Location	Latitude	Longitude
NW Australia	17.5°S	117.5°E
Western Indian Ocean	15.0°S	50°E
NE Pacific	15°N	250°E
NW Pacific	15°N	130°E
Bay of Bengal	17.5°N	90°E
Gulf of Mexico	25°N	275°E
N Atlantic	30°N	285°E

Table 2 WCRP–CMIP3 model results in use in this study. Model results are derived from the WCRP/WGCM archive at PCMDI. For more details of the model simulations, details of the model components and an explanation of the acronyms employed, see www-pcmdi.llnl.gov/ipcc/about_ipcc.php.

		Model years analysed		
Model	**Details**	**Pre-industrial control (PICTRL)**	**Climate of the 20th century (C20C)**	**550 ppm stabilisation experiment (SRES B1)**
CSIRO Mk 3	CSIRO Mk3.0 (2000): atmosphere: spectral (T63L18); ocean: MOM2.2 (1.875 x 0.925L31)	1871–2250	1871–2000	2001–2150
HADCM3	HadCM3 (1998); atmosphere: (2.5 x 3.75); ocean: (1.25 x 1.25); land: MOSES1	1870–2100	1870–1999	2000–2150
GISS-ER	GISS ModelE/Russell atmosphere ModelE (E3, 4 x 5 x L20); ocean: Russell (E3,4 x 5 x L13); sea ice: ModelE (E3); land: ModelE (E3)	1900–2399	1880–2003	2004–2150
NCAR	CCSM3.0, version beta19 (2004); atmosphere: CAM3.0, T85L26; ocean: POP1.4.3 (modified), gx1v3; sea ice: CSIM5.0, T85; land: CLM3.0, gx1v3\0	1870–2099	1870–2000	2001–2150
GFDL	CM2.0 – AOGCM; AM2P13 – atmosphere; OM3P4 – ocean LM2 – land; SIS – sea ice	1870–2100	1870–2000	2001–2150

Table 3 Changes in mean cyclone season intensity (positive indicates more severe season). Values are scaled to indicate mean change in hPa for each month. Response varies between model and basin, but is largely indicative of more severe cyclone seasons.

	CSIRO	GISS	UKMO	GFDL	NCAR
North Atlantic	8	1	26	10	6
Gulf of Mexico	3	4	21	7	6
Bay of Bengal	10	6	17	7	8
NE Pacific	11	6	-1	10	11
NW Pacific	2	12	25	4	8
SW Indian	10	1	8	4	6
NW Australia	11	24	12	16	11
SW Pacific	1	4	3	4	8

3. Analysis

3.1 Season severity

The computed maximum potential intensity (MPI) (Emanuel 1988; Holland 1997) does not offer a direct indication of the characteristics of individual storms. Other factors affect the nature of a storm, such as lower level vorticity and wind shear. Here, the problem is addressed in terms of characteristics of the cyclone season for a given basin. Since the MPI can reflect both number and intensity of storms, the resulting MPI computed for each month, basin and model and the area between the mean sea level pressure curve and the MPI curve is integrated (Figure 3) to give an index of the season severity under the assumption that a longer season and/or a potential for more intense storms are likely to be more severe seasons. This 'season severity index' is re-computed as anomalies from the long term mean of the pre-industrial simulation. Clearly, changes in the length of season and intensity of season need not occur together or even with the same sign and may not directly compensate in terms of impacts, but the index presents a first order attempt to quantify the severity of a season. The basins identified in WMO (1993) have been characterised by single points at the peak of the cyclone occurrence maps.

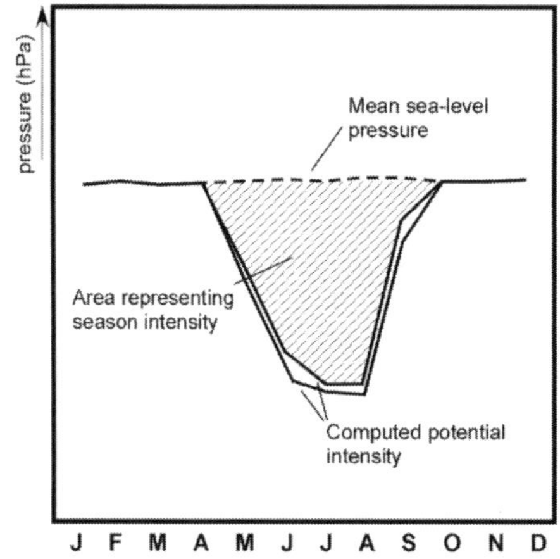

Figure 2 Representation of how the 'cyclone severity index' is computed. The length of the season and the maximum MPI (minimum pressure) reached in each month affect the total shaded area. In addition to mean differences in the season, the maximum increase in intensity for any particular month is also calculated.

3.2 Changes in enhanced greenhouse conditions

Changes in the mean cyclone season severity have been computed for the 50-year period 2100–2150 and expressed as a difference from the mean of the pre-industrial control for that model and basin. The changes are summarised in Table 2. Almost all model and region combinations show an increase in season severity when viewed on a monthly mean timescale. The changes are around half a category (when converted to approximate cyclone intensity) in the mean case. Some models show much greater changes than others: some modelled changes are only a few hPa. Some idea of the overall picture can be gained from Figure 3, which shows the mean time series and range of anomalies from the mean for the control simulations (dotted lines) and anomalies from the pre-industrial control mean for the SRESB1 550 ppmv simulation (solid lines).

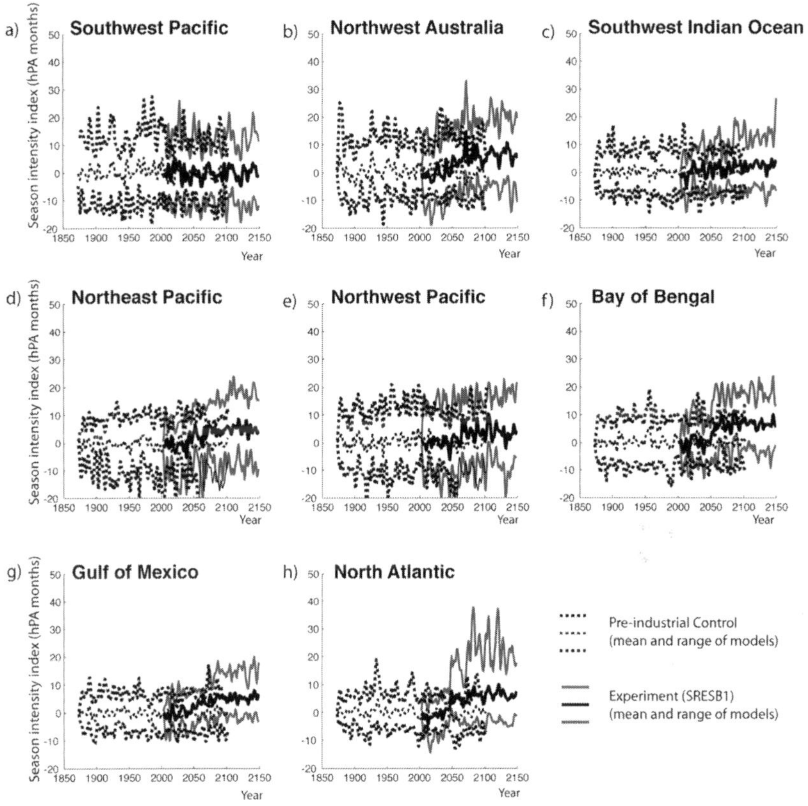

Figure 3 Time series of changes in cyclone season severity index for the eight tropical cyclone basins in Table 1. The mean and range of model results is shown for both control and experiment runs. A five-year [1 3 4 3 1] weighted filter has been applied to these results.

The general increase (indicating a more severe cyclone season) is apparent in most basins with the notable exception of the Southwest Pacific. This is consistent with the changes observed by Elsner et al. (2008) from the historical record. The small changes noted in Table 2 are difficult to see against the interannual variability of the model mean and the inter-model range (indicated by the grey lines).

In order to take some steps to quantifying a worst-case scenario from these analyses, the mean and range of models is calculated for the worst-case monthly change in MPI. Figure 4 shows peak monthly average changes in MPI (SRESB1–PICNTRL) for the eight basins studied here. Error bars show the range of model simulations for each basin. All basins show an increase in intensity, with peak increases in intensity of between roughly 15 and 30 hPa although some model-basin combinations give much higher changes in the worst case. These numbers correspond to changes in the MPI in the worst-case month, indicating storms that could by more intense by about one category.

4. Discussion

The changes in potential tropical cyclone intensity in enhanced CO_2 conditions have been examined using output from five coupled ocean atmosphere models from the WCRP/CMIP3 archive and a diagnostic model for computing the maximum potential intensity of tropical cyclones. There is a wide range in modelled intensities of season between models and some models are more sensitive than others to changes in the greenhouse forcing (as indicated by this particular measure). Although the methodology has been shown to have some promise, further work is needed to develop the technique further. In a calibration study, Camargo et al. (2007) noted that each model had a specific relationship between cyclone genesis potential and cyclone number. However, such calibrations have not been done here, and this may improve the model estimates. Such studies require good observation records of cyclone activity and this is lacking in many places of interest.

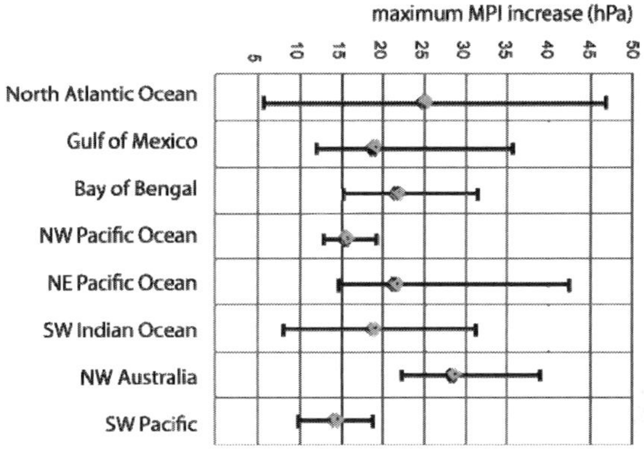

Figure 4 Maximum changes in monthly MPI for the eight tropical cyclone basins, showing the mean change and the range of model differences.

The results for the Australian region show significant differences between changes in the Indian Ocean and the Southwest Pacific. Based on these results it is suggested that it is inappropriate to make conclusions on the future tropical cyclone climate by referring to 'Australian Region' or 'Southern Hemisphere' storms. The study could be further enhanced by using more models (and/or CO_2 scenarios) and the nature of the changes in season could be further examined (changes in length and intensity have different implications). Overall, the role of scalable diagnostic studies such as this is shown to have potential to contribute to the ongoing study of the likely effects of climate change on tropical cyclone activity.

Acknowledgements

The modelling groups are acknowledged for providing their CMIP3 data for analysis, the Program for Climate Model Diagnosis and Intercomparison (PCMDI) for collecting and archiving the model output, and the WCRP Working Group on Coupled Modelling (WGCM) for organising the model data analysis activity. The multi-model data archive is supported by the Office of Science, US Department of Energy.

References

Bengtsson L, Hodges KI, Esch M, Keenlyside N, Kornblueh L, Luo J-J & Yamagata T (2007). How may tropical cyclones change in a warmer climate? *Tellus*, 59, 539–61.

Broadbridge, LW & Hanstrum BN (1998). The relationship between tropical cyclones near Western Australia and the Southern Oscillation Index. *Aust Meteor Mag*, 47, 183–89.

Camargo SJ, Emanuel KA & Sobel AH (2007). Use of a genesis potential index to diagnose ENSO effects on tropical cyclone genesis. *J Climate*, 20, 4819–34.

Camp JP & Montgomery M (2001). Hurricane maximum intensity: past and present, *Mon Weath Rev*, 129, 1704–17.

Curry JA, Webster PJ & Holland GJ (2006). Mixing politics and science in testing the hypothesis that greenhouse warming is causing a global increase in hurricane intensity. *Bull Amer Met Soc*, 87, 1025–37.

Donnelly JP & Woodruff JD (2007). Intense hurricane activity over the past 5000 years controlled by El Niño and the West African monsoon, *Nature*, 447, 465–68 doi: 10.1038/nature05834.

Elsner JB, Kossin JP & Jagger TH (2008). The increasing intensity of the strongest tropical cyclones. *Nature*, 455, 92–95, doi: 10.1038/nature07234;

Emanuel KA (1988). The maximum intensity of tropical cyclones, *J Atmos Sci*, 45, 1143–55.

Emanuel KA (1999). Thermodynamic control of hurricane intensity, *Nature*, 401, 665–69.

Emanuel KA (2005). Increasing destructiveness of tropical cyclones over the past 30 years, *Science*, 436, 686–88.

Emanuel K (2007a). Comment on "Sea surface temperatures and tropical cyclones in the Atlantic basin" by Patrick J Michaels, Paul C Knappenberger & Robert E Davis. *Geophys Res Lett*, 34, doi: 10.1029/2006GL026942.

Emanuel K (2007b). Environmental factors affecting tropical cyclone power dissipation. *J Climate*, 20, 5497–509.

Emanuel K, Sundararajan R & Williams J (2008). Hurricanes and global warming: Results from downscaling IPCC AR4 simulations. *Bull Amer Meteor Soc,* 89, 347–67.

Haak H et al. (2003). Formation and propagation of great salinity anomalies, *Geophys Res Lett,* 30, 1473, doi: 10.1029/2003GL17065.

Henderson-Sellers A, Zhang H, Berz G, Emanuel K, Gray W, Landsea C et al. (1998). Tropical cyclones and global climate change: a post-IPCC assessment. *Bull Amer Meteor Soc,* 79, 19–38.

Holland GJ (1997). The maximum potential intensity of tropical cyclones, *J Atmos Sci,* 54, 2519–41.

Holland GJ (2007). Misuse of landfall as a proxy for Atlantic tropical cyclone activity. *Eos, Trans Amer Geophys Union*, 88, 349–50.

Holland GJ & Webster PJ (2007). Heightened tropical cyclone activity in the North Atlantic: Natural variability or climate trend. *Proceedings of the Royal Society* A, 365, 2695–716, doi: 10.1098/rsta.2007.2083.

Hoyos CD, Agudelo PA, Webster PJ & Curry JA (2006). Deconvolution of the Factora contributing to the increase in global hurricane intensity, *Science*, 312, 94–97.

Knutson TR, Tuleya RE & Kurihara Y (1998). Simulated increase of hurricane intensities in a CO_2-warmed climate. *Science*, 279, 1018–20.

Knutson TR & Tuleya RE (1999). Increased hurricane intensities with CO_2 induced warming as simulated using the GFDL hurricane prediction system, *Climate Dynamics*, 15, 503–19.

Knutson TR & Tuleya RE (2004). Impact of CO_2-induced warming on simulated hurricane intensity and precipitation: Sensitivity to the choice of climate model and convective parameterization. *J Climate*, 17, 3477–95.

Knutson TR, Sirutis JJ, Garner ST, Vecchi GA & Held IM (2008). Simulated reduction in Atlantic hurricane frequency under twenty-first-century warming conditions. *Nature Geoscience*, 1, 359–64.

Knutson TR, Sirutis JJ, Garner ST, Held IM & Tuleya RE (2007). Simulation of the recent multi-decadal increase of Atlantic hurricane activity using an 18-km-Grid Regional Model. *Bull of the Amer Meteor Soc*, 88, 1549–65.

Kossin JP (2008). Is the North Atlantic hurricane season getting longer? *Geophys Res Lett*, L23705, doi: 10.1029/2008GL036012.

Kossin JP, Knapp KR, Vimont DJ, Murnane RJ & Harper BA (2007). A globally consistent reanalysis of hurricane variability and trends, *Geophys Res Lett*, 34, L04815, doi: 10.1029/2006GL028836.

Landsea CW, Anderson C, Charles N, Clark G, Dunion J, Fernandez-Partagas J et al. (2004). The Atlantic hurricane database re-analysis project: Documentation for the 1851–1910 alterations and additions to the HURDAT database. In RJ Murnane & K-B Liu (Eds). *Hurricanes and typhoons: past, present and future*, (pp177–221). New York: Columbia University Press.

Landsea CW, Harper BA, Hoarau K & Knaff JA (2006). Can we detect trends in extreme tropical cyclones? *Science*, 313, 452–54.

Liu K & Fearn M (2000). Reconstruction of prehistoric landfall frequencies of catastrophic hurricanes in Northwestern Florida from lake sediment records. *Quaternary Research*, 54, 238–45.

Mann ME & Emanuel K (2006). Atlantic hurricane trends linked to climate change. *Eos, Trans Amer Geophys Union*, 87, 233–41.

McDonald RE, Bleaken DG, Cresswell DR, Pope VD & Senior CA (2005). Tropical storms: representation and diagnosis in climate models and the impacts of climate change. *Climate Dynamics*, 25, 19–36.

Meehl GA, Covey C, Delworth T, Latif M, McAvaney B, Mitchell JFB et al. (2007). The WCRP CMIP3 multi-model dataset: a new era in climate change research. *Bulletin of the American Meteorological Society*, 88, 1383–94.

Michaud LM (2001). Total energy equation method for calculating hurricane intensity, *Meteorol Atmos Phys* 78, 35–43.

Miller BI (1957). Tropical cyclone intensity and sea surface temperature, *J Meteor*, 15, 184–95.

Nicholls N, Landsea C & Gill J (1998). Recent trends in Australian region tropical cyclone activity. *Meteorol Atmos Phys*, 65, 197–205.

Oouchi K, Yoshimura J, Yoshimura H, Mizuta R, Kusunoki S & Noda A (2006). Tropical cyclone climatology in a global-warming climate as simulated in a 20 km-mesh global atmospheric model: frequency and wind intensity analyses. *Journal of the Meteorological Society of Japan*, 84, 259–76.

Schade LR (2000). Tropical cyclone intensity and sea surface temperature. *J Atmos Sci*, 57, 3122–3130.

Tonkin H, Holland GJ, Holbrook N & Henderson-Sellers A (2000). An evaluation of thermodynamic estimates of climatological maximum potential tropical cyclone intensity. *Mon Weath Rev*, 128, 746–62.

Vecchi GA & Knutson TR (2008). On estimates of historical North Atlantic tropical cyclone activity. *Journal of Climate*, 21, 3580–600.

Vecchi GA & Soden BJ (2008). Increased tropical atlantic wind shear in model projections of global warming. *Geophys Res Lett*, 34, L08702, doi: 10.1029/2006GL028905.

Walsh KJE & Ryan BF (2000). Tropical cyclone intensity increase near Australia as a result of climate change. *J Climate*, 13, 3029–36.

Webster PJ, Holland GJ, Curry JA & Chang H-R (2005). Changes in tropical cyclone number, duration and intensity in a warming environment. *Science*, 309, 1844–46.

WMO, 1993: *Global guide to tropical cyclone forecasting*. World Meteorological Organization Technical Document, WMO/TD No. 560, Tropical Cyclone Programme, Report No. TCP-31.

Woodruff JD, Donnelly JP, Emanuel K & Lane P (2008). Assessing sedimentary records of paleohurricane activity using modeled hurricane climatology. *Geochem Geophys Geosyst*, 9, Q09V10, doi: 10.1029/2008GC002043.

12
Assessing climate change adaptation options for local government

S Trück, W Bradford, A Henderson-Sellers, S Mathew, J Scott, M Street and R Taplin

*Abstract**

Uncertainty in the prediction of low probability high-impact climate events makes it difficult for local governments to foresee locality specific climate risks, prioritise policy options and implement suitable actions for the future. Climate projections and economic evaluation need to be available at the local levels (scales of a few tens of kilometres at most) to ensure local climate investment actions. Observations of the historical climate impacts at local scales could give a probability distribution that could be projected into the future using new information and expert opinions. This research describes the early stages of a study of the adaptation options for local government authority, based on available climate information and perceptions of the local community. In our overview we concentrate on the following key issues and questions:

i. Review of the literature on economic modelling of adaptation strategies at the local scale
ii. A description of the legal background and the quadruple bottom line (QBL) approach that can be used to evaluate potential adaptation strategies for local governments
iii. How to quantify and evaluate losses from climate change impact related hazards at the local scale

---

* Chapter originally submitted in March 2009.

iv. The adequate choice of discount rate for economic analysis.

Our findings are that to date there have been surprisingly few studies at the local level on the economic cost of mitigation and adaptation to climate change effects. We suggest that valuing the non-monetary impacts of climate change via Quadruple Bottom Line analysis considering the social, economic, environment and governance dimensions could be a method to identify the appropriate importance of each impact category. We further suggest the use of the so-called Loss Distribution Approach (LDA) in combination with expert judgements and discounted present value analysis as a potential framework for modelling the losses from climate change impacted hazards at the local scale. For the choice of the optimal discount rate we suggest for the financial dimension to use the cost of capital for local governments while for the social discount rate it might be appropriate to use clearly lower discount rates as they are suggested – for example in the reports by Stern (2006) and Garnaut (2008).

1. Introduction

Climate change mitigation and adaptation are challenging tasks and require input from all levels of government. Generally, mitigation and adaptation have not been handled on a comparable basis to date even though they are closely interrelated. Mitigation efforts are aimed at reducing the likelihood of bad climate outcomes, while adaptation efforts are aimed at reducing the severity of any bad climate outcomes that may be faced. As such, they are closely related, and decisions about the best policy or mix of policies depend upon complex and sometimes contradictory factors. These include: international treaties like the United Nations Framework Convention on Climate Change's (UNFCCC) Kyoto Protocol which came into force in 2005 and was ratified by Australia in December 2007; national legislation and targets (e.g. the National Pollution Reduction Scheme proposed for Australia for 2010); regional requirements (e.g. NSW Climate Change Action Plan); and local community desires and demands.

At the local level, where action is often the most effective, local government is confronted with the complex and problematic task of planning and implementing mitigation and adaptation actions within existing budgetary constraints. While in New South Wales (NSW), the state government has provided funding for some initial climate mitigation and adaptation initiatives such as energy and water savings, the effects of altered weather patterns associated with a changing climate are generating a plethora of interconnected impacts that demand a multi-disciplinary, multi-faceted response.

Ku-ring-gai Council, a local government authority, administers an area located on the North Shore of Sydney that may be very vulnerable to increased storm and bushfire activity as predicted by regional climate change models (DECC 2008). Significant costs and losses have already been experienced within the council area – a 1991 storm that cost around $670M and residential home losses that were incurred during the 1994 bushfires. Climate change modelling predicts future climate shifts that highlight significant caveats regarding uncertainty which creates dilemmas for determining appropriate responses. Ku-ring-gai Council has determined that the development of any climate change response strategy should incorporate the Principles of Ecologically Sustainable Development (ESD) to comply with the requirements of the *NSW Local Government Act 1993* within a Quadruple Bottom Line reporting framework (GRI 2006) that is used for the council's existing management and reporting system. Overall, the following key issues with respect to climate change have been identified through a community consultation process:

- Increasing frequency and intensity of bushfires
- Increasing frequency and intensity of storms
- Water security – decreasing potable supply per capita
- Biodiversity: loss of Critically Endangered Vegetation communities
- Increasing heat stress mortality rates.

With respect to these identified issues, for a local government (in this case Ku-ring-gai Council) it is particularly important to investigate the costs and benefits of potential adaptation strategies to such climate change hazards at a local scale. Unfortunately, in the literature there seems to be an obvious lack of economic modelling of adaptation strategies at the local level. The few exceptions include the recent work by Gardner and Stern (2008), Hayhoe et al. (2008) and Kirshen et al. (2008).

Gardner and Stern (2008) consider a number of energy retrofit options for homes. They find that although financial incentives are indeed a motivating factor, public information campaigns and government commitment to facilitating the energy-saving actions are by far the most effective at invoking private adaptation investment. Hayhoe et al. (2008) aim to downscale results of different global circulation models to northeast regions in the United States. They illustrate how statistical and dynamical downscaling methods are able to improve simulation of spatial and temporal variability in temperature and precipitation across the region. They further develop high-resolution projections of future climate change across the northeast USA. Kirshen et al. (2008) investigate interdependencies of urban climate change impacts and adaptation strategies in the USA for metropolitan Boston. In their approach they use climate change scenarios developed for the American New England region and argue that due to the interrelations among infrastructure systems, it is critical to take account of the impacts that adaptation actions have on each other and other systems.

While important issues for finding optimal adaptation strategies are provided in these studies, none of them delivers results that can be used in particular for investigating optimal adaptation or mitigation strategies with respect to identified key issues mentioned above. Therefore, the aim of the research reported in this chapter was to develop a framework that will be helpful for the evaluation of greenhouse gas mitigation and climate change adaptation strategies for local government decision makers. With respect to these tasks the following aspects were considered:

1. Results of the economic and climate change literature on economic modelling of adaptation strategies focusing on the local scale
2. The legal background for local governments in NSW for decision-making with respect to different adaptation and mitigation strategies
3. How to include not only financial or economic but also non-monetary impacts of climate change into the analysis such that

next to the economic also social, environmental and governance aspects can be considered

4. The quantification of potential risks and losses from climate change impacted hazards at the local scale
5. A framework that enables determination of the costs and benefits of different adaptation strategies with respect to climate change impacted hazards at the local scale
6. The adequate choice of a discount rate for the financial, economic and social analysis.

The remainder of this chapter proceeds as follows. Section 2 provides a literature review on results from other studies focusing on economic modelling of mitigation and adaptation to climate change. Section 3 briefly describes the legal background for local government in NSW and the use of a quadruple bottom line (QBL) approach for policy development. Section 4 suggests a framework for economic comparison of different adaptation strategies that includes modelling the frequency and severity of natural hazards at the local level. Section 5 illustrates the controversy in the literature concerning the choice of discount rate to be used in project evaluation with focus on the optimal choice of the discount rate within a QBL framework. Section 6 concludes and provides recommendations for future research.

2. Results from other studies

While there is great interest by stakeholders in the economic costs of mitigation and adaptation to climate change, there seems to be an obvious lack of economic modelling of adaptation strategies at the local level. As pointed out by Kuik et al. (2006), impacts of climatic change have been studied extensively by the natural sciences, but research on the social and economic consequences is far more limited. Kuik et al. (2006) also note that insufficient work has been done on the costs of adaptation versus no adaptation as well as on what changes there might be to adaptation costs if there were some level of mitigation. The importance of such work re-emerges in Garnaut's (2008) conclusion where he implies that the high costs of climate change do not in themselves make a case for any level of mitigation, rather a comparison of the costs of mitigation must be weighed against the costs saved from avoided climate change. Agrawala et al. (2008) note that the literature concerning mitigation is comprehensive and demonstrates clear boundaries to the concept of mitigation, while the literature for adaptation remains sparse and contested and without clear boundaries or a metric for assessing the effectiveness of adaptation measures. Nevertheless, governments at local and national levels will implement adaptation action because of its public-good characteristics. Studies at the local level have been restricted to certain sectors such as public health and tourism. Also, there has been very little attention to government's role in facilitating private adaptation investment decisions. This review first looks at economic models and approaches with respect to climate change adaptation strategies reported in the recent literature. The review then summarises methods for valuing non-market goods, considers the controversial topic of discount rates, and finally refers to downscaling, before providing an evaluation of the existing literature with respect to the aims of this chapter.

2.1 Economic models and uncertainty

Many authors advocate an *integrated approach* or *an integrated model* to link economic analysis and climate change projections, although the term *integrated* is used with a variety of meanings. Among the authors advocating *integration* in one form or another are Ermolieva and Sergienko (2008), Jones et al. (2007), Kirshen et al. (2008), Kuik et al. (2006) and Godard (2008). Kuik et al. (2006) distinguish integrated approaches from exogenous or externally influenced approaches. Fully dynamic integrated approaches, linking long-term socio-economic feedback and climate scenarios to the costs of inaction are rare. Godard (2008) points out that integrated assessment models are intended to capture interactions between natural and economic systems and the aim should be 'projected equilibrium', where achievements match expectations, and should include intergenerational equity considerations.

Ermolieva and Sergienko (2008) review models and approaches taken in managing catastrophic risk. Planning for natural risks requires a strict risk-specific methodology. They acknowledge that mathematical models for quantitative assessment and prediction that can deal with incomplete information are necessary. The fact that a one-in-500-years event may occur tomorrow is all too often overlooked by traditional deterministic models. They also point out that catastrophes have the characteristic of being infrequent in any given locality; they may not resemble one another, meaning that past experience is not a reliable guide; and there are few or no real observations to facilitate traditional statistical modelling based on the law of averages. Ermolieva and Sergienko (2008) conclude that stochastic optimisation models are valuable in taking into account the goals and constraints of the different agents, and that an integrated approach including preventive measures to reduce catastrophe probabilities is necessary.

Calzadilla et al. (2005) use a dynamic computable general equilibrium model to assess the impact of climate change on El Niño – Southern Oscillation and the North Atlantic Oscillation cycles and expected losses at the regional level. They recognise that local impacts of economic loss due to extreme weather events propagate within the globalised world economy. They remark, however, that although climate scientists are studying these relationships, no model is available that links climate change to local extreme weather conditions. Their paper does not address climate change adaptation or mitigation efforts.

Kuik et al. (2006) use the term *general equilibrium effect* to refer to the economy as a whole. Indirect effects of climate change, having been spread throughout the economy by markets, can amplify and diminish the direct economic impacts of climate change. According to them, most studies to date have underestimated the indirect economic effects on the general economic equilibrium.

Jones et al. (2007) explore the critical complementarity of mitigation and adaptation and the need for an *integrated approach* by managers. They describe the case of two Australian regional bodies – the Central Victorian Greenhouse Alliance (CVGA) and the North Central Catchment Management Authority (NCCMA) – that maximise the benefits of mitigation and adaptation actions through integrating such actions at the regional scale. They believe that Australian policy affecting adaptation and mitigation is likely to be 'top down' from the state or national level, while the actors are local. The CVGA and NCCMA both recognise that regional reductions in greenhouse gases will not directly benefit the local area but rather will contribute to the global common good. However, they do believe that a local area being an early adopter of mitigative technologies and actions will provide economic benefits. White et al. (2008) share this optimism in considering policy options for English Regional Development agencies and local authorities who seek opportunities for business and the economy as a whole while responding to the government's carbon reduction pathway obligations.

In a similar stream of work as Jones et al. (2007), Kirshen et al. (2008) provide a case study on integrated urban impacts and adaptation strategies for the city of Boston. They analyse Boston's urban infrastructure systems and recognise the interdependencies of climate change impacts and adaptation actions. The authors summarise the implications regarding energy use, coastal and river flooding, transport, water supply and quality, public health, tall buildings, and bridge scouring infrastructure systems. The authors argue that

i. Structural actions taken before full climate change impacts occur will result in fewer total expected negative impacts,

ii. Actions taken soon will result in less negative weather-related impact even without climate change

iii. While climate change will magnify negative impacts due to demographic changes, such impacts and largely complementary adaptations will not only affect target infrastructure systems but also interrelated systems

iv. Adaptation actions must be integrated with land-use planning, environmental impact assessments, and socio-economic impact assessments, for example, of disrupted supply chains.

With respect to the choice of an appropriate model, Ganderton (2005) asserts that Benefit-Cost Analysis (BCA)/ Cost-Benefit Analysis (CBA) is the most appropriate method for assessing hazard mitigation actions as it takes into account all the benefits that may be achieved by a mitigation action. He warns against keeping benefits constant as with Cost-Effectiveness Analysis, where benefits, often non-monetary, are stated and fixed at a particular level thereby ignoring other non-specified benefits that result from the action. He comments that although projects should be evaluated from society's point of view, decision-makers often have to confine themselves to one sector or level of government and must also recognise that not all people see an action as a benefit; others may see it as a cost. Unfortunately BCA has not produced a definitive prescription for dealing with uncertainty, a major issue in anticipating risk induced by climate change (Ganderton 2005).

On the other hand, Norman et al. (2007) doubt that CBA is an appropriate tool for analysis of global problems that extend over centuries and impact non-human species. They find that *ex-ante* estimations of costs may be higher than later, realised *ex-post* estimations, owing to research breakthroughs that lowered cost or improved quality.

Finally, Prato (2008) shows how fuzzy set logic can be applied when there is uncertainty about the nature and extent of climate change impacts. Hereby, the author distinguishes between two case approaches: first, the case of climate risk and second, the uncertainty case. The former assumes that the decision-maker is able to assign subjective probabilities to future climate change scenarios while in the uncertainty case the decision-maker is unable to assign these probabilities. Further, both cases assume there is uncertainty regarding conditional outcomes of adaptive strategies. In the so-called climate risk case, the author suggests the determination of the preferred overall adaptive strategy for the climate risk case by selecting the adaptive strategy that maximises expected performance over all future climate change scenarios. The suggested procedure involves specifying subjective probabilities for future climate change scenarios and then ranking the adaptive strategies for each climate change scenario using a fuzzy set operation. On the other hand, the preferred overall adaptive strategy for the climate uncertainty case is determined by applying the minimax regret criterion to the maximum losses in performance for all combinations of adaptation strategies and climate change scenarios. The model suggested in Prato (2008) involves the potential inclusion of expert opinions and updating of strategies when new information (additional observations or knowledge) becomes available.

2.2 Valuing non-market goods

Economists employ various methods for assigning monetary values to non-market goods such as environmental values. Ganderton (2005) describes revealed preference methods, where behaviour is actually observed, such as Value of Intermediate Goods, Hedonic Price Models and Travel Cost Models. He also describes stated, or hypothetical, preference models such as the Contingent Valuation Method and Conjoint Analysis. In many applications Benefit Transfer, using existing data from other projects, is also implemented in order to reduce the costs. Kuik et al. (2006) believe that although all current studies use a mix of valuation techniques, a rough version of Benefit Transfer is generally applied in most of the techniques. Unfortunately studies into the merits of one valuation method over another are lacking.

Ganderton (2005) reminds us that environmental economists find that the wider society places very high values on the existence of ecosystems; furthermore economists cannot measure the emotional and psychological dimensions of disaster mitigation. Rehdanz (2007) investigates how EU citizens value species preservation, biodiversity and ecosystem services. Humans place value on the proximity of ecosystems that conserve biodiversity and reduce numbers of threatened species. He links human life-satisfaction with biodiversity and numbers of threatened species by using spatial econometric techniques to determine whether spatial relationships exist and to what extent.

A summary of studies since 2000 involved with key methodological issues such as valuation of non-market goods, adaptation costs and uncertainty is provided in Kuik et al. (2006). Unfortunately, none of these studies were focused particularly on the local government level.

2.3 The choice of the discount rate

As it is pointed out by Karp (2005), the choice of the discount rate has substantial impact on analysis and may well ultimately determine the cost-benefit ratio for investments related to climate change. Even moderate discounting discourages small present-day investments to avoid large damages in the distant future.

Kuik et al. (2006) observe that very few studies incorporate a measure of statistical risk with respect to the discount rate. There is uncertainty regarding future discount rates; in fact there is no agreement in the empirical literature as to how a discount rate should be chosen. In their overview of the literature Kuik et al. (2006) present the following points of view on discounting:

i. Discounting is equivalent to the present generation owning all future resources. Although objectionable from a moral standpoint, it seems to reflect current reality. This line of research has not led to practical alternatives to discounting.

ii. A constant rate of pure time preference between 1 and 3% is employed by most studies.

iii. Conventional exponential discounting accords the same difference between two years whether they fall in the present decade or in a century from now. That is, using the discount factor $(1+r)^{-t}$, where r = the rate of discount and t = time, leads to that effect. Hyperbolic discounting could be implemented where one year in the present is equivalent to ten years in one hundred years' time. Present discount rates would be little affected and would mainly affect long-term decisions.

iv. A low discount rate puts the cost of inaction with regard to the environment and climate change much higher than does a high discount rate.

v. A hyperbolic discount rate (becoming lower over time) assigns higher marginal damage costs as the discount declines.

Azar and Sterner (1996) point out that assuming constant discount rates over longer time scales might not be appropriate. They suggest basing the discount rate on the social rate of time preference which consists of two components: the expectation that we will be richer in the future, and pure time preference. Karp (2005) confirms other researchers' views that hyperbolic discounting tends to conform to psychological perceptions of value over time and reduces the problems associated with constant discounting.

Summers and Zeckhauser (2008) take the view that market rates of discount are too high for projects benefiting society as a whole, and that social investments should give the future far more weight than do private projects. They say that humans, particularly those alive today, have despoiled the environment, and future generations have as much right to inherit a hospitable earth as we did. They point out that humans are shown to have a strong aversion to causing suffering or loss, and their valuing of non-monetary goods rises disproportionately with rising income; while standard discounting implies that an increment in consumption will have less marginal utility among richer future generations.

In contrast, Kirshen et al. (2008) avoid the ethical arguments surrounding the discount rate by not discounting any impacts of climate change, thereby assuming that property values and adaptation costs appreciate at the same rate.

Ismail-Zadeh and Takeuchi (2007) find the standard discount rate approach inadequate for making decisions about when to take preventative measures in relation to an analysis of disasters that have already occurred in the 21st century. On the other hand, Winkler's (2006) analysis finds that the outcome of hyperbolic discounting is not only unsatisfactory from the present generation's point of view, but may also be inefficient. This is because hyperbolic discounting assumes that the present generation faces the costs of investment while the benefits are spread over all subsequent generations. Winkler (2006) thus challenges the use of hyperbolic discounting in long-term decision making and argues that success in hyperbolic discounting requires a commitment mechanism.

Overall, it appears that contributions to the literature are clearly aware of the substantial impact discounting may have with respect to identifying optimal adaptation and mitigation strategies. However, there seems to be no common agreement on the adequate choice of a discount rate for evaluating such strategies. We refer to Section 5 for a further discussion on the choice of discount rates for assessing climate change adaptation options for local governments.

2.4 Conclusions regarding research to date

Our survey of the literature reveals that economic studies to date have tended to focus on particular economic sectors such as agriculture, tourism or health. The proposed models in the literature that address climate change often do not adequately integrate feedback from socio-economic mitigation and adaptation efforts. Also, there has not been enough work on comparing the costs and benefits of adaptation strategies with respect to the costs of the 'reference case' of no adaptation effort. Further, assigning costs and benefits to present-versus-future generations in the face of uncertainty enmeshes ethical, and therefore controversial, considerations. In particular there is no common agreement on the adequate choice of the discount rate for evaluating future costs and benefits of adaptation or mitigation strategies.

Finally, so far no satisfactory method for assigning dollar values to non-market environmental services exists despite human attempts to approximate a value for what is essentially priceless.

3. Legal and sustainability issues for local government in NSW

Local government in NSW has a duty to manage land, infrastructure and natural assets in a manner consistent with the Precautionary Principle (*Local Government Amendment (Ecologically Sustainable Development) Act 1997*). Recent court judgements in NSW and elsewhere (Scott et al. 2008) confirm the need for governments to ensure responses to issues are proportionate to the scale and magnitude of the risk. This section provides a brief overview of the legal setting for local governments in NSW and explores the use of the Quadruple Bottom Line (QBL) framework for considering climate change mitigation and adaptation strategies at the local level.

3.1 Legal background for local governments

In planning for climate change, it is worth noting that there have been a growing number of successful administrative challenges involving climate change in the NSW Land & Environment Court and other jurisdictions around Australia (NSW LEC 2006). The NSW Land & Environment Court has jurisdiction over local government matters and has adopted a receptive disposition to the consideration of future climate change. The outcomes of recent cases considered by the court have hinged upon the application of the ESD Principles as defined in the *Environmental Planning and Assessment Act 1979* (NSW). These principles are:

i. The precautionary principle
ii. Inter-generational equity
iii. Conservation of biodiversity and ecological integrity
iv. Improved valuation, pricing and incentive mechanisms.

These principles also appear in the *Local Government Act 1993* (NSW). As a result it is clear that local government in NSW must have regard to these principles when managing the environment.

Litigation regarding climate change issues may occur via the common law of tort. While the likelihood of successful litigation appears low due to difficulty in proving specific causation, this may change as McDonald (2007) notes:

> Like the tail-effect of greenhouse gas emissions, legal claims may be slow to gestate. But the law has a long memory, so courts of the future will reflect on the state of knowledge currently at hand to determine whether decision-makers of today did enough to avoid or minimise the worst exposures to climate change.

The need for due diligence on behalf of local government is outlined by England (2007) who points out that local governments should "take care to ensure their actions, decisions and policy responses to matters that may either contribute to, or be affected by, climate change remain current and reasonable in what is a rapidly evolving policy context."

Overall, to reduce the risk of exposure to litigation, local councils must take into account the future effects of climate change in the approval of new developments and management of their assets. This can be achieved by adhering to the ESD Principles as they have been applied to climate change in the courts. Councils must also recognise the need to fulfil their duty of care to the community. Whilst the scope for actions in negligence and nuisance arising from climate change appears limited at present, the legal concepts of reasonableness and causation are evolving at a rapid rate. It would be prudent that local government authorities act with due diligence in a manner that is consistent with shifting legal and community expectations.

In the case of local government and climate change, it may be argued specific causation exists between a failure to act (to reduce emissions or implement adaptation strategies) and a consequence (the magnification of the risk due to increasing greenhouse gas (GHG) levels in the atmosphere and lack of preparedness for foreseeable impacts). Local government contributes to the cause of climate change if it fails to reduce GHG emissions and so assist efforts to offset risks and/or does not recognise the consequences of climate change by failure to identify risks and prepare infrastructure under their care and control.

Decision-making under uncertainty is difficult but almost always an unavoidable constant where environmental change is concerned. Pindyck (2007) notes we cannot know with much precision the effects created by environmental damage even though we may spend considerable resources on trying to find out. With regard to climate change, Pindyck claims the very long time horizons and the ability of humans to adapt makes it very difficult to establish any firm grasp of the impacts. Pindyck (2007) claims there will inevitably be hidden or unanticipated costs and benefits that emerge as a result of policy change due to the dynamism and complexity of climate change scenarios and their associated uncertainties.

The challenge is to identify the significance of the predicted effects of climate change and realise local consequences in relation to future liability. Such decisions must also consider the benefits lost and the cost of failure to take pre-emptive action to mitigate and adapt to the more extreme impacts of climate change. This has the potential to cover many broad areas and responsibilities of the local government sector including but not restricted to the area of Ku-ring-gai.

3.2 The Quadruple Bottom Line (QBL) framework

Climate change is a typical sustainability issue. In terms of general sustainability, planning and reporting frameworks have progressed well beyond simply interpreting sustainability through the exclusive lens of the environment. Initially the notion of sustainability was embedded within the limits to growth debate (Smith & Scott 2006), resource depletion and impacts of human activity on natural cycles. In more recent times the complexity of the sustainability agenda has become apparent. Change cannot occur in isolation and ramifications emerge across all sectors and ecosystems including human health and wellbeing. Equity, risk analysis, cost benefit estimates and biodiversity resilience assessment are all key tools used in sustainability planning and reporting.

Ku-ring-gai is bounded by National Parks with substantial forest conservation areas located throughout the municipality. The bushland/urban interface extends 89 km and the local government area is listed as the third most vulnerable community to bushfire in the Greater Sydney Region (Chen 2005) with 13,000 homes on the bushland interface. Planning responses to climate change impacts require an understanding of the ramifications of change across the Quadruple Bottom Line (QBL). The QBL framework, in accord with the Global Reporting Initiative (GRI 2006), consists of the following elements:

(i) Environment – biodiversity, ecosystem health (including air and water), natural resource conservation

(ii) Social – urban planning equity in distribution of costs and benefits between current and future generations and all sectors of the current Ku-ring-gai community

(iii) Economic – identification of the true cost of activity including non-monetary and indirect costs including pollution of environment and consumption of non-renewable resources

(iv) Governance – transparency, accountability, risk management, and preparedness.

These four elements correspond with the ESD Principles mentioned above. In broad terms they can be considered to mirror one another and as such the QBL framework permits the ESD Principles to be transferred from rhetoric to reality. As illustrated by Table 1, each aspect of the QBL will potentially contain a combination of the ESD Principles depending upon the nature of the specific issue under examination. It is rarely possible to exclusively assign an impact to a single element of the QBL and it is critical to sustainability to recognise and deal with both intended and unintended consequences of any response.

Integration of the QBL into a council's planning and reporting system requires structuring management and administrative systems around the QBL from the annual report through to each staff member's work plan. This framework ensures that the traditional 'silos' – so often a problem in past local government structures – are dismantled as council officers consider policy development, projects and procurement and work plans from a multidisciplinary perspective. By doing so, council can demonstrate that the non-monetary and indirect positives and negatives of change are factored into any decision or plan.

Table 1 The interdisciplinary relationship between ESD and QBL.

	Precautionary principle	**Inter/intra generational equity**	**Biodiversity conservation**	**Valuation and pricing**
Governance	*√	√	√	√
Environment	√	√	*√	√
Social	√	*√	√	√
Economic	√	√	√	*√

* *Key ESD Principle for each QBL element.*

Climate change will generate consequences across all dimensions of the QBL and there will be consequences for both council and the community. Planning responses to these impacts developed within the QBL framework should ensure that the full range of potential costs and benefits are included as well as highlighting any 'trade-offs' that may result from investment responses. For example, consider the risk of bushfires that is expected to increase as a consequence of increasing temperatures and drought periods. Council faces the task to investigate the scale and nature of risk arising from these changes. Any response applied to offset that risk will have a combination of financial, social and ecological consequences. The beneficiaries of risk abatement ideally need to be the same people as those bearing the cost of risk abatement. For example implementations of more stringent building codes to reduce property vulnerability will incur a cost to the resident and will also generate a benefit to the insurers of residents' homes. Therefore the home owner could expect a reduced premium to share the benefit with the insurer. This kind of equity issue is becoming recognised and is not always a trade-off between social and economic costs and benefits.

Risk reduction associated with potential climate change impacts may cause negative consequences to biological assets – for example, the clearing of fire trails. Here the costs are bequeathed to the environment as has often been the case in the past. However, decision-makers using QBL judgement endeavour to design a strategy that attempts to 'balance' the costs with benefits; that is, if there are environmental costs there should be environmental benefits as well. Equity across the QBL is considered a benchmark of sustainable practice in the 21st century (GRI 2006).

4. Quantifying losses and discounted present values

In this section, we will suggest an approach for quantifying potential losses from extreme events like storms, droughts or bushfires that might be further increased in frequency and severity due to climate change impacts. We suggest the use of the so-called Loss Distribution Approach (LDA) that is quite popular in the financial industry for modelling insurance claims and losses arising from operational risks in the banking industry, see e.g. Klugman et al. (1998) or Bank of International Settlement (2001). The Loss Distribution Approach involves the estimation of an adequate frequency and severity distribution for the considered extreme events. The aggregate loss distribution for the hazard is then computed by combining these two distributions such that expected annual loss and the loss at any arbitrary confidence level α can be computed. Often, simulations also are used to derive higher quantiles of the aggregate loss distributions. Using the calculated figures of the aggregate loss distribution for each year and cost estimates for adaptation and mitigation strategies it is then possible to apply discounted present value (DPV) analysis to compare the costs and benefits of the individual strategies with respect to a no-action scenario. In the following, we will discuss first the LDA for modelling risks from natural and climate change impacted hazards, second how expert opinions can be included in the estimation procedure and finally the use of DPV analysis to compare different strategies.

4.1 The Loss Distribution Approach

The LDA is a statistical approach for generating aggregate loss distribution. In this section, we provide the algorithms that can be used to compute the aggregate loss distribution and show how to calculate extreme quantiles for the losses based on the generated aggregate loss distribution. As mentioned above this approach has become particularly popular in the financial and insurance industry. To compute the probability distribution of the aggregate loss over a one year time horizon, we need to estimate the probability distribution function of the single event loss and its frequency. For a natural or climate-impact related hazard, this means that we have to determine a probability distribution for the number of events per time period as well as a severity distribution for potential losses from the events. Then it is possible to compute the cumulative losses for each time period.

In the LDA, the loss severity and frequency distribution are assumed to be independent. Thus, the cumulative loss G for one year is expressed as

$$G = \sum_{i=1}^{N} X_i \qquad (1)$$

where N is the annual number of events modelled as a random variable from a discrete distribution and $X_i, i = 1, \ldots, N$, are severities of the event modelled as independent random variables from a continuous distribution. Modelling frequency and severity of the losses is a well-known actuarial technique (Klugman et al. 1998). For the modelling of losses arising from low frequency high severity hazards typically a Poisson distribution is chosen to model the frequency while a heavy-tailed distribution like the Lognormal, Weibull or Pareto distribution is suggested for severity. Figure 1 gives illustrative examples of a Poisson distribution with parameter $\lambda=2$ (where λ is the mean rate of occurrence of the low frequency high severity events) and the typical shape of a Lognormal distribution.

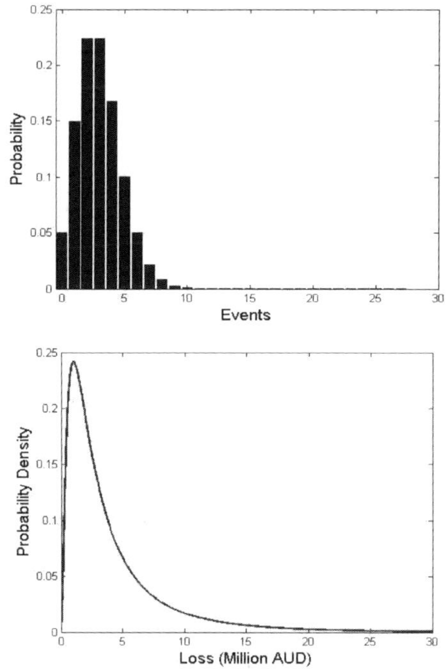

Figure 1: Example of the Poisson distribution (upper panel) for modelling the frequency of events and a Lognormal distribution (lower panel) for the severity.

Usually, Monte Carlo simulation is used to compound the severity and frequency distribution and calculate the aggregate losses from an event type, see e.g. Fishman (1996). The simulation algorithm to generate an annual loss distribution is as follows:

1. Take a random draw from the frequency distribution: suppose this simulates *N* events per year.
2. Take *N* random draws from the severity distribution: denote these simulated losses by $X_1, X_2, \ldots, X_N$
3. Build the sum of the *N* simulated losses to obtain an annual loss $G = X_1 + X_2 + \ldots + X_N$

4. Return to step 1, and repeat k times. Usually, with respect to the low frequency high severity nature of natural or climate-impacted hazards, the number of simulation runs is chosen to be a rather large number, for example k=5000, k=10,000, k=50,000 or even higher. For each run a simulated number for the annual loss G_1, $G_2,\ldots,G_k$ is obtained that enables derivation of a distribution for the aggregate losses.

Therefore, using the LDA, based on the simulated aggregate loss distribution *G*, we can compute the average or so-called expected loss but also the loss at a confidence level α of the event for one year or also longer time horizons. The expected loss *EL* and the loss at a confidence level α, $L(\alpha)$ are defined by

$$EL = \int_0^\infty x dG(x) \qquad (2)$$

$$L(\alpha) = \inf\{x \mid G(x) \geq \alpha\} \qquad (3)$$

The expected loss is the expected value of the aggregate loss distribution function *G* whereas the loss at confidence α is the quantile for the level α. By using Monte Carlo simulation, we can generate the aggregate loss distribution of the event and for example obtain the quantiles at the 90%, 95% and 99% confidence levels. The accuracy of the estimation depends on the adequacy of the parameter estimates but also on the number of simulations, see e.g. Fishman (1996). Overall, it is important to run a high number of simulations, in particular if probabilities for the extreme events and thus the tail of the distribution need to be considered.

4.2 Including expert opinions into the analysis

The algorithm for Monte Carlo simulation can be applied to different kinds of extreme risks or losses and be altered to include climate change scenarios or adaptation scenarios. For example, increasing or decreasing frequencies or severities can be included in the simulation procedure based on the assumptions about the effects of climate change or adaptation strategies on the distributions. However, we point out that the estimation and adjustment of frequency and severity distribution is a challenging task, especially for low frequency high impact hazards when only a small number of observations are available. Note that generally sufficient sample data is required for producing meaningful parameter estimates for the considered probability distributions. Therefore, in order to estimate an adequate probability distribution for the potential losses from extreme events at a local scale, it might be necessary to take into account information beyond the (often rare) historical data. This could include relevant external data from other areas and expert opinions on the probability and severity of such events. Considering the usually very small number of observations that will be available it could be very important to have such expert judgements that are often also referred to as scenario analysis, incorporated into the model.

Therefore, in the following we will provide a brief outline on how different sources of information – in particular historical data on losses from extreme events and expert judgements can be combined. Hereby, we follow an approach using Bayesian estimation as was initially suggested in Shevchenko and Wüthrich (2006) for the quantification of operational risks in the banking industry. As mentioned above in the so-called loss distribution approach suggested by the Basel Committee of Banking Supervision, a very similar problem of estimation of frequency and severity distributions often based on low-frequency high-severity data is required. To overcome these difficulties, according to Bank of International Settlement (2001), it is mandatory to include scenario analysis into the model for risk quantification to meet regulatory requirements. Hereby, scenario analysis gives some rough quantitative assessment of risk frequency and severity distributions based on expert opinions. It is important to point out that for any application, by itself, scenario analysis is rather subjective and should if possible always be combined with actually observed loss data. For combining such sources of information, in recent years particularly Bayesian inference has gained some popularity in the insurance and financial industry. For an introduction to Bayesian inference methods and their application to insurance and finance, we refer, for example, to Berger (1985), Bühlmann and Gisler (2005) or Rachev et al. (2008).

Generally, Bayesian techniques allow for structural modelling where expert opinions are incorporated into the analysis via the specification of so-called prior distributions for the model parameters. The original parameter estimates are then updated by the data as they become available. Additionally, the expert may reassess the prior distributions at any point in time if new information becomes available (e.g. additional information on climate change, effects of climate changes on the frequency and severity of extreme events like storms, droughts or bushfires etc.). To our knowledge this technique has not been used yet to quantify the risk of such losses – in particular not at the local scale. We recommend the application of the approach for determining more appropriate estimates of frequency and severity distributions on the local scale and better quantification of risks from natural and climate change impacted hazards.

4.3 Discounted Present Value (DPV) analysis

Once the potential losses have been determined using the appropriate discount rate the discounted present value (DPV) of the expected losses and costs for a considered time horizon can be calculated. Since adaptation and mitigation strategies at the local level might only pay off after several years, it is important to take into account these longer time horizons in the analysis.

The examination of adaptation strategies to climate change, costs and benefits for the effects of different strategies can be compared with respect to their DPV of total costs. The following formula illustrates the necessary calculations for a considered time horizon up to period T:

$$DPV = Inv_0 + \sum_{t=1}^{T} \frac{Costs_t + Loss_t}{(1+i)^t} \qquad (4)$$

Here, Inv_0 denotes the cost of the initial investment at time 0 for an adaptation strategy, $Costs_t$ are the further costs in each time period related to the adaptation strategy, $Loss_t$ is the loss due to a catastrophic event in period t and i is the appropriate discount rate. In financial applications the discount rate is usually chosen as the cost of capital. However, see the discussions in Section 5 about the differences between economic and financial modelling and the appropriate choice of the discount rate. It should be pointed out that in this approach it is assumed that the chosen strategies will have different effects on the parameters of the frequency and severity distribution which will lead to differences in the calculated or simulated loss figures for each year. Further, the approach also provides the possibility to include the effects of climate change by adjusting the parameters of the frequency and severity distribution in the model.

Once the DPVs of the costs for different adaptation strategies have been calculated, they can be compared with respect to their net benefit or to the business-as-usual scenario that involves no investment for adaptation strategies, but will also most likely yield higher figures for the losses arising from catastrophic events. So, in general we expect a mitigation strategy to have the following effects: increased costs in comparison to a no-action scenario, such that Inv_0 and $Costs_t$ will increase, while $Loss_t$ will decrease. A mitigation or adaptation strategy is then considered to be beneficial from a financial point of view if the overall discounted present value of costs and losses from natural hazards is less than for the no-action scenario

$$DPV_{strategy} < DPV_{no\ action} \qquad (5)$$

while it might be considered to be non-beneficial if its DPV of costs and benefits is higher than for the no-action scenario:

$$DPV_{strategy} > DPV_{no\ action} \qquad (6)$$

As mentioned above, if we are interested in the comparison of different adaptation strategies over a longer period of time, it will be necessary to simulate not just for onc year but over a longer period of time. However, the described algorithm for the loss distribution approach can easily be adjusted for this purpose and the simulated losses for each year $t=1...T$ can then be discounted using the specified discount rate i. Note that both the considered time horizon and the choice of the discount rate may have substantial impact on the classification of whether a strategy is beneficial or not. For considering longer time periods, it might also be necessary to adjust the model parameters of the frequency and severity distribution through time, if for example an increase in the frequency and/or severity of bushfires is expected through time. Assuming that the effects on parameter estimates and changes through time can be quantified correctly, it is then possible to gain more realistic figures for the costs or benefit of different strategies.

5. Choice of the discount rate

Discounting as a procedure involves expressing future and present values of particular variables in a common unit so that decisions regarding possible courses of action in the present can be made rationally. There is considerable ambiguity and indeed controversy in the literature concerning the choice of discount rate to be used in project evaluation. The basis of this controversy is twofold:

i. The relationship between discounting in a financial sense and discounting in an economic sense
ii. In the context of discounting in an economic sense, the magnitude of the discount rate that should be used.

In particular for evaluating climate change adaptation or mitigation strategies, there is a wide discussion on what should be used as the appropriate discount rate. While the reports by Stern (2006) and Garnaut (2008) recommend the use of rather low social discount rates, their arguments have been criticised by various authors. Recent publications on the issue include Carter et al. (2006) and Nordhaus (2006), Quiggin (2006) and Tol and Yohe (2007), just to mention a few.

5.1 Discounting: financial vs. economic

The key difference between discounting in a financial sense and discounting in an economic sense is the variable that is being compared (future vs. present) in each case.

Financial discounting involves the comparison of explicit monetary flows – hence discounting is necessary to convert future dollars (expected net earnings) into present dollars so that a valid comparison can be made with known costs that must be incurred in present dollar terms. It is only through appropriate discounting of future net earnings that, for example, the question of the maximum price that should be paid for an asset can be answered (for a given earnings profile the maximum justifiable price is higher the lower the discount rate applied and vice versa).

Clearly it is desirable in a financial context to use a discount rate that reflects the cost of capital (or the cost of acquiring funds) in some way – using a rate lower than the cost of capital associated with the project implies that avoidable losses will be incurred. In the context of an organisation with a binding budget constraint (such as Ku-ring-gai Council) the use of artificially low discount rates (from a financial perspective) for some projects will necessitate reallocation of expenditure away from other uses.

Discounting in the economic sense involves comparisons between future and present *welfare*, and so raises a different set of questions.[1] The paradigmatic case in economics (relevant to the current project) is the choice among available public investment projects, each of which involves the use of resources in the present and the generation thereby of a path of welfare. The optimal investment is that which maximises welfare, and the comparison of projects necessitates the conversion of future welfare (i.e. the welfare of future agents) into present-welfare equivalents i.e. discounting.

Unlike the financial case, there is no obvious 'commonsense' value of the discount rate to use in comparing welfare streams. The choice of discount rate is a choice between the weight given to the interests of agents in the present and near future, and the interests of agents further separated in time from the project's implementation. High discount rates favour projects with immediate benefits and/or deferred costs, and imply that the justifiable present cost (in terms of welfare foregone) of a given future increase in welfare is low. Low discount rates render large present costs justifiable in return for future welfare gains and so favour projects with deferred benefits and immediate costs. Clearly the choice of discount rate in economic terms has a significant, if not decisive, ethical component which has been the source of much debate in the literature and, obviously, in response to the Stern (2006) and Garnaut (2008) reviews.

---

[1] Of course, welfare cannot be measured directly (if indeed it can be defined unambiguously!) and so it is standard practice in economics to measure it by proxy in terms of monetary values such as Willingness To Pay and so on. The superficial similarity between the two cases should not obscure the fundamental difference between them in terms of purpose and interpretation.

5.2 Choosing a discount rate from an economic perspective

In economic theory, welfare means the satisfaction of preferences – by definition an increase in welfare for an agent entails moving to a more preferred outcome. Preferences are defined over consumption, so an increase in welfare is defined as a move to a more preferred combination of goods consumed.

Economists represent preferences for technical purposes through the use of a *utility function*

$$U = u(c) \qquad (7)$$

where U is utility, an index that indicates the place of a given consumption combination *(c)* relative to others (higher utility numbers indicating more preferred combinations). The typical investment project will imply a consumption trajectory over the given time horizon (possibly infinite) giving rise to a total welfare effect of

$$U_0 = \int_0^{\infty} u(c_t)\, dt \qquad (8)$$

where U_0 is the total utility of the stream of *c*s over time, evaluated in period 0 (i.e. the present, in which the project will be implemented).

As was recognised by Ramsay (1928) there is a *technical* need for discounting (which he decried on ethical grounds) if a choice is to be made among possible consumption paths (i.e. projects). The zero discount rate case depicted involves summing finite quantities over an infinite horizon; project choice would then require an illegitimate comparison of infinites. Discounting of utilities converts the stream of utilities into a declining geometric series which therefore has a finite sum

$$U_0 = \int_0^{\infty} u(c_t)\, e^{-\theta t}\, dt \qquad (9)$$

Given the differing consumption paths, the choice of project will now be determined by the value taken by the discount rate applied to utilities (θ).

It is at this point that the ethical/governance aspects of the discount rate choice begin to bite. The first issue concerns the role of *pure time preference* or *intrinsic discounting*. This involves assigning different weights to consumption baskets *purely* on the basis that they will occur at different points in time. In other words, a positive rate of pure time preference implies that a given quantity of consumption to be experienced *n* periods in the future would be considered inferior to the same quantity experienced now, *ceteris paribus*.

Philosophers such as Rawls (1971) and Parfit (1986) hold pure time preference to be inconsistent with individual rationality, a fairly standard position in the political philosophy and distributive justice literatures. Economists are more circumspect on the issue (the basic textbook model of the intertemporally optimising consumer does not assume zero subjective time preference) but generally concur with the philosophers that pure time preference is unjustifiable in terms of *collective choice*. That is, a government considering a choice among investment projects *ought not* treat the welfare of future generations as less important than that of those currently alive *simply* because of the displacement in time.

This position is reflected in the Garnaut (2008) Report in which a baseline value of zero is chosen for pure time preference, with only a slight adjustment made to allow for the possible occurrence of an extinction event. A discount rate of zero would imply that the welfare of the current generation is weighted equally to that of a billion years in the future; the 0.05% rate set by Garnaut means that the welfare of people around 1400 years in the future is worth half of that of people alive today. For more relevant timeframes (such as 100 years) the 0.05% rate implies only a marginal decrease (around 5%) in the importance of future welfare.[2]

However, where future generations are expected to be better off than present agents there is scope for discounting on grounds other than pure time preference. In other words, if future agents will consume more than current agents then the sacrifices of current agents on behalf of future agents ought to be lower notwithstanding that the welfare of agents is considered to be of equal importance regardless of the generation in which they live.

---

[2] By way of comparison, with a 1% rate of discount the welfare of current agents is worth twice that of those in 69 years time and nearly three times that of agents a century hence.

The degree to which future utility should be discounted depends on the relationship between the growth rate of consumption and the growth rate of utility. A standard assumption in the literature – also followed by Garnaut (2008) – is that the underlying utility function is drawn from the class of *constant relative risk aversion* (CRRA) functions which have a constant elasticity of marginal utility. In other words the percentage change in marginal utility resulting from a given percentage change in consumption is the same no matter when (generationally speaking) the increase occurs. Furthermore the elasticity of marginal utility is assumed to be 1 (so-called log utility, again typical) so that the change in marginal utility is the same as the change in consumption, in percentage terms. As the increment to welfare while positive is falling at the same rate as consumption is rising, it is defensible to discount utility at a rate equal to the growth rate of consumption.

As the Garnaut (2008) modelling assumes a per capita income growth rate of 1.3% p.a. out to 2100, the rate of discount employed is 1.35% (the sum of the pure time preference and consumption growth rates). For sensitivity analysis Garnaut (2008) also considers the case where the elasticity of intertemporal substitution is 2 (i.e. the utility function is $U=(1/2)(c\ )^{1/2}$) so that the discount rate used is 2.65% (i.e. two times the growth rate plus the 0.05% pure time preference rate).[3]

5.3 Choice of rate for the considered key issues

The key question is how far, if at all, we wish to diverge from the Garnaut assumptions. Specifically:

i. Are there grounds for assuming that incomes in Ku-ring-gai will rise more quickly or more slowly than the national average out to 2100?

[3] The intuition behind the higher discount rate is straightforward. For the non-log utility function, marginal utility declines at a slower rate as consumption grows. Hence, future generations are made better off by any given increment to consumption relative to the log utility case. Thus future utility must be discounted more strongly for any rate of consumption growth.

ii. Are there grounds for assuming different forms for the utility function?

iii. Are there grounds for assuming a higher or lower rate of pure time preference than is used in the Garnaut Report?

The pure time preference question is involved and complicated. On the one hand the principle of non-discrimination among generations has intuitive appeal, and yet on the other pure time preference does appear to be an identifiable aspect of individual behaviour (and hence a characteristic of the preferences of each generation) and so raises the question of whether the 'authoritarian' response (Marglin 1963) of overriding those preferences in a democratic context is justifiable. Furthermore, the argument is based on the assumption of the homogeneity of preferences across time, the coherence of which is questioned by some commentators (e.g. Ball 1985). Finally, we might consider whether Ku-ring-gai Council functions as the guardian of future generations of ratepayers in the same way that a national government does with regard to future citizens according to the advocates of low social discount rates.

The choice of the form of utility function is typically driven by analytical convenience as noted, and the empirical evidence, such as it is, is mixed regarding the 'fit' of log utility. The choice of a different form could not really be justified empirically for Ku-ring-gai simply due to the difficulty of testing involved. In any case the decision to use a different form of the utility function and the decision to use a higher social discount rate are effectively one and the same.

Finally the issue of growth rates is also difficult. Note that it is important not to confuse levels and growth rates. Ku-ring-gai almost certainly has an above average per capita income but are there grounds for predicting that income will *grow* faster than average over time? In the absence of such grounds it might be sensible to stick with low discount rates as they are suggested in the reports by Stern (2006) and Garnaut (2008).

On the other hand, there is an extensive discussion on the choice of the adequate social discount rate in the literature, see e.g. Carter et al. (2006), Nordhaus (2006), Quiggin (2006) or Tol and Yohe (2007). Carter et al. (2006) claim that the very low discount rates derived from the use of social time preference theory and in particular the numbers chosen by the Stern review are all open to question. They further illustrate that the results of the review are not robust and take no account of the opportunity cost of crowding out other forms of future-directed expenditure. Nordhaus (2006) argues that the actual social discount rate used in the Stern Review is only vaguely justified by extinction estimates and that the logic behind the *review's* social welfare function is not as conclusive as it claims. He suggests that based on an argument that each generation should leave at least as much total societal capital – tangible, natural, human or technological – as it inherited, quite a wide array of social discount rates would be admitted (Nordhaus 2006). Tol and Yohe (2007) summarise the critiques of the Stern estimates with focus on the discount rate by the following statements: (i) Stern (2006) uses an extraordinarily low discount rate without reporting a sensitivity analysis. (ii) The Stern Review's low discount rate does not match the equally low assumed rate of risk aversion. (iii) The time horizon in the Stern Review is too short for the chosen discount rate. On the other hand, Quiggin (2006) argues that low discount rates are primarily the result of applying the standard utilitarian view that all people count equally. He suggests that if this view is accepted, the pure rate of time discount has to be close to zero. Under these assumptions, there is nothing remarkable about extremely low discount rates.

In terms of the conflict between the social discount rate and the rate to be used in a financial sense, there are few concrete indications to be gleaned from either theory or practice. The Garnaut Report notes the distinction (using the positive-normative distinction) and argues that it is appropriate to use market rates of interest for discounting when analysing private market activity (such as pricing emissions permits) but social discount rates (used when comparing utility outcomes) should be lower. In terms of the considered issues there might be grounds to argue that appropriate market rates are used in assessing the financial viability of climate change adaptation projects at the local government level, with appropriate non-market rates used to decide among financially viable adaptation options. Overall, discounting should still embody the opportunity cost of the funds employed and/or the consumption foregone. Investigating local government decision making, the cost of funds will be higher in most relevant cases than for national governments. Moreover, many adaptation investments may be pushed into the private sector, so the social discount rate is not applicable. If the discount rate used is lower than the cost of funds, government will have to raise taxes, which will reduce current disposable income and thus consumption. Overall, the mechanics certainly remain unclear, but that is true of the approach taken to date by government agencies in prioritising adaptation measures.

6. Discussion and conclusions

6.1 Discussion

This chapter describes the results of the early stages of a study directed towards assessing climate change adaptation options for Ku-ring-gai Council, a local government on the North Shore of Sydney, based on available extreme event and climate information together with perceptions of the local community. Climate change mitigation and adaptation has to be considered as one of the most challenging tasks in environmental management and requires input from all levels of government. At the local level, where action is often the most effective, local government is confronted with the complex and problematic task of investigating the costs and benefits of potential adaptation strategies to climate change hazards in order to implement the most efficient actions within existing budgetary constraints. This task requires an appropriate economic framework for decision-making to justify implemented actions.

Unfortunately, in the literature there seems to be an obvious lack of economic modelling of adaptation strategies focusing at the local level. Therefore, in this chapter we have tackled several key issues in order to develop an economic framework that will be helpful for the evaluation of climate change adaptation strategies for local government decision makers.

In a first step, the legal background for local governments in NSW for decision making with respect to different adaptation and mitigation strategies was reviewed. We suggest that via Quadruple Bottom Line analysis the social, economic, environment and governance dimensions also the non-monetary impacts can and should be identified and included into the analysis.

To quantify the potential risks and losses from climate change impacted hazards on the local scale, we suggest the use of the so-called Loss Distribution Approach. The LDA is a statistical approach for generating aggregate loss distributions initially suggested for the financial and insurance industry but also appropriate for adaptation in order to model losses from natural or climate change impacted hazards. The approach generally involves the determination of a probability distribution for the number of events per time period as well as a severity distribution for potential losses from the events. With respect to the small number of historical observations available on the local level, we suggest that the reliability of the approach can further be improved by combining observed data with expert opinions on frequency and severity distributions. To do this we suggest the use of a Bayesian estimation framework where expert opinions are incorporated into the analysis via the specification of so-called prior distributions for the model parameters. The original parameter estimates are then updated by the data as they become available. Once the probability distributions have been specified we recommend the use of Monte Carlo simulations and discounted present value (DPV) analysis for comparison of different adaptation strategies to climate change impacted hazards on the local scale.

For the choice of the optimal discount rate we distinguish between the different dimensions that are considered in a QBL analysis framework. For evaluating the financial costs and benefits of different adaptation options the use of the cost of capital for local governments seems to be the most appropriate approach. On the other hand, there is a discussion in the literature on the appropriate choice of social discount rates. While Stern (2006) and Garnaut (2008) suggest extremely low discount rates derived from the use of social time preference theory, other authors question the adequacy of these rates (Carter et al. 2006; Nordhaus 2006; Tol & Yohe 2007).

6.2 Conclusions

Overall, this chapter extends the literature on prioritising adaptation strategies to climate change impacted hazards by focusing in particular on the local scale. Several key questions determining the decision process of local governments have been investigated. We have provided suggestions on how these issues can be tackled with respect to considering the social, economic, environment and governance dimensions of adaptation, a quantitative framework for evaluating losses and different adaptation strategies and finally the choice of the discount rate. With respect to the application of the suggested frameworks, further empirical work will be required. Applying the framework using empirical data will also help to determine potential necessary model extensions that could also include, for example, dependencies between different hazards and loss categories, sensitivity analysis of the results to risk factors and most importantly finding optimal adaptation strategies in real world applications.

References

Agrawala S & Fankhauser S (Eds) (2008). Economic aspects of adaptation to climate change: costs, benefits and policy instruments, *OECD Governance;* 11 July 2008, Organisation for Economic Cooperation and Development.

Azar C & Sterner T (1996). Discounting and distributional considerations in the context of global warming. *Ecological Economics,* 19, 169–84.

Ball T (1985). The incoherence of intergenerational justice. *Inquiry* 28, 339–57.

Bank of International Settlement (2001). Consultative document: operational risk.

Berger, JO (1985). *Statistical decision theory and Bayesian analysis.* New York: Springer.

Bühlmann H & Gisler A (2005). *A course in credibility theory and its applications.* Berlin: Springer.

Calzadilla A, Pauli F & Roson R (2005). Climate change and extreme events: an assessment of economic implications. *Computing in Economics and Finance,* 49.

Carter R, de Freitas CR, Goklany IM, Holland D & Lindzen R (2006). The Stern Review: a dual critique. *World Economics,* 7(4).

Chen, K (2005). Counting bushfire prone addresses in the greater Sydney region.Working Papers, Risk Frontiers – Natural Hazards Research Centre, Macquarie University, Sydney.

DECC (2008). Summary of climate change impacts Central Coast region, NSW climate change action plan. NSW Department of Environment and Climate Change.

England P (2007). *Climate change: what are local governments liable for?* Brisbane: Griffith University.

Ermolieva T & Sergienko I (2008). Catastrophe risk management for sustainable development of regions under risks of natural disasters. *Cybernetics and Systems Analysis,* 44(3), 405–17.

Fishman G (1996). *Monte Carlo: concepts, algorithms and applications.* Springer.

Ganderton PT (2005). 'Benefit-cost analysis' of disaster mitigation: application as a policy and decision-making tool. *Mitigation and Adaptation Strategies for Global Change,* 10, 445–65.

Gardner GT & Stern PC (2008). The most effective actions US households can take to curb climate change. *Environment* 50(5).

Garnaut R (2008). *The Garnaut climate change review: final report.* Cambridge, UK: Cambridge University Press.

Godard O (2008). Climate modelling for policy-making: how to represent freedom of choice and concern for future generations? *Interdisciplinary Science Reviews,* 33(1), 51–69.

GRI (2006). Global Reporting Initiative: G3 guidelines. Amsterdam: Global Reporting Initiative.

Hayhoe K, Wake C, Anderson B, Liang X, Maurer E, Zhu J et al. (2008). An integrated climate change assessment for the Northeast United States, *Mitigation and Adaptation Strategies for Global Change,* 13, 419–23.

Ismail-Zadeh A & Takeuchi K (2007). Preventive disaster management of extreme natural events. *Natural Hazards,* 42, 459–67.

Jones RN, Dettmann P, Park G, Rogers M & White T (2007). The relationship between adaptation and mitigation in managing climate change risks: a regional response from North Central Victoria, *Australia Mitigation and Adaptation Strategies for Global Change,* 12, 685–712.

Karp L (2005). Global warming and hyperbolic discounting. *Journal of Public Economics,* 89, 261–82.

Kirshen P, Ruth M & Anderson W (2008). Interdependencies of urban climate change impacts and adaptation strategies: a case study of Metropolitan Boston USA. *Climatic Change,* 86, 105–22.

Klugman SA, Panjer H & Willmot G (1998). *Loss models.* New York and Chichester: Wiley.

Kuik OJ, Buchner B, Catenacci M, Goria A, Karakaya E & Tol RSG (2006). *Methodological aspects of recent climate change damage cost studies,* Working Paper Fridtjof Nansen Institute. Oslo and Research Unit for Sustainability and Global Change, Hamburg University, revised December 2006.

Marglin S (1963). The social rate of discount and the optimal rate of investment. *Quarterly Journal of Economics,* 77, 95–111.

McDonald J (2007). A risky climate for decision-making: the liability of development authorities for climate change impacts. *Environmental Planning and Law Journal,* 24, 405–07.

Nordhaus W (2006). The Stern Review on the economics of climate change. NBER Working Paper Series, WP12741, December 2006.

Norman CS, DeCanio SJ & Fan, L (2007). Opportunities and challenges for the 20th anniversary of the Montreal Protocol, University of California at Santa Barbara. Economics Working Paper Series 12–07.

NSW Government Climate Change Action Plan (2008).Draft. Available at: www.environment.nsw.gov.au/climateChange/actionplan.htm (20/2/2009).

NSW LEC (2006). Gray v Minister for Planning & Ors [2006] NSWLEC 720; (2006) 152 LGERA 258; Walker v Minister for Planning [2007] NSWLEC 741; Taralga Landscape Guardians v Minister for Planning & Anor [2007] NSWLEC 59.

Parfit D (1986). *Reasons and persons.* Oxford: Oxford University Press.

Pindyck RS (2007). *Uncertainty in environmental economics.* Association of Environmental and Resource Economists. Oxford: Oxford University Press.
Prato T (2008). Accounting for risk and uncertainty in determining preferred strategies for adapting to future climate change. *Mitigation and Adaptation Strategies for Global Change*, 13, 47–69.
Quiggin J (2006). Stern and the critics on discounting, Working Paper, School of Economics and School of Political Science and International Studies, University of Queensland.
Rachev ST, Hsu J, Bagasheva B & Fabozzi F (2008). *Bayesian methods in finance*. New York: Wiley.
Ramsay F (1928). A mathematical theory of saving. *Economic Journal*, 38, 543–59.
Rawls J (1971). *Theory of justice.* Cambridge: Belknap.
Rehdanz K (2007). Species diversity and human well-being: a spatial econometric approach, Working Paper, Fridtjof Nansen Institute, Oslo.
Scott J, Hayward L & Joyce A (2008). Climate change adaptation: socialising the science. *Local Government Law Journal*, 14, 52–68.
Shevchenko PV & Wüthrich MV (2006). The structural modelling of operational risk via Bayesian inference: combining loss data with expert opinions. *Journal of Operational Risk,* 1(3), 3–26.
Smith G & Scott J (2006). *Living cities: an urban myth? The government and sustainability in Australia.* Sydney: Rosenberg Publications.
Stern N (2006). *The economics of climate change: the Stern Review.* Cambridge: Cambridge University Press.
Summers LH & Zeckhauser RJ (2008). Policymaking for posterity, Working paper 14359, National Bureau of Economic Research.
Tol R & Yohe G (2007). The Stern Review: a deconstruction, Working Paper, Economic and Social Research Institute, Dublin, Ireland.
White G, Gardner N & Swift J (2008). Carbon reduction: obligation and opportunity at regional and local levels. *Local Economy,* 23(3), 235–41.

Winkler R (2006). Now or never: environmental protection under hyperbolic discounting. Center of Economic Research at ETH Zurich, Working Paper No. 06/60, Zurich.

Index

Printed in Dunstable, United Kingdom

91223678R00239